Prophet—or Professor?
The Life and Work of Lewis Fry Richardson

Prophet—or Professor?
The Life and Work of Lewis Fry Richardson

Oliver M Ashford

Adam Hilger Ltd, Bristol and Boston

©Adam Hilger Ltd 1985

All rights reserved. No part of this publication may be reproduced, stored in a retrieval system or transmitted in any form or by any means, electronic, mechanical, photocopying, recording or otherwise, without prior permission of the publisher.

British Library Cataloguing in Publication Data

Ashford, Oliver M.
  Prophet—or Professor?
  1. Richardson, Lewis Fry   2. Meteorologists
  —Great Britain—Biography
  I. Title
  551.5'092'4   QC858.R5

  ISBN 0-85274-774-8

Consultant Editor: Professor A J Meadows
University of Leicester

Published by Adam Hilger Ltd
Techno House, Redcliffe Way, Bristol BS1 6NX, England.
PO Box 230, Accord, MA 02018, USA.

Typeset by Mathematical Composition Setters Ltd, Salisbury, Wilts, UK in 10/12 point Palacio.
Printed in Great Britain by J W Arrowsmith Ltd, Bristol.

With thanks for her help and encouragement,
this book is dedicated to Lilias.

Nullum quod tetigit non ornavit
He touched nothing that he did not adorn

(Samuel Johnson, of Oliver Goldsmith)

... a mind for ever
Voyaging through strange seas of thought, alone

(William Wordsworth, of Isaac Newton)

Contents

Preface   xi

1   Family Background   1
2   Early Life and Education   9
3   Early Career 1903–13   19
4   Eskdalemuir and France 1913–19   42
5   Benson 1919–20   72
6   Westminster Training College 1920–29   108
7   Paisley Technical College 1929–40   142
8   Retirement: Paisley 1940–43   171
9   Retirement Kilmun 1943–53   193
10 Epilogue   237
Appendix A   Published Works of L F Richardson   266
Appendix B   Unpublished Papers by L F Richardson   271
Appendix C   Published Works about L F Richardson   272
Appendix D   Other References   274
Appendix E   Archives Containing Material Relating to
             L F Richardson   278
Appendix F   Notes, Including Sources of Quotations   280
Appendix G   Acknowledgments   292

Index   294

Preface

When Lewis Fry Richardson died in 1953, the event was reported briefly in *The Times* and somewhat more fully in *The Friend* (he was a Quaker) and in the local press. In due course, obituary notices appeared in the journals of most of the scientific societies to which he had belonged. His former meteorological colleague, Ernest Gold, wrote a 19-page appreciation of his life and work for publication by the Royal Society. Richardson's demise thus received the kind of recognition that might be expected for any scientist who had achieved the distinction of being elected as a Fellow of the Royal Society. It seemed, at the time, that his name would be remembered by future generations chiefly in connection with the *Richardson number*, a criterion for determining whether atmospheric turbulence will increase or decrease.

In the course of the next twenty years or so, it became increasingly evident that Richardson's contributions to the sciences, both physical and social, had not been fully appreciated during his lifetime. In 1957, for example, the *Journal of Conflict Resolution* devoted a whole issue to his mathematical theory of war; the main contribution was a 50-page article by Anatol Rapoport, which is still mandatory reading for anybody interested in this aspect of Richardson's work. In 1960, the Royal Meteorological Society instituted the annual L F Richardson Prize. In the same year, his two books on the causes of war, for which he had never been able to find a publisher, finally appeared in print. In 1965 his *Weather Prediction by Numerical Process*, copies of which were still obtainable from the publishers 30 years after its original publication in 1922, was re-issued as a paperback; this inspired George Platzman to write a scholarly 36-page essay review of the book for the *Bulletin of the American Meteorological Society*. In 1969, a centre for conflict resolution in London was named the Richardson Institute. In 1972, the British Prime

Minister opened a new building for the Central Forecasting Office (with its giant computer) of the Meteorological Office, and named it the Richardson Wing. In 1980, a book by David Wilkinson, dealing solely with Richardson's statistical study of war, was published. And so it goes on. It is hoped that this first full-length biography will help to keep up the momentum and to stimulate further interest in this fascinating man and in his many pioneering works.

The origin of this book dates back to 1955, when Dorothy Richardson expressed the hope that some day I would write a biography of her husband. The feeling that I really ought to take up this challenge remained with me over the years; it was usually fairly dormant but occasionally it would surge into my consciousness, as, for example, whenever I heard the name Richardson. The main reason for my hesitation was the thought that to do an adequate job would require more time and effort than I could spare from my professional work. I did, however, make a few notes and carefully retained any relevant material which came my way.

When, after my retirement, I finally started to tackle the job seriously, I soon discovered that during the intervening quarter of a century many of the possible sources of information had disappeared; files of correspondence and other records had been destroyed and most of Richardson's close friends had died. It seemed, at first, unlikely that I would be able to add much of significance to what had already been written by Gold, Rapoport, Platzman, Wilkinson and others. Then I made two remarkable discoveries. I found that my true vocation was to be a detective, and that librarians and archivists are the most helpful people in the world. I also received encouragement from members of the Richardson family and from friends in many different walks of life. Within a few months, I had unearthed some unpublished articles by Richardson which had not been available to Gold. Then I found a few of his letters which had been carefully kept either by the addressee or in a university or other archive. To my great joy, I established contact with four of his colleagues from his days in the Friends Ambulance Unit; they were all in their mid-80s but were able to share with me their recollections of the 'Prof' and of their harrowing experiences together during the First World War. Gradually, I filled in the gaps in the Richardson story. Each new piece in the jigsaw puzzle seemed to lead me on to another; this in turn stimulated me to intensify my efforts, and before long I found myself thinking about Richardson in all my spare moments. As my knowledge and understanding of his life and work deepened, something that had started as a desire to express a debt of gratitude to a family friend became more and more exciting and absorbing.

In all his writings, Richardson's literary style was distinctive.

Sometimes rather introspective, sometimes more extrovert, he frequently interrupted his otherwise terse narrative by thought-provoking remarks and witty comments. In attempting to summarise, for the first time, all his work in a single book, I have therefore felt it best to make liberal use of quotations, especially from his unpublished articles and correspondence.

As I approach the end of this undertaking, I now wonder in what ways the book would have been different had I started writing it in 1955. Would the material which has since been lost have added much of value? Or has time acted as a beneficial filter, retaining the really important and discarding the trivial, and thereby spared me from having to wade through mountains of paper, mostly irrelevant? I shall never know the answer, but I like to think that time has done at least as good a job of selection as I could have done myself.

As the biography is intended for a wide variety of readers, I have refrained from entering into highly technical details and, to the extent possible, from using mathematics. My aim is not to explain to professional meteorologists the significance of Richardson's contributions to meteorology, nor to psychologists the way in which he helped to advance their science. I have only discussed the history of peace research to the extent that it is relevant to my main theme. As a result, I may well have lapsed into some over-simplifications which may not please the specialists in these various fields. My hope is that meteorologists will understand what I have written about psychology and peace research, and that those engaged in these latter disciplines will find something of interest about meteorology. I would like to think that Quakers will be happy to read about Richardson's work in all these fields; if they are disappointed that I have not said more about his religious beliefs, I can only reply that in this respect I have made full use of the limited archival material at my disposal.

This book could not have been written without the help of numerous institutions and individuals, as indicated in Appendix G. Their assistance has been so valuable and generous that it seems almost invidious to pick out anybody for special thanks. I must, however, mention a few outstanding contributions. Stephen Richardson not only provided me with much of the new information about his father but also made many helpful suggestions at various stages in the preparation of the book. Henry Charnock, who has long shared my admiration for Richardson, was at one time planning to write a biography himself; he nobly shared all his material with me and reviewed my typescript. George Platzman analysed some of the correspondence between Richardson and Quincy Wright and gave me ready access to the information he had collected for his article on Richardson's work on weather prediction. Ralph Jewell shared with me his vast store of knowledge about Richardson's association with V Bjerknes and commented very constructively on my manuscript. But, above all, my indebtedness is to Lewis and Dorothy Richardson; they befriended me while I was still at school and were a constant source of inspiration. When I visited their home in Kilmun to gather material for this biography, somebody who remembered the 'Doctor' told me that it would be of great interest to a few—and of no interest to many. I shall be satisfied if the first half of this prophecy is fulfilled.

I was also told in Kilmun of an incident which occurred while Richardson's body was being taken to the crematorium. The driver of the hearse stopped at an inn to have a 'dram' and was invited by a friend to have another. He refused on the grounds that he had a passenger waiting. 'That's fine,' said the friend, 'ask him to come and join us.' The driver replied 'I couldna' dae that. You see, its Doctor Richardson and he doesna' drink—and in any case he's deid!' Richardson has now been dead for over thirty years but his life and work will surely continue to be a source of inspiration for many years to come.

Oliver M Ashford, 1984

CHAPTER 1

Family Background

'A persistent influence in my life has been that of the Society of Friends [Quakers] with its solemn emphasis on public and private duty.' Thus wrote Lewis Fry Richardson in 1953, just three months before his death. Regarding his ancestors he commented: 'It is difficult to say who were noteworthy as some of the best work passes unpraised'. He was more specific about his parents: 'My father was affectionate and constructive. His arguments were confused. My mother, on the contrary, was unemotional, clear-minded, and critical. I seem to have a bit of both'[1]. By including these remarks in response to a request for some biographical information, he clearly indicated his support for the view that in order to understand a person's life fully it is essential to know something about his family background.

Richardson's father, David, belonged to a well known family from the north-east of England whose contacts with Quakers have been traced back to the time of the founder of the Society of Friends, George Fox, in the middle of the 17th century. Contemporary accounts of Fox and his associates describe how they were 'called in scorn' Quakers because of the manner in which they sometimes trembled when moved to speak at a Meeting for Worship. Quakers are perhaps best known nowadays for their strong stance against war or any other form of violence, based on their interpretation of the teaching of Jesus Christ. They try to follow Fox's example by living 'in the virtue of that life and power that takes away the occasion of all wars'. In earlier times they were equally well known for their objection to taking judicial oaths; many Friends were sent to prison and some even died there in support of this testimony. In refusing to swear, they were following the command of Christ: 'Swear not at all', and were also avoiding any implication of a double standard of truth, one in court and the other outside. Quakers were also known for their simple style of living. They were enjoined not to indulge in the arts, sports, games and other forms of recreation which might be incompatible with the Quaker standards of silence, gravity and sobriety. Their homes and Meeting Houses were plainly furnished, music was banned, and brightly coloured clothes were frowned upon. The custom in those days was to address superiors as 'you' and everybody else as 'thou'; Quakers objected to all marks of distinction and rank and called everybody 'thou'.

One of the earliest Richardson Quakers—a collateral of Lewis' direct ancestors—was John (1667–1753), whose father had joined the Society of Friends immediately after hearing George Fox preach. John himself became a Friend at the age of 16 and started preaching when he was only 18. He travelled widely in the Ministry and twice visited the 'English Plantations' in America. On his first visit between 1700 and 1703, he stayed two or three days with William Penn at his country house in Pennsbury, Pennsylvania. In his diary *An account of the Life of that Ancient Servant of Jesus Christ, John Richardson* (1774), he describes a meeting between Penn and the Indians. At the end of the meeting Penn gave them 'Brandy or Rum, or both', a practice which would hardly be approved by modern Quakers! But of course there are many other differences between Friends of the 20th century and the early Friends. It is doubtful if many visitors to a Meeting for Worship nowadays would feel that the nickname Quaker is still appropriate. Nor would they hear Friends addressing each other as 'Thee' or 'Thou'. They would have difficulty to find any marked difference of dress between Friends and others, whereas in earlier times the sober Quaker grey of the ladies would have been all too apparent.

The story of Lewis' branch of the family is related in *Records of a Quaker family: The Richardsons of Cleveland* by Anne Ogden Boyce (1889). This charming book was originally intended to be a biography of three sisters, Elizabeth, Mary and Hannah, daughters of Henry Richardson, a brother of Lewis' great great grandfather, but it proved to be as changeable as Austin Dobson's Ode which turned to a Sonnet, and it developed into a family history with many delightful digressions. The story begins with William Richardson, a tanner, who lived in the Cleveland district of Yorkshire from about 1660 to 1740. He and his wife Elizabeth Wilson were brought up in the Church of England, but became devout members of the Society of Friends. William was fined many times for refusing to take the oath that was required for calculating the duty payable on all the leather he handled. The fines became so onerous that he considered giving up the tanning business, but this became unnecessary when the Government wisely decided that Quakers could make a simple affirmation instead of taking an oath.

The tanning business stayed in the Richardson family until recent times, but not always in the same location. One of William's grandsons, John Richardson (1733–1800), moved north from Cleveland and founded another tannery on the River Tyne in 1765. His house was known as the Low Lights, from its proximity to the lighthouse nearest to the harbour mouth of South Shields. Other members of the family were successful in diverse branches of industry, including chocolate manufacture; some were well known scientists, writers and politicians.

The youngest of the heroines of Anne Ogden Boyce's book, Hannah Richardson, was Governess of the Quaker school at Ackworth, near Pontefract in Yorkshire, from 1836 to 1846. It was here that the Quaker chemist, Luke Howard, had a country home to which pupils at the school were often invited to tea. He was of course famous as an amateur meteorologist; his system for classifying and naming the clouds (cumulus, nimbus, cirrus etc) is still in use with but minor changes to this day and his book on the climate of London is a classic. Anne Boyce recalls an occasion when many of the pupils had fallen prey to an epidemic and Howard broke up a weekday Meeting for Worship after only half an hour with the remark that 'Under present circumstances, I think the children ought to have shorter meetings and more generous diet'. The latter proposal was duly carried out, no doubt helping to increase Luke Howard's popularity.

Ackworth is proud of its famous student, John Bright, remembered especially for his outstanding oratory as a Member of Parliament, his major role in helping to repeal the Corn Laws and his fight to obtain the franchise for working men. He was a friend of the Richardson family and married Elizabeth Priestman, a second cousin of Hannah. After their wedding at the Quaker Meeting in North Shields, the happy couple went to the Richardson home at the Low Lights for dinner. Anne Boyce gives an interesting account of this ceremony and also of a meeting at Ackworth School in 1834, when John Bright seconded a resolution expressing gratitude that slavery had just been abolished throughout the British Empire.

The Richardsons of Cleveland also tells the story of the banking firm Richardson, Overend and Gurney, founded towards the end of the 18th century (Quakers had a large share in building up the British banking system). One of the founders, Thomas Richardson, became a great benefactor to the Quaker schools at Wigton in Cumberland and at Great Ayton in Cleveland; his first cousin, Edward Pease, with whom he was closely associated, was the moving spirit in the development of the English railways. Thomas retired from the bank at a relatively early age; a few years later the directors were tried for fraud but were found not guilty.

Another famous member of the family was George Richardson (1773–1862). He kept a record of his extensive travels as a Minister in the Society of Friends; this was published in 1864, shortly after his death, together with a biographical sketch by his daughter. He also wrote *The Annals of the Cleveland Richardsons* (1850) which proved to be a valuable source book for Anne Boyce. Like his relative, Thomas, he was interested in education and helped to found the Royal Jubilee Schools at Newcastle upon Tyne.

Lewis' father, David Richardson, belonged to the tanning branch of the family, the Richardsons of the Low Lights in Anne Boyce's book. He was born in Newcastle upon Tyne, where his grandfather had set up home. He was educated at Bootham, the famous Quaker boarding school in York about which we shall hear more later. He then studied chemistry and engineering without, however, going to university. His first business venture was as a manufacturer of agricultural machinery, but this was not successful and he therefore joined his father's leather firm. He had plenty of opportunity of displaying his engineering skills in the construction of a new factory in the Newcastle suburb of Elswick and in the design of the machinery which was to be installed there. Under the direction of David and his brother James, the business prospered. They developed improved methods of tanning for a variety of animal skins, concentrating on the production of high quality leather for shoes, upholstery and bookbinding. They also manufactured driving belts and glue. Their Oasis leather is still being produced for binding fine books and it is reported that they supplied leather for the seats in the House of Commons. They refused however to accept work for the military: an order for army belts was dropped into the wastepaper basket.

In 1861 David Richardson married Catherine Fry, daughter of Robert and Jane Fry of Wellington in Somerset[2]. Robert was a corn merchant but his forefathers had been engaged in wool stapling and spinning. His grandmother Jane (née Coleman) was a remarkable person; having been left with seven young children when her husband died at an early age, she not only brought them up very successfully but also carried on the business by herself. The Fry family had been members of the Society of Friends for three generations but were apparently not related to the well known chocolate manufacturers nor to the in-laws of Elizabeth Fry, the great prison reformer. Catherine's brother Lewis Fry worked as an accountant at the Richardson leather factory; he and his wife Mary (Cruikshank) were devout Quakers and kept a most hospitable and friendly home in Wensleydale after his retirement.

Constructing the Elswick factory was not sufficient outlet for David Richardson's passion for building. He also erected a large family house 'The Gables' in Elswick Road, at that time a fashionable part of Newcastle (in more recent years the house became a maternity home and then a Salvation Army hostel). Six years later, in 1882, he bought

Wheelbirks Farm, a 300 acre estate in the rolling countryside near the village of Stocksfield, some 10 miles west of Newcastle. It was in a somewhat derelict condition, which perhaps explains the very modest price of £5500, and David improved it gradually over the next 30 years, including an extension to the existing farmhouse to provide adequate accommodation for his growing family during their holidays. He kept several diaries, one devoted mainly to his business transactions[3]. Here we read of how he invested his quite considerable savings in a variety of companies, not all as profitable as his own. He had a special fondness for rather speculative mining enterprises and railways. On attaining the age of 21, each of his sons received a very handsome gift of some of these investments.

Figure 1.1 Lewis Richardson's parents, David and Catherine.

Outside the factory, David Richardson was well known for maintaining the family tradition of service to the community based on profound religious beliefs. He was active with the Boys Brigade and enjoyed having them camp in the woods which formed about a third of Wheelbirks. He was an ardent teetotaller and often addressed public meetings on the benefits of total abstinence from alcohol. In April 1875, when he was not enjoying the best of health, he had a letter from a business acquaintance:

> You sometimes send me one side of an interesting question—do you ever read the other? I am sorry your health is poor. What else can you expect if you go so contrary to nature's laws? I am expecting a cask of Burgundy

from the Cote d'or—may I bottle you off a dozen and put some sunshine into your washed out tissues? Do let me. It would be all right for you to live on anchorite's fare did you sit on the top of a pillar and do nothing, but hard work (and *brain* work) requires something very different. Were I to do like you I should be afraid of my boys taking to drinking from want of stamina.

David was not converted, but he did paste the letter into his diary.

David Richardson was a highly respected member of the leather industry and was chosen to represent the British leather manufacturers at the Vienna International Exhibition in 1873 and at the Paris Universal Exhibition in 1878. He wrote several pamphlets on economics in which, according to an obituary in a leather journal, his ideas were 'original and full of thought'. Anybody visiting Wheelbirks today can still see the inscriptions on the estate wall which he built; one of them 'Be sure your work is better than what you work to get' seems particularly apt.

Like many Victorian couples, David and Catherine were blessed with a large family. The first child, Hugh, was born in 1864 and was educated at Bootham and at King's College, Cambridge, where he obtained a first in the Natural Science Tripos. He was science master at Bootham from 1897 to 1914 and then took possession of the Wheelbirks estate on the death of his father; here he was able to indulge in his hobbies of forestry and gardening. During the First World War he visited many military prisons as a friend of the prisoners. He was joint author of *An Introduction to Practical Geography* (1905) and editor of the Cambridge Nature Study Series from 1911 to 1919.

The second son, Arthur, was born in 1865. He was a very delicate sensitive youngster and displayed great artistic ability from his earliest years. Quakers of a former generation would probably not have encouraged him to develop this gift but by that time the Society of Friends had adopted a more liberal attitude to the arts and his parents accordingly allowed him to follow his natural bent. It was while studying painting in Paris that he met his future wife, Lydia Susie Russell (often called Lillian), a fellow student. For a time he gave life classes at a studio in Newcastle but in 1893 he moved to Cheltenham to take up a post as art master at the renowned Cheltenham Ladies College. Some of his water colours were exhibited in Newcastle and Cheltenham and at least two were selected for the annual exhibition of the Royal Academy in London. His marriage proved to be unhappy and his wife ultimately ran off, taking with her their youngest son, Ralph, who became one of the greatest actors of his generation[4].

The third child, born in 1867, was the Richardsons' first daughter, Edith. She was sent to the Mount School, the sister school to Bootham. After a spell at home, she followed the footsteps of her brother Arthur by studying art, at Bushey in Hertfordshire and later in Paris. In 1900, she bought a studio in Bushey and then moved to a studio which she had built herself near Hertford in 1914. By nature Edith was not very sociable; she never married, disliked visitors and was happiest when on her own. At the Mount she had a tendency to turn faint and once had to go out in the middle of a Quaker Meeting in York on this account; thereafter she had an aversion to Meetings for Worship. She used some of her paintings to illustrate a story for children, *Doors* (1909). It relates the fantastic adventures of a boy who succeeded in climbing a mountain after overcoming a formidable series of obstacles. Edith also published several volumes of poetry. Her philosophy of life is well expressed in her poem *Silence and Solitude*[5], which I suspect appealed strongly to Lewis:

> Beyond the whirl of all the world
> Two things abide most good
> Two for the soul most needful are
> —Silence and solitude.
>
> Silence to hear the still small voice
> That tells of things divine
> And solitude to climb the heights
> Towards where the heavens shine.

The third son, Lawrence, arrived in 1869. Although he was not so delicate as his brother Arthur, he had a great deal of illness as a child, which perhaps explains why he was backward in his studies at Bootham. On the death of his Uncle James in 1890, he was pressured into the family leather business. His real interests were, however, elsewhere. While at Bootham he had already shown enthusiasm for astronomy and is reputed to have found a new star at the same time as the official discovery by a German astronomer. This proved to be his life passion and he became an amateur astronomer of some note; at the advanced age of 79 he was awarded an honorary degree by the University of Durham for his contributions in this field. Like many other members of the Richardson family, he was an active member of the Society of Friends; he visited South Africa twice immediately after the Boer War on behalf of the Friends Relief Committee to investigate conditions, provide relief, renew links with Friends and make recommendations for the continuation of Quaker humanitarian work. His letters and diaries relating to this period of reconstruction in South Africa are of such historical interest that an edited version was published by the Van Riebeck Society in 1977.

The fourth son, Gilbert Hancock Richardson (1871–1950) was another fascinating character. He too was educated at Bootham School and at Armstrong College, Newcastle, before joining the family business. By nature, though, he was more of a scholar than an industrialist, and he derived his greatest pleasure from studying languages. Not satisfied by learning Latin and Greek, he also became knowledgeable in Sanskrit and then specialised in international languages. He preferred Ido to its parent Esperanto and helped to publish Ido–English and English–Ido dictionaries (which Lewis later used in his book on numerical weather prediction). To meet the criticism that there was no literature in the international languages, he himself wrote several poems in Ido. Gilbert was an active Quaker and served on several Friends' Committees, including those for Ackworth and Ayton 'Scools' (he was a strong advocate of simplified spelling). He was also a talented artist and produced some sketches which have been compared favourably with those of his brother Arthur.

The next to arrive, in 1874, was a second daughter, Catherine. She was educated at Cheltenham Ladies College and Edinburgh University where she graduated as a doctor in 1900—a rather unusual accomplishment for a young lady in those days. She worked in hospitals for five years and then set up in private practice after taking her MD. In 1913, she was appointed Assistant Medical Officer for Council Schools at Stockton. She did not share her sister's love of solitude but never married. She retired to a small house in Hexham, not far from the family home at Wheelbirks; there she bred terriers and grew fruit and flowers in her garden. By all accounts she was the most accomplished musician of the family and in her retirement played in an amateur orchestra which gave concerts in Hexham Abbey.

By any standard, David and Catherine Richardson and their six oldest children were a very talented group. There was no lack of intellectual or artistic ability, nor of the Quaker tradition of service to the community. But, perhaps above all, this remarkable family displayed signs of originality, not to say eccentric genius, which were to find full expression in the life of the youngest child, Lewis.

CHAPTER 2

Early Life and Education

Lewis Fry Richardson, named after his mother's brother, was born on 11 October 1881 at 'The Gables', the house his father had built in Newcastle upon Tyne. He was very much the baby of the family, being seven years younger than Catherine and no less than 17 years younger than Hugh. Apart from his comments on the influence of his parents, he recorded very little about his family life. The chief source of information about his early years is his mother's diary, but some of the background can also be gleaned from one of his father's notebooks, which combined the functions of a personal diary with those of a guest book.

The atmosphere at home was doubtless that of a rather strict but loving Quaker family, with prayers and Bible reading every day at meal-times and regular attendance at Meeting for Worship on Sundays. According to his mother, Lewis began life as 'a vigorous infant, with keen senses, taking notice very early'. At six months he was 'over 20 lb in weight, 27 inches in length, very vigorous, and given to growling'. Two weeks later he 'cut his first teeth without any difficulty' and 'began to sit up by himself on the floor'. At nine months, he began to stand and by his first birthday he could 'walk across a room though not very steady' and had become a 'very fine boy', but 'rather fierce in his desires from tooth irritation'. He had already started to talk, his first word being 'Tiss' for cat, followed a few months later by 'naughty poserman' when he found no letters in the post box. At about the age of 18 months Lewis became very interested in maps and could locate 'Turtee' (Turkey) and 'Paul' (Nepal). He was also very fond of "am', his name for any kind of cold meat, and 'appils'. 'Altogether' his mother wrote, 'he is a child of great size, strength and energy, and naturally enough of great hunger'[1].

By the age of two, Lewis could say anything he needed to express himself, and was in the habit of asking many questions of the kind 'what dat called?' Six months later he had an attack of scarlatina but recovered quickly. In his fifth year he was 'much interested in electricity'. Lewis himself wrote of an incident from about this time: 'My earliest memory of a sceptical and experimental attitude, that later became habitual, dates from when I was about four. My elder sister Catherine and her nurse came into the nursery wearing their hats.

LFR	Where are you going?
They	We're going to the bank.
LFR	What for?
They	To put money in.
LFR	What for?
They	To make it grow larger.

I was left alone to think about this remarkable way of "making money grow larger". I took my money—a silver threepenny bit and a new farthing—found a trowel, dug a hole in the steep bank in the garden, put the money in and covered it over with earth. About three days later my sister and I together dug up the money. It was not any larger. I was grievously disappointed, and felt that the words of grown-ups were not to be trusted'[2].

Just before his sixth birthday, Lewis went to a kindergarten school run by a Miss Betts. His mother felt that he enjoyed this greatly but noted that 'the preparations for the Christmas fête and examination were too much of a strain for so young a child: he was too anxious and eager and then very cross'. He left the kindergarten after only one year and was taught at home, mainly by his sister Edith, for the next three years. At the age of 10 he was 'extraordinarily given to chemistry'.

Lewis then spent three years at Mr Sawer's day school, where his chief enjoyment was Euclid 'as taught by Mr. Wilkinson'. He won the school prize for science in July 1892 and suffered from chicken pox and a slight attack of mumps.

Among the visitors to the Richardson home during this period were Dr William Garnett and his wife Rebecca. Garnett was then Principal of the Durham College of Science at Newcastle upon Tyne: he probably met the Richardsons through their relative Robert Spence Watson who was a member of the Council of the College. The continuing friendship between the Garnetts and Richardsons was of great importance for Lewis, as he later married one of the Garnett daughters, Dorothy.

Early Life and Education

Figure 2.1 Lewis (aged about four) with his sister, Catherine.

Another visitor was Henry Richardson Procter, who from 1888 to 1890 worked as a chemist at the Elswick leather works. He exerted a 'potent scientific influence' on Lewis and gave him when he was barely 10 a balance and some other scientific apparatus. Procter left to take up a newly created post as head of the applied chemistry department at Leeds University and became famous for his research on the chemistry of the tanning process. He was greatly liked, not only by the Richardsons—to whom he was related through both of his parents—but also by his students and colleagues. He was elected Fellow of the Royal Society fairly late in life and died in 1927.

In January 1894, at the tender age of 12, Lewis was sent as a boarder to Bootham School, where the Richardson family was already well known from his father and his brothers, Hugh, Lawrence and Gilbert. In later years he looked back on his five years at Bootham with great affection and recognised the influence which some of the masters had had on his attitude to science. One of them, Alfred Neave Brayshaw, convinced him for example that 'science ought to be subservient to morals' while another, James Edmund Clark, gave him 'glimpses of the marvels of science'[3]. A brief account of the lives of these two men may help to show the high quality of the Bootham staff in those days.

Figure 2.2 Lewis, aged about nine.

Neave Brayshaw, affectionately known as 'Puddles', was a solicitor by profession but gave this up to follow a call to the service of teaching in his late twenties. He was greatly inspired by John Wilhelm Rowntree's vision of what the Society of Friends might do and be and at the age of 45 he decided to devote the rest of his life to the Society. He wrote several books on Quakerism and travelled widely to visit Friends, to attend meetings and give lectures. He was very interested in Gothic architecture and for more than 40 years he led parties of boys from Friends' Schools on holiday tours to Normandy. He was greatly loved by all who met him. He died in 1940 at the age of 79 shortly after being struck by a car during the blackout[4].

James Edmund Clark came from Street, Somerset, where his family ran the well known shoe manufacturing business. He was also a man of many interests from astronomy and meteorology to sport and service for the Society of Friends. He was an active member of the Royal Meteorological Society and was joint editor of its annual phenological report from 1911 to 1925 (phenology is the study of seasonal changes in nature, meteorologists being especially interested in the relation between the weather and the dates of such events as the budding of different species of flowers and the arrival of migrant birds). While at Bootham, Clark helped to found the *Natural History Journal* which served all the Quaker schools in England.

When Clark left Bootham in 1897 owing to his increasing deafness, he was replaced by Lewis' oldest brother, Hugh who, in Lewis' words, 'taught us how to observe and describe, while he supplied us with very little information'[5]. Hugh apparently had the ability to improvise convincing illustrations from simple materials but was at times dazzled by his own ingenuity; boys preparing for their examinations did not always appreciate how the experiments with flowerpots, hay, string, confetti and the like served to bring home the points which he was trying to make.

Lewis' sense of humour manifested itself while he was at Bootham. He himself related the story of a practical joke which he played on one of his schoolmates. On the way to school by train this unfortunate lad fell asleep and Lewis took the opportunity of hiding his rail ticket. When he woke up he could not find his ticket and asked Lewis what to do when the train stopped before arriving in York for the ticket collector to make his rounds. Lewis suggested that he should hide under the seat. When the collector entered the compartment Lewis handed over two tickets and explained that the second one belonged to his friend who preferred to travel under the seat!

There is a more serious anecdote about Lewis at Bootham. The story goes that while he was heating some bones in the natural history room the supper bell rang. He left the bunsen burner alight, intending to return after the meal. Unfortunately he forgot and went to bed, with the result that a large part of the school building was destroyed by fire. One task of a biographer is to verify such stories, which is sometimes quite easy and in other cases impossible. Investigation of the fire at Bootham showed that it occurred in 1899[6], the year after Lewis left! I have nevertheless included the anecdote as a warning that some of the subsequent stories in this book would have had to be omitted if I had to swear that they were true in every detail.

An outstanding feature of Bootham was the encouragement of leisure hour pursuits—all the prizes were given for these rather than for academic achievements. The Bootham Natural History Society is claimed to be the oldest school society of its kind in Europe. Like many other Bootham scholars, Lewis kept a natural history diary in which he recorded his observations of birds, insects, flowers and the weather. The volume which he wrote at the age of 13 has been preserved in the Bootham archives. It contains some quite pleasing water-colour illustrations of butterflies, birds etc, but is in no way outstanding on Bootham standards. Towards the end of the diary he frequently mentions the weather, with such phrases as 'the glass is falling' or 'overcast and rainy'[7].

By this time the *Natural History Journal* had been widened in scope to include sports and other activities in Friends Schools. Edmund Clark continued for several years as editor and was doubtless responsible for providing tabular records of weather observations from most of the schools. The first entry in the *Journal* about Lewis was in January 1897, when he displayed an insect collection of 167 species at the annual Bootham winter exhibition. The judges considered that the specimens were 'beautifully set and labelled' and awarded him first prize both for this collection and for his natural history diary. In March 1897, he was reported to be studying bacteria in various putrefying solutions while in July he discoursed on entomological specimens to the Natural History Club, of which he became assistant secretary at the beginning of the next term.

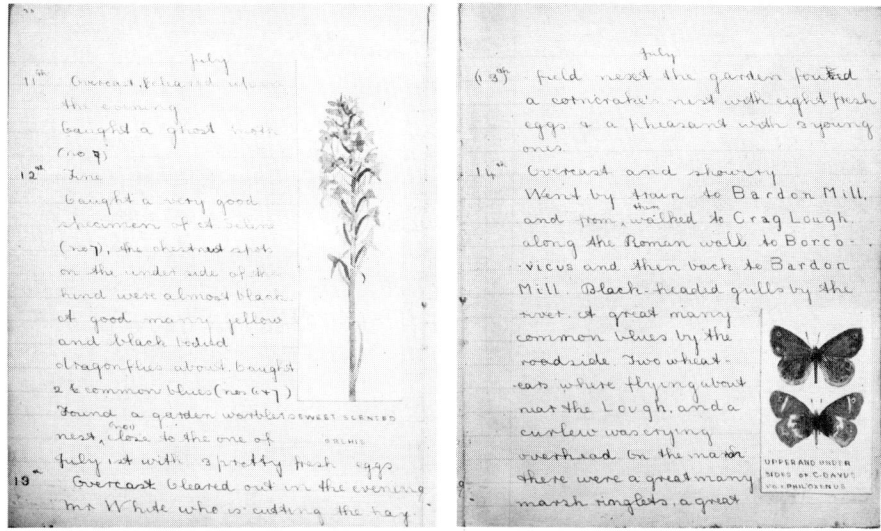

Figure 2.3 A page from Lewis' natural history diary (at age 13).

In January 1898, Lewis exhibited a set of plaster casts of footmarks of animals and birds and was again awarded first prize for his natural history diary. This time the judges commented that 'his diary reveals careful scientific investigation, while his attempts at bacteriology study stamp it as a very valuable production. He has not made many collections this year, having gone beyond the collecting stage'. Next month he was elected president of the Natural History Club, an office which more often than not was held by an adult. He reported to the Club that after a 'series of interesting experiments to ascertain whether a bean's rootlet grew downwards because of its own weight or because it shrank from light' he had 'decided on the former'. As further evidence of his interest in different branches of nature, Lewis gave a report on the wonderful aurora borealis of 15 March 1898. Two months later he was awarded the Bootham natural history scholarship.

Figure 2.4 Senior Class, Bootham School, 1898. *Middle row* (from centre to right): John Firth Fryer (with beard—Headmaster), Lewis Richardson, Francis Knight, Hugh Richardson (with moustache). The future Headmaster, Arthur Rowntree, is also in the middle row (with bow tie).

All this leisure hour activity did not mean that Lewis was neglecting his academic studies. In June 1898 he passed the London University Matriculation Examination, being placed 19th in the national honours list (his classmate, Francis H Knight, who was appointed mathematics master at Bootham in 1904, came 9th). Details of his achievements in different subjects are not available, but it seems safe to assume that he excelled in answering questions which required deductive abilities rather than straightforward memory. Lewis himself later wrote that 'In comparison with the average university graduate, my retentivity was low. I disliked, or shied away from, memory-subjects such as history at school, and organic chemistry and human anatomy at Cambridge. I was incapable of becoming learned on any subject to the extent of the average university professor. It was easier and pleasanter for me to explore new subjects than to try to maintain a reputation for knowledge which I was rapidly forgetting'.

In his last year at Bootham, Lewis was also Chairman of the Reading and Discussion Group and an active member of the Senior Essay Society. His essay on a visit to a North Durham Colliery is all that survives. It is concise, factual and free from social comment, but does not display any exceptional literary gifts. Perhaps the most remarkable thing about the essay is the handwriting, which is childish yet full of character; clearly he made a big effort to ensure legibility. Throughout his life he wrote very slowly and deliberately, and always used an old-fashion stylograph; more modern types of pen, and especially ball-points, were quite unsuitable for his painstaking style. When asked many years later if it had been a handicap in sitting an examination to be such a slow writer he replied that he had always had difficulty in finding time to answer all the questions but that this had become less important in the more advanced examinations when quality became more important than quantity.

Lewis' distinctive style of handwriting may also reflect the slight hand tremor which affected him from childhood. He wrote of this: 'The tremor is presumably congenital; for I do not know of any experience that could have caused it; and my elder brother Lawrence has it worse. This defect ruled out surgery as a career, and hindered manipulation in physics'[8]. In spite of this, he was a skilled practical experimenter and made much of his own apparatus. On one occasion at Paisley Technical College, when a laboratory technician was unable to repair a galvanometer because the quartz suspension which had to be fitted was too fine for him to see, Lewis did the job himself.

On leaving Bootham at the age of 17, Lewis went to the Durham College of Science in Newcastle for two years (his parents' friend, Dr William Garnett, had left the College in 1893 to take up an appointment with the newly constituted Technical Educational Board of the London County Council). Here he studied mathematics, physics, chemistry, botany and geology and came second in the final Associate of Science (ASc) examination. He observed that 'the social and ethical discipline of Bootham School was lacking'[9].

Evidence of Lewis' inquiring mind is provided by an entry he made in his father's notebook in September 1898 'testing some waters with potassium permanganate for oxydizable matters (i.e. organic matter?)'.

As it was not possible in those days to take a full university degree at Newcastle (the ASc was roughly equivalent to a present-day A-level), Richardson followed in his brother Hugh's footsteps by going to King's College, Cambridge, which he entered in 1900 as a Minor Scholar. It is of interest that until 1871 entry to King's was limited to Anglicans—prior to 1861 only Etonians were eligible! When these conditions were relaxed, the Quakers were the first outside the Church of England to go there and the history of King's is rich in the names of such well known Quaker families as Cadbury, Clark, Rowntree and Fry[10]. Richardson wrote that at King's he 'continued to acquire information and skill from more brilliant teachers, notably in physics, from Professor J J Thomson and Dr G F C Searle'[11]. Thomson was of course one of the greatest physicists of all time, best known perhaps as the man who discovered the electron. Evidence of his vast influence as a teacher is provided by the fact that no fewer than 25 of his students were subsequently elected Fellows of the Royal Society. One of these was the New Zealand physicist, Ernest Rutherford (later Lord Rutherford of Nelson) popularly regarded as the first to split the atom. He received the Nobel Prize for Chemistry in 1908 and succeeded Thomson as Cavendish Professor of Physics in 1919. Collectively, Thomson and Rutherford trained 17 future Nobel laureates, an achievement without parallel in the academic world. Searle wrote several papers jointly with Thomson and was the author of numerous books on experimental physics. He served as a demonstrator in experimental physics in the Cavendish Laboratory from 1890 to 1935.

Figure 2.5 Lewis, aged 19, ready to leave for Cambridge University.

While at Cambridge, Richardson did not specialise in any one branch of science. He left in 1903 after obtaining a First in the Natural Science Tripos Part I, which included chemistry, zoology, physics, botany and geology; there was no separate paper on mathematics and the level of knowledge required in this subject for the physics paper was very elementary. One might well wonder why a student of his calibre did not choose to take Part II of the examination, which demands a much higher level in a more limited range of subjects. This may have been partly because he did not enjoy the best of health at Cambridge. He had measles in 1901 and was ill again for several weeks early in 1902. I suspect, however, that the main reason was that at that stage he had not decided what career to follow. He later wrote that 'In the midst of my Cambridge course on natural science I was hesitating whether to specialize on the physical or on the biological when someone, probably my friend W.E. Agar, told me that Helmholtz had been a medical doctor before he became a physicist. It thereupon occurred to me that Helmholtz had eaten the meal of life in the wrong order, and that I would like to spend the first half of my life under the strict discipline of physics, and afterwards to apply that training to researches on living things. I kept this programme a secret'[12].

One of Richardson's close friends at Cambridge was Stuart Garnett, the second son of William Garnett. There is a brief account in David Richardson's notebook of an occasion in 1901 when Lewis and Stuart set off in foggy weather for a walk together over the moors. They lost their way and had to spend the night in the open. Stuart was brilliant academically and a very fine athlete—as we shall see, this was typical of the Garnett family. In spite of his indifferent health, Lewis also engaged in sporting activities and became a good oarsman. He may have overdone this for it has been suggested that the heart condition which affected him more as he grew older originated from a strain dating from his Cambridge days.

Another friend, as already mentioned, was Wilfred Eade Agar, who took his Natural Science Tripos Part I at the same time as Richardson, but went on to take Part II in zoology the following year. He had a fine academic career, finishing up as Professor of Zoology in Melbourne University, Australia. He was elected to the Royal Society in 1921.

At King's, Richardson already gained a reputation for being somewhat unconventional. On one occasion, his bedmaker found him, in a state of nature, painting his room in colours he preferred to the official bursar's green. On another, a fellow student heard a tap at the window; it was Richardson paying a call having climbed up the drain-pipe in preference to the normal approach. Another student recalled that one of Richardson's inventions was a method of recording talk with friends in a sort of musical notation, with signs to indicate pace, pitch and quality of voice. In reminiscing on these incidents[13], a contemporary of Richardson's at King's wrote that 'he was much liked for his strong independence, his quaint but stimulating opinions and his very kind heart.... He already had the pioneering spirit which distinguished him in later life'.

CHAPTER 3

Early Career 1903–13

The pioneering spirit mentioned by Richardson's Cambridge contemporary was not apparent in his achievements in the years immediately following his graduation. We shall see in this chapter how he drifted from job to job, with little sense of continuity; his stated desire was to carry out research, but the subject of the research was at this stage of his career not of primary importance. Reflecting on this many years later, he wrote: 'At times I have been paid to do research: by the National Physical Laboratory, by Karl Pearson, by the Sunbeam Lamp Company, and by the Meteorological Office. I thoroughly enjoyed being paid; and indeed have often dreamt of large sums of money, and what I would do with them. Yet, paid or not, I have gone on with research; so money cannot have been the controlling motive'[1].

His first appointment was as a Student Assistant in the Metallurgy Division of the Physics Department at the National Physical Laboratory, where he worked under Dr H C M Carpenter (later Sir Harold Carpenter). As part of a project on the properties of different steel alloys, Richardson was put to work on testing the tensile strengths of steel plates. In a paper published in the *Proceedings of the Institution of Civil Engineers*, a Professor Unwin had referred to differences in tensile strength between thick and thin steel plates. Carpenter and Richardson reported to the Institution in 1903 that their observations had shown that part of these differences could be accounted for by differences in the composition of the plates. This first scientific note under Richardson's name[2] contained no indication of the originality which was to mark his later work. Richardson himself did not consider it to be of sufficient importance for inclusion in the bibliography of his writings that he prepared some 40 years later for the Royal Society.

In his historical account of the National Physical Laboratory, Pyatt

(1983) describes 1903 as a bumper year when the work of this two-year old institution really got into its stride and several new assistants were recruited. Even so, there were still fewer than 20 university graduates on the staff and Richardson must have known them all personally. He became especially friendly with the Director, R T (later Sir Richard) Glazebrook, a senior member of the Physics Department, J A Harker, and the Superintendent of the Observatory Department, Charles Chree. Chree and his staff were based at Kew Observatory, a few miles from the main building at Teddington; they all came under the Meteorological Office in 1912, when the observatories at both Kew and Eskdalemuir (about which we shall hear more later) were transferred from the NPL. The friendships which Richardson made at the NPL did not compensate for the nature of the work, which to him was rather dull. Nor was the salary of £50 per annum very attractive. It is not surprising therefore that after serving for just over a year he left to take up an appointment as a junior demonstrator at University College, Aberystwyth where he was paid £100 per annum.

At that time, Aberystwyth, which had always claimed to be a college of the people, had just achieved a genuine university standard[3]. Even so, the teaching staff in physics consisted solely of the Professor (D Morgan Lewis), the Assistant Lecturer and Demonstrator (G A Schott) and the Junior Demonstrator. The course was at the intermediate level but covered a very wide range of theory supplemented by practical work. The staff would clearly have had very little time for research and it is hardly surprising that Richardson did not publish any papers during his stay, which in any case lasted only a year—he resigned in October 1905. The only mention of Morgan Lewis in Richardson's notes relates to the reported disorderliness of his classes, apparently because his mathematics were incomprehensible and he had not a strong will. He had a closer relationship with Schott, through whom he became interested in mathematical physics. Schott was later elected Fellow of the Royal Society for his contributions to the mathematical theory of electromagnetic radiation. Richardson wrote some humorous poems about him but, as can be seen from the following example, they are of no great literary merit[4]:

> 'Well then' says Schott 'of course you see
> That this and that amount to three
> Times $y^2x + k^2g$.'

One of the students at Aberystwyth at this time was David Brunt, who knew Richardson and his work in later life when they were both serving in the Meteorological Office. He recorded that Richardson 'was keenly interested in designing a planimeter. He carried his model in his pocket,

Early Career 1903–13

and would take it out and modify it in any free moment. He remained in my memory as a quiet friendly man, always ready to help in any difficulty with an experiment'[5].

Only one other anecdote of Richardson's spell in Aberystwyth appears to have survived; it was told to me by Professor Mansel Davies of the chemistry department[6]. 'Like many young scientists he was a keen conscientious worker. Finding himself locked out from College, and more particularly from the physics laboratory, Richardson one evening climbed in through a window, only to be confronted by one of the senior members of the staff (almost certainly G A Schott) who berated him for his irregular entry. This, apparently, he took badly and flew into a temper, a most un-Quaker and un-pacifist reaction.' As we have seen, this was not the first time that Richardson had climbed in through a window—and it was not the last occasion on which he lost his temper.

Richardson only stayed at Aberystwyth for a year. He left in 1906 and was replaced by a laboratory assistant whose commencing salary was only 25 shillings a week; it seems that even in those days universities had to economise. His next move was into industry as a chemist with National Peat Industries Ltd, a small and very young company based in Newcastle upon Tyne. The company prospectus, issued in August 1905, stated that its object was 'to deal in a scientific way with peat and its various products and uses, and for that purpose to acquire, combine, and consolidate certain businesses, properties and interests, and to acquire peat mosses, patent rights, licenses, plant and machinery'[7]. It went on to state that the series of failures in the peat industry had been due to lack of scientific knowledge of the properties of peat and of engineering skill, mainly in the drying process. It then claimed that these difficulties had been overcome by processes patented by A Clark Kennedy and others. The solicitor for registering the company was Herbert J Richardson (apparently no relation). The total capital was £70 000 and the share capital open for public subscription was £37 500, although it was decided that in the first instance this would not be publicly advertised. One of the major initial shareholders was none other than L F Richardson: he started with £1000 and increased his holding to £1550 in May 1907. He was listed in the company books as a 'scientist'. His father also purchased some shares in 1906 and again in 1907.

Richardson's only reference to his first job in industry is a sentence in his biographical notes to the effect that he was employed from 1906 to 1907 by National Peat Industries Ltd 'whose managing director stole a large sum and fled abroad'. From this it has generally been deduced that Richardson had to leave the company because of its financial difficulties. Let us see, for example, what was said in the authoritative obituary written for the Royal Society by his contemporary, Ernest Gold (another student of J J Thomson who was for many years a prominent figure in the meteorological world). Gold wrote: 'one is inclined to speculate on the course of his life if the managing director had not defalcated—would Richardson perhaps have put Britain in the forefront of hydrological research?'[8]. The main difficulty in accepting this interpretation of Richardson's remark is that the defalcation did not apparently occur until several years after he had left the firm. The first indication in the company records that everything was not as it should be was a reference in the auditors' report to the shareholders in 1911 to 'defalcations of £6000 by Mr. Kennedy' (who had been appointed managing director shortly after the formation of the company) and of £1159 by two ex-officials. Moreover, there was a loss in excess of £515 000 on the sale of some of the firm's Irish property at Umeras, about forty miles from Dublin. Richardson's father promptly sold his shares, but Richardson himself hung on, hoping no doubt for better times. This proved to be unwise, for in 1913 the company was wound up; the business was ultimately taken over by Umeras Peat Ltd in 1916.

I am led to conclude that Richardson remembered the defalcation not because of its impact on his job but because he had been one of those who suffered financial loss. In any case, the importance of his work for National Peat Industries was not its effects on his finances; it was, in Gold's words, that 'having been brought face to face with the practical necessity of solving differential equations which were not formally soluble, he was led to study approximate methods'[8]. Let us pause for a moment to reflect on what this means.

We are all familiar with the concept of an average speed. If, for example, we drive 70 kilometres in an hour, our average speed was 70 kilometres per hour, usually written 70 km h^{-1}. But in the course of the hour our speed may have varied considerably; during one ten-minute period we may have averaged 80 km h^{-1}, during another, only 60 km h^{-1}. During one particular minute our average speed could have been as much as 100 km h^{-1}. If we continue to reduce the time interval over which we measure the distance travelled we arrive ultimately at the concept of instantaneous speed, the limit reached when the time interval and the distance travelled both become infinitesimal. The branch of mathematics dealing with such infinitesimals is the calculus, and the relationship between the variables is expressed in differential equations. If a differential equation is not too complicated, it can be used for calculating the exact values of the variables under given circumstances; the equation can, in mathematical language, be solved. For many equations, however, an exact solution is not possible and it becomes necessary to resort to approximate methods, one of which is to replace the infinitesimals by small but finite differences. Richardson was by no means the first to use these 'finite-difference' methods; his major contribution was to apply them to certain types of equations and to introduce a quick and accurate way of calculating the answer to the required degree of accuracy.

The problem which Richardson had to tackle for National Peat Industries was, in his own words, 'Given the annual rainfall, how must the drains [in a peat-moss] be cut in order to remove just the right amount of water?'[9]. By a simple experiment he confirmed that some already known equations could be applied with appropriate modifications, and he then went on to solve the equations approximately by a freehand graphic method. He reported the results at a meeting of the Royal Dublin Society in January 1908 and they were published by the Society in May. (In passing, one may well wonder why he submitted his paper to Dublin. He had already presented a fuller account of the purely mathematical aspects of his work to the Physical Society of London[10].)

Once his interest in approximate methods had been aroused, Richardson was not content solely with the use of graphical methods. His further work on arithmetical methods led to his first important scientific paper, published in 1910 by the Royal Society of London. Its full title was impressive: *The approximate arithmetical solution by finite differences of physical problems involving differential equations, with an application to the stresses in a masonry dam*[11]. In spite of support from Sir Richard Glazebrook, who had of course been Richardson's director during his spell at the NPL, the paper was not accepted by the Society without some difficulty. It had been written in two parts; part A was general while part B dealt with the specific case of a dam. In correspondence with a mathematical friend 40 years later, Richardson recalled that he had been advised by the Society's Secretary to act on comments made by two referees. 'Probably he had never read them', he wrote, 'for I was appalled to find that the first referee recommended that part A should be omitted and B condensed while the second referee recommended that B should be omitted and A condensed! Perceiving that even referees were not infallible, I decided to persist, and after a lot of bother to myself and to other referees I got both parts published.'[12]

In Gold's obituary of Richardson, the last half of the title of this paper is omitted. One might in fact wonder why Richardson selected the stresses in a masonry dam to demonstrate how his method could be applied. In the paper he gives two reasons for his choice, namely that it is obviously a problem of great practical interest, bearing in mind the potentially disastrous results of a dam failure, and that the equations involved are relatively simple. But there was another reason. In 1907 Karl Pearson and A F Campbell Pollard had published a paper *An experimental study of the stresses in masonry dams* in the Drapers Company

Research Memoirs and Richardson is listed as one of the two people who assisted in the project[13].

It has not been possible to ascertain exactly when Richardson left National Peat Industries Ltd, or what he did during the rest of 1907. His mother's diary does not help as she makes no mention of any employment in the year after he left Aberystwyth; she refers simply to 'a rather vacant year with disappointments'. Richardson himself wrote that this was the year in which his distant aim of applying his training 'to researches on living things' broke out temporarily when he sold his physics books 'in order to raise money to go and see Karl Pearson and learn about statistical proof'[14]. Elsewhere, as already mentioned, he listed Pearson as one of the people who had paid him to do research and there is yet another reference to this outstanding man in the section of Richardson's biographical notes dealing with his temperamental peculiarities. Here he describes how he avoided obvious leaders but, by exception, attached himself to Karl Pearson and later to Sir Napier Shaw because his programme temporarily overlapped theirs.

More will be said about Sir Napier Shaw (a student of James Clerk Maxwell) when we come to Richardson's work in the Meteorological Office. Let us concentrate for the moment on Karl Pearson, who will always be remembered as a pioneer in the application of statistics to biology. Richardson was by no means alone in having fallen under his influence, for a whole generation of young scientists had been deeply impressed by his *Grammar of Science*, published in 1892[15]. Pearson himself had been stimulated by Francis Galton, a scientist of Quaker origin who founded the school of eugenics—the study of how to produce fine offspring. Galton had become interested in this after reading the *Origin of Species* by his cousin, Charles Darwin; he was also a distinguished meteorologist, still remembered as the inventor of a method of mapping the weather and as the man who introduced the word 'anticyclone'. When Galton died in 1911, he left a substantial sum to the University of London for founding a chair of eugenics and Pearson became the first professor. A few years earlier the two men had collaborated with W F R Weldon in founding the journal *Biometrika* for statistical studies of biological problems. In its early days, this journal contained many contributions by biologists, but in recent years it has become almost entirely concerned with advances in statistics. Pearson himself wrote many papers for *Biometrika* and for other journals, for he had a prolific scientific output. His research was supported very handsomely by an annual grant from the Drapers Company and he was thereby enabled to hire mathematical assistants who helped in solving mathematical problems and in carrying out the calculations.

As far as I have been able to piece together the fragmentary evidence about Richardson's 'rather vacant year', it seems that he was employed as one of the mathematical assistants by Pearson for several months in the first half of 1907. This may seem surprising in view of his lack of formal training in mathematics, but Pearson had been favourably impressed by his report on the flow of water in peat, which Richardson had prepared in the form of a dissertation for an application for a fellowship at King's College. We have already mentioned one of Richardson's tasks while working for Pearson, namely assistance in the work on the stresses in masonry dams. The only other significant task appears to have been the preparation of an index for the first five volumes of *Biometrika*, which was published with the 1907 volume[16]. This index is a fairly straightforward compilation with no comments or explanations by the author. Quakers might be amused to find themselves listed in the section of the index which deals with 'Local races of man'! As with his joint paper with Dr Carpenter, Richardson made no mention of this index or of the paper on stresses in masonry dams in his own bibliography of publications.

Richardson was clearly very happy to be working with Pearson and was in no hurry to leave. In a letter to Pearson in April 1907 he wrote[17]: 'I would prefer to stay somewhat longer at University College if this is convenient to you, but understand that it is a question of ways and means and that you badly need someone to take part of the lecturing. Of course I would be delighted to take on the elementary lectures, if you would let me'. Richardson had been giving more thought to the further development of his graphic method for solving differential equations and to other practical applications. In the letter to Pearson he gave the first pointer to his future work on weather prediction with the remark that 'There should be applications to meteorology one would think'. He also wrote about his desire to be 'free to get on with the heredity work', but unfortunately did not elaborate on what he had in mind. We shall see how this interest in heredity, and in the closely related topic of eugenics, recurred spasmodically throughout his life.

By this time Richardson's dissertation had been returned from Cambridge and he knew that his submission for a fellowship had not been successful. It was later reported that although his work had been highly praised he had just failed to be elected in the face of stiff competition. The successful candidates were a historian, Esmé Wingfield-Stratford, and Richardson's old friend W E Agar. Ernest Gold speculated about how Richardson's life might have been different if he had not had to leave National Peat Industries Ltd, but the difference would surely have been even more dramatic if he had become a Fellow at King's; this would have enabled him to continue his research according to his own inclinations, with an assured income and the possibility of supplementing it by undertaking some teaching. At a party with Cambridge friends at about this time somebody asked Richardson

'Are you writing for a Fellowship?' 'Yes', he answered, 'something about peat'. To the questioner's astonished cry, 'Dried peat?' the only answer was 'Just peat'[18].

Richardson's next appointment was once again in the National Physical Laboratory, this time in the Metrology Division (this is not a mis-spelling of meteorology!). There is some doubt about the actual date of the appointment for although his name appears on the staff list in the NPL annual report for 1907, the 1908 report announces that he started work in 1908. However that may be, it seems that metrology had no greater appeal for him than metallurgy and he left after less than two years.

Richardson had continued his friendship with his Cambridge colleague, Stuart Garnett, and it was reputedly during an Easter holiday with him on the Isle of Wight that he met Stuart's sister, Dorothy. It seems more likely that they had already met as young children at the Richardson home in Newcastle and that it was on the Isle of Wight that they fell in love. They became engaged in April 1908 and in August joined in the annual vacation of the Garnett family in Switzerland. The Garnett boys were enthusiastic mountaineers and Kenneth later climbed the Matterhorn.

As mentioned earlier, Dorothy's parents William and Rebecca Garnett had known the Richardson family from the late 1880s. Prior to that, William Garnett had served for several years at the Cavendish Laboratory in Cambridge as assistant to the mathematical physicist James Clerk Maxwell. He was an ardent admirer of Maxwell and was joint author with Lewis Campbell of *The Life of James Clerk Maxwell* (1882). Maxwell was succeeded in 1879 by Lord Rayleigh, an equally eminent scientist. Garnett decided that this would be a suitable time to make a move, and in 1881 he accepted the chair of mathematics and physics at the new University College of Nottingham; his place at the Cavendish was taken jointly by R T Glazebrook and W Napier Shaw. Three years later Garnett moved further north to become Principal of the Durham College of Science in Newcastle, where he succeeded against great obstacles in having a new building constructed. He loved working with his hands and earned the respect of those engaged in the construction. In 1893 he was appointed secretary to the new Technical Education Board of the London County Council and he continued in this position until his retirement in 1915. He is especially remembered for his work in the development of secondary education and for his prominent role in the legal disputes about the educational responsibilities of the Council. One of his main interests was the education of artisans, and in 1919 he became involved in higher education for ex-servicemen. The Garnett College of Technology in Roehampton is named after him[19].

William and Rebecca Garnett had three sons and two daughters. The first son was named James Clerk Maxwell Garnett but was usually called Maxwell. His life has some interesting parallels with that of Lewis Richardson. He too was a brilliant mathematician and gained a First in Part II of the Mathematical Tripos at Trinity College, Cambridge in 1903. One of his first posts was with Karl Pearson who described the appointment in his 1904 report on the Drapers Company grant: 'I have been able to procure the services of two distinguished Cambridge wranglers, first, Mr Clerk Maxwell Garnett of Trinity .... They have assisted those working in the department, revised for press mathematical and other memoirs, and solved mathematical problems arising in the course of our statistical work'[20]. Maxwell Garnett spent the next 16 years in the world of education, first as an examiner at the Board of Education and then as Principal of the College of Technology, Manchester, where Lewis would also work for a time. Here he tried to expand work at university level but failed to obtain the support of the education committee. This difference of opinion led to his resignation in 1920 when he was appointed secretary of the League of Nations Union. After 18 years he resigned once again, this time because he could not agree to the Union being used for political propaganda. He has been described as a controversial figure with a strong personality and not easy to work with. Athough they had so much in common—mathematical ability, educational experience (including being principal of a technical college) and working for peace—it seems that he and Lewis never became close friends. Lewis felt more at ease in the company of Maxwell's wife, Margaret Poulton, who came from a wealthy family with Quaker connections.

We have already mentioned the second of the Garnett sons, Stuart. He was two years younger than Maxwell (and hence a year younger than Lewis) and also took a first in the Mathematical Tripos at Trinity. He was a keen sailor and in the early days of the First World War he assembled a crew from amongst his friends to man a yacht which he offered to the Admiralty as a minesweeper. He later joined the Royal Flying Corps and was killed in active service in 1916.

The youngest son, Kenneth, followed in the footsteps of his two brothers by gaining a first in the Mathematical Tripos at Trinity. He also rowed both for his college and for the university. He was a member of his brother's minesweeper crew at the beginning of the war and then took a commission in the Royal Field Artillery. He invented several mechanical devices to improve the efficiency of his guns and also worked on an instrument to enable an observer in a gun pit to measure the direction and speed of the wind at a height of several metres. In August 1916 he was hit in the neck by a piece of shrapnel and he died from his wounds a year later[21].

In the Garnett family the sons were all-important; daughters tended to be treated as second-class citizens. Although both Dorothy and her older sister, Hilda, had no lack of intellectual ability, neither went to university. They were educated at North London Collegiate School and then at Westfield College, a small establishment in Hampstead which had been founded in 1882 for the improvement of education of women, especially for teachers and social workers. Although the College is now part of London University, very few of the students in those days took the university entrance examination. Hilda never married and devoted most of her life to looking after her parents; her nieces and nephews looked on her as a much-loved archetypal spinster aunt. She and Dorothy were intimate friends as youngsters and kept in close touch throughout their lives. In her younger days at least Dorothy was something of a tomboy; she really enjoyed trying to keep up with her three talented athletic brothers. At the Garnett summer house which her father built at Seaview on the Isle of Wight, she took part enthusiastically in all the swimming, boating and other outdoor activities. It was here that both Maxwell and Stuart met their future wives. The Garnetts were just as religious as the Richardsons, initially as Congregationalists and later as Anglicans. Dorothy joined the Society of Friends after her marriage and became more Quakerly than many birthright Friends; she wore very simple clothes and always addressed Lewis as thee.

Figure 3.1 The Garnett holiday home, Horestone Point, Seaview, Isle of Wight. It was built by William Garnett in 1902, in spite of being warned that, owing to the instability of the land, it would stand for only 40 years. The house was washed away in 1942.

Figure 3.2 Lewis and Dorothy Richardson on their wedding day, 1909.

Dorothy and Lewis were married on 9 January 1909 at the Congregational Church in Lyndhurst Road, Hampstead. It was quite a social occasion and was written up in the *Court Journal*. More than 300 guests were present at the reception in the nearby Garnett home—one wonders how Lewis, with his preference for solitude, felt about all the pomp and ceremony. Even the Garnett house was not big enough to accommodate everybody; the members of Dorothy's Sunday school class who sat with their parents in the gallery of the church during the marriage were afterwards entertained to tea in the church hall. Dorothy wore an Empire gown of cream satin charmeuse, with a train embroidered in silver and arranged with Irish lace. Her veil, lent by an aunt, was arranged over a coronet of orange blossom. Little wonder that Lewis' father noted in his diary that the wedding was going to be rather formidable. It doubtless reminded him of his son Hugh's wedding in 1896 which was reported in the local press as having been witnessed by a large and fashionable audience who admired the very numerous and costly collection of presents.

For some years after their marriage Lewis and Dorothy continued to spend their holidays at the Garnett home at Seaview and would attend the local Church of England with Dorothy's parents. Lewis had a reputation for not suffering fools gladly, and Anglican ministers were not exempt from his scorn. On one occasion, whilst attending a church service where members of the Garnett family read the lessons, the minister had unfortunately chosen to illustrate his sermon with some comparisons from science—a field in which he was woefully ignorant. After several totally inappropriate remarks, such as 'the centrifugal force which drives everything inward to our hearts' Lewis could stand it no longer and with an almighty slap on his thigh and an exclamation 'The man's a fool' rose to his full six feet and stalked out of the church. A sequel was that the minister happened to be the Garnett's guest to lunch after the service. The two men spent an uncomfortable forty minutes glowering at each other after which Lewis felt that good manners called for an apology, however tentative. But he qualified it by adding 'I can recommend you some good books for beginners in the study of science'. To which Rebecca Garnett, who was supervising a table of twenty, was heard to say 'Cream in your coffee, Vicar?'[22].

After his second spell at the National Physical Laboratory, Richardson went back north again in October 1909, this time to serve as head of the chemical and physical laboratory of the Sunbeam Lamp Company in Gateshead[23]. This was an older established firm than National Peat Industries, having been founded in 1887 with a capital of £25 000 to manufacture incandescent electric lamps. The initial shareholders included William Garnett and a well known Gateshead figure, Robert Spence Watson, father-in-law of Lewis Richardson's brother Hugh. The managing director was John Wigham Edmundson whose daughter Gertrude Mary was married to another of Lewis' brothers, Lawrence. Lewis was therefore in a good position to find out all he wanted to know about the firm before accepting the position. He was sufficiently impressed with the company's fortunes to buy 500 of its £1 shares two months before taking up his new job; this time he was described in the shareholders list as 'chemist'. Earlier in the year his father had also taken £500 shares in the company 'in expectation of their requiring Lewis' help in making their new lamp[24].

Early Career 1903–13

Figure 3.3 The Richardson brothers at their parents' home, The Gables, on New Year's Day, 1911. *Standing* (left to right): Lewis, Gilbert and Lawrence. *Seated*: Arthur and Hugh.

As Gateshead is so close to Newcastle upon Tyne, Lewis was able to stay with his parents at The Gables while another house, also belonging to his father, was being prepared. It was here that Dorothy had the first of what was to prove to be a disastrous series of miscarriages—she had no less than seven. More than 30 years later they discovered the cause of the trouble after donating blood to the blood transfusion service during the Second World War. They were then told of their rhesus incompatibility. If only our understanding of the rhesus factor and our ability to deal with it had been developed while the Richardsons were young enough, they would doubtless have been able to have a family of their own, something which they strongly desired.

From the visitors' book which Lewis and Dorothy maintained from January 1909 onwards, it can be seen that their home in Newcastle was open to a constant stream of visitors. Prominent among them are members of the extended Richardson and Garnett families and friends of Lewis from his Bootham, Cambridge and NPL days (such as the Glazebrooks and the Harkers). There are records of parties given to employees of the Sunbeam Lamp Company, to people from the Adult School in which David Richardson was still active, and to members of the 5th Elswick troop of Boy Scouts which Lewis had started. There are tantalising references to visits by Lewis to Vienna in November 1909 and to New York in August 1910. Unfortunately, this visitors' book appears to be the only surviving account of Lewis' social life from this period.

Figure 3.4 Lewis and Dorothy Richardson at The Gables, 1911.

Information about his activities at work is equally scant. Two patents for improvements to lamp filaments were taken out in his name in June 1911 but they were not followed up and copies are not available in the Patent Office. Evidence of his continuing sense of humour is provided by a set of cartoons which Richardson drew about life in the Moonbeam Lamp Company. The cartoons, some of which are reproduced in figure 3.5, do not reveal much artistic ability, although the sketch of three men drawing tungsten wire does give a fair impression of the human body under stress: according to the caption, the wire itself is too fine to be seen. Richardson let his imagination run riot in his description of the manager's office:

> This is on the top floor of the factory. The telephones, and the periscopes which look into the various floors of the factory are attached to the central pillar H. The chair is marked A. The desk CCCC is ring-shaped. It is set revolving by pressing the pedal M. By this means a large number of papers can be laid out together. For signing letters it is set agoing at slow speed and the letters are fed in by a boy at one side. When desiring to work in private the manager presses another pedal and a cylindrical aluminium shield (shown dotted in the figure) descends from the ceiling and completely encloses the ring shaped desk. Visitors are shown into the part of the floor marked V and, though it is not generally known, there is a pedal (marked D) which if pressed causes the floor of the visitors' compartment to descend until it reaches the ground level, and then to tip up thus conveying a hint that it is time to leave. We are glad to hear, however, that it has been used but once. When desiring refreshment he presses a fourth pedal. The chair A then rises on the end of a ram passes

through an opening in the roof and the weary manager finds himself in a little roof garden from which the surrounding country may be seen through the tall spikes of larkspur and hollyhocks with which it is bordered. A fountain plays in the centre in summer.

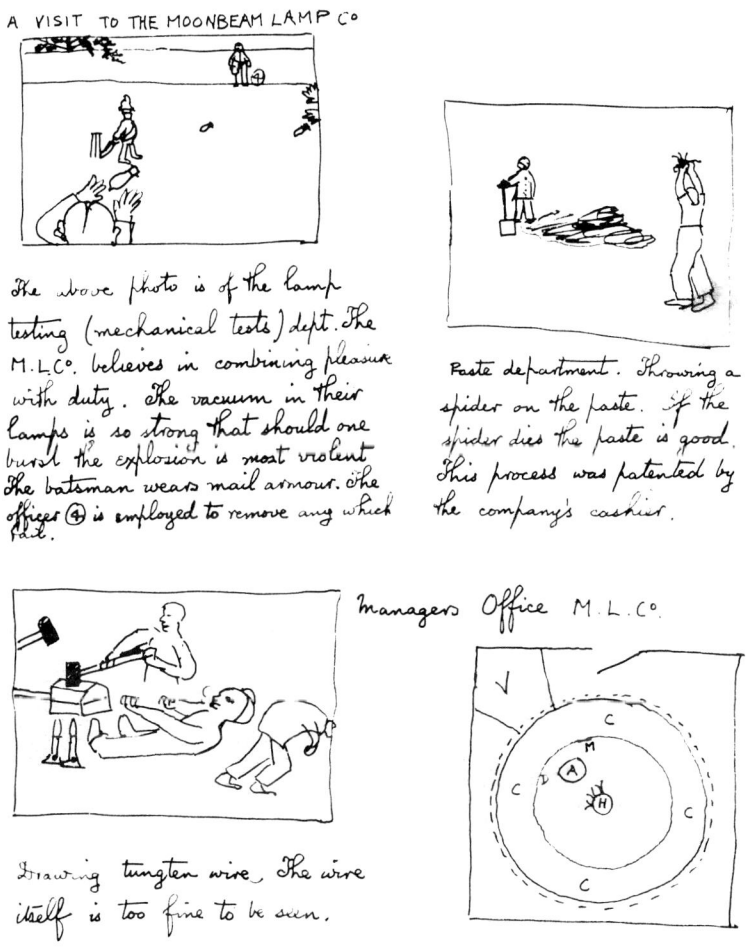

Figure 3.5 A visit to the Moonbeam Lamp Company (about 1912, when Richardson was working with the Sunbeam Lamp Co).

It is not known whether these cartoons were simply for his own amusement or whether they were reproduced in any way, for example in a house magazine.

In its early years, the Sunbeam Lamp Company had been very successful. They were the first to make high wattage lamps in the 300–1000 W range and pioneered the production of lamps with the relatively complex internal structure needed to carry the heavy currents involved. In this they were involved in a legal dispute over an Ediswan patent for a filament lamp but the court ruled that their elements were too large to be classified as filaments. Over the years their own patents ran out and they were faced with increasingly fierce competition. It seems that there was insufficient capital to exploit the new techniques which Richardson and his colleagues were developing. Furthermore, the managing director was not a very tough businessman: he was reputedly too kind-hearted, especially with the female employees who were able to twist him round their little fingers[25]. The optimism displayed by Richardson and his father in buying shares in the Company proved to be unwise and in 1912 it was decided that the firm should be wound up voluntarily. After three years of congenial employment—the longest period that he had ever stayed in one post—Richardson was given notice. Before leaving, he was sent round the country for several weeks arranging for the sale of the Company scientific equipment—and looking for a good permanent job.

His natural inclination was towards a university research post, preferably in his beloved Cambridge. His former professor, J J Thomson, responded favourably to an enquiry about the possibility of doing research at the Cavendish Laboratory[26], but it can only be presumed that Richardson was unable to obtain the necessary financial support, for in the end he accepted a position as demonstrator and lecturer in physics at the Municipal School of Technology in Manchester. Although this college is now part of the university—its present title is the University of Manchester Institute of Science and Technology (UMIST)—in those days it came under the local government authority; less than a tenth of the students were reading full-time for a degree, the remainder being only in attendance for evening classes[27]. We have already seen how Maxwell Garnett would later become involved in the process of raising the academic level of the school during his spell as Principal when it became known as the College of Technology. As at Aberystwyth, Richardson found himself with a heavy teaching load which left him little time for research.

From a notebook entitled *Reflections on discipline in schools and colleges* which was found amongst his papers after his death, it is clear that Richardson had difficulty in maintaining discipline in his classes at Manchester[28]. This led him to wonder how some teachers become more successful than others in this respect. 'Why' he wrote, 'was Jim Clark so much "fooled" at Bootham when he was science master? Nearly every boy liked him, for he was gentle and kindly in private life, even quaintly entertaining sometimes, and a good swimmer, skater and slider. The chief attraction in fooling Jim was first that it was possible to do quite bold facetious things without him dropping on you, second that his own behaviour was a little erratic and so had a pleasing element of surprise in it.' Other masters at Bootham kept perfect order yet one was grave, one inspiring, one hasty tempered and one crude and apt to tease. 'But the general characteristic of their discipline was that it was prompt and orderly, rather than severe. The same punishment for the same offence, inflicted with almost mechanical precision.' Richardson then analysed the situation which he had experienced at the Durham College of Science.

> The same students went to Stroud for Physics and Bedson for Chemistry. Both lectures were experimentally well illustrated, and there was little or nothing to choose between them in point of interest, as I saw it. Perhaps Bedson was a trifle more cogent. But from the point of view of order there was no comparison. There was once a disturbance in Bedson's class,—so faint that some of us hardly noticed it; but Bedson pounced on it and said the most stinging things: 'Persons, I will not call them gentlemen, in the 3rd row, who apparently do not know how to conduct themselves in a lecture. Unless they can regulate their conduct, I shall have to ask them to withdraw from the room', or something to that effect, and this was the only occasion on which I remember any disturbance. In Stroud's class however they used to stamp, play toy musical instruments. On one occasion a man played a concertina and another whistled in tune with it for several minutes while Stroud stood perfectly silent and motionless apparently looking in vain for the offender. The seats had solid backs so that he had no chance of seeing the concertina. It was mostly the stupid students who made a row. This contrast between Stroud and Bedson shows that interest alone will not keep order.

At Cambridge Richardson had not observed any disorder. 'It would have been bad form to have made a row while so distinguished a man as J J Thomson was lecturing.'

Richardson then recorded his own experiences as a lecturer in Manchester. At his first lecture to second year engineering students, 'they were as quiet as mice but said they hadn't understood it'. Trouble started during the third lecture when, after some initial disturbance which he thought he had dealt with successfully, 'the whole class took to talking, and some to humming, making farmyard noises etc. I could not locate them exactly, but made sarcastic remarks about pigs etc, which did not have much effect'. Later on

> some took to throwing shot and chalk when I was writing on the blackboard; and, finding that I could not spot the offenders, this practice increased reaching a climax on the last lecture of the autumn term when nearly everybody came in with two or three pieces of chalk. This last hour before this Christmas vac. is well known to be the worst of the session for disorder. I turned out about six people. Gee says that 'on seeing the chalk he would have left the lecture table and moved about among the class

dictating'. I took this affair too much as a joke and its bad effects were seen next term. In lent term throwing of shot recommenced. I discussed how to deal with it with several lecturers. Barrett never has any trouble of this kind now. Prescott said he had on occasion asked who'd done it. And once when no one owned up he asked every man in the class in turn whether he had done it. Then if no one owns up someone is a liar and you say so and rub it in. Prescott says be severe, be as severe as you like.

I did not much like this scheme but it seemed the only way; so next time things were thrown or noises deliberately made I asked everyone. Clare and Horrocks and Whitrow owned up. I turned them out and suspended the first two.

Next lecture we were getting on happily, Macorra, Schiele and some others arrived late, immediately clicking began. I said I hoped I shouldn't have to turn anybody out. It recurred. I asked each in turn who'd done it. I said someone added to his other ungentlemanly qualities that of being a liar; that I knew pretty well who'd done it and if it recurred certain persons would have to go out. It recurred twice. The noise came from near where Macorra and Clare were sitting so I said 'Macorra and Clare you will go out. If you think it unjust you may appeal to the Professor but you must go now. You are also suspended'.

Circumstantial evidence only. Macorra came to me afterwards and said it wasn't him. Clare had gone home. I wouldn't believe him (Fool, he spoke the truth and Clare had done it).

A few weeks later, Richardson wrote

This day I was tired and overstrained. The air was cold and raw and several students sneezed. Then more sneezed. Then Clare sneezed ostentatiously. Now the distinction between a natural and a forced sneeze is rather a difficult one. Indeed it is questionable if many people can force a sneeze. I think they must have taken snuff. So if I had only been sensible I should have smiled and said nothing. People get tired of sneezing. But like a fool I said, 'Clare if you do it on *purpose* you'll have to go out'. The class, realizing that they could plead that they couldn't help it, sneezed the more vigorously. I got on with the lecture punctuated by sneezes, for about three quarters of an hour. Then Macorra who was just in front sneezed as loud as possible. I told him to go out. This was unjust for Duguid, Leonido Watts etc. had all been as bad. He said 'I won't go out, I can't help it'. In a moment of madness I seized him and began dragging him towards the door. He hung on to the desk dragging it off the raised step with a loud bang. When about half way to the door, I came to my senses, let go of Macorra, helped him back to his desk and papers, saying angrily 'I'll report you all the same'. Then looking at the class I said 'I think I ought to apologise to Mr. Macorra for handling him; a lecturer ought not to do that'. Kilner said 'Oh it doesn't matter'. The rest sat quiet. Macorra complained 'I didn't do it on purpose, I can't sneeze on purpose if I try'.

'Alright' I said, 'I'll take your word for it, and we'll say no more about

Early Career 1903–13

it.' Macorra blushed and looked down at his book. After that there was much less sneezing. Degiarde still kept it up. Degiarde is a boxer and would no doubt have enjoyed a fight.

I made two serious mistakes here. First to say anything about sneezing. It is remarkable that I am never troubled with stamping or humming now, although students usually employ these methods of annoying a lecturer, and that I have never said anything about either of these. Also eating in class has gone out of fashion and I've never said anything about that. The second mistake was to threaten Clare and Macorra when lots of others were just or nearly as bad.

In April 1913 Richardson went to hear Rutherford (then a professor at Manchester University) lecture on the electrical state of the atmosphere. He wrote 'A clear orderly account of experimental work, classified and illustrated. No mathematical connections beyond such as current = charge per time etc. Spoken extempore in a clear flowing manner with a certain amount of errr-r. Afterwards Lustgarten and I talking to him about university lectures. Rutherford agreeing technical terms too long. He said too much teaching bad. Men must learn to think for themselves. Knowledge no use unless man can apply it. "If teaching what is wanted we might as well hire a lot of schoolmasters at £100 a year to give them definitions etc. over and over again until they know them. I mean it is absolutely wasting my time to do that sort of thing." I brought up Max's suggestion that men might read text book in lecture time. Rutherford said "Well, my idea of course of a lecture is to get them interested"'.

His final entry while at Manchester was 'Have had no trouble worth mentioning for several weeks now'. Evidently his enquiries and introspection had produced results.

The notebook concludes with a discussion on the art of lecturing based on questions which Richardson posed to a number of people. Frank Knight, who had been placed ahead of Richardson at Bootham in the matriculation examination, commented that 'an air of supreme indifference, coupled with the suggestion of heavy artillery in reserve, is the most effective. If your voice suggests irritation, the boys are pleased'. Richardson's brother Hugh, who it will be recalled had preceded Knight as the science master at Bootham, said that a science lecture should give 'enough idea of subject to enable (the students) to get on in the laboratory'. Maxwell Garnett remarked that at St Paul's, the senior mathematics students each read their book and worked the examples. If any difficulty arose they took it to the master. He thought that this method could be used at technical colleges.

These extracts from Richardson's notebook, written at the age of 31, show him to be conscientious, introspective, interested in psychology — and not particularly gifted as a teacher. Although he subsequently spent 20 years in the education world, it seems that he never mastered the art of lecturing to students at technical colleges and it is probable that he would not have been more successful in this respect had he achieved his ambition of being a university professor.

Richardson's notebook on discipline in schools also reflects his continued interest in eugenics. The entry for 14 March 1913 begins: 'This day I was tired and overstrained as a result of decision about Eugenics Education Society ...'. A few months earlier he had in fact published a note in the Society's *Eugenics Review* under the title *The measurement of mental 'Nature' and the study of adopted children*, which is also missing from the bibliography of his publications which he prepared for the Royal Society. It related to the question of the extent to which the intelligence of children of different social classes depends respectively on 'nature' and on 'nurture'. In those days, the customary definition of nature was 'that which is contained in the fertilized ovum' or 'that which is inborn'[29]. Richardson argued that from the point of view of scientific measurements a more satisfactory definition would be 'the growth attained under standard conditions of nurture and personal effort'. He supported the belief of Francis Galton that for each individual 'there is a limit to his general ability, beyond which no effort, learning, or enthusiasm will carry him' and he therefore proposed as a measure of nature 'the growth attained when conditions of environment and personal effort have been so favourable, throughout the whole time of growth, that a limit is reached'. In other words, nature could be considered as 'saturated growth'. As an example of saturating conditions he considered the case of the sons of an eminent man who 'usually have the benefit of his experience, and in addition a good deal of stimulus to emulate him; so that if they show no capacity for their father's work it is a pretty clear proof that their nature differs from him'. From this Richardson concluded that 'difference between brothers is a surer indication of the strength of heredity than is resemblance'. Such arguments led him to propose that members of the Eugenics Education Society should co-operate in a study of a hundred cases where children of poor parents have been adopted 'when newly born into well-to-do and well-educated families'. If these children 'turn out markedly different from the birthright members of those families', this would support the presumption that 'the dullness, of whichever is the duller, is a saturated growth'. If, on the other hand, 'they all turn out much alike there is no proof that growth is saturated for any of them'. On the assumption that the conditions are alike for all the members of one family, Richardson suggested that such a study would give a useful comparison of native worth.

The following issue of the *Eugenics Review* contained an appeal for 'anyone in a position to furnish information about an adopted person'[30] to communicate with Richardson. His intention was to invite

Early Career 1903-13

those who responded—provided they were sufficiently numerous—to complete a questionnaire about each adopted person. I have not been able to ascertain the nature of the response to this appeal nor whether Richardson pursued the investigation.

The question of the relative importance of genetic and environmental factors in determining a child's intelligence is still the subject of much research. All the reliable data from studies based on the method proposed by Richardson were reviewed by Harry Munsinger (1975). To him, the results showed that 'The adopted parents' home environment only has a modest effect on their adopted children's intellectual growth, while the heredity and environment of the biological parents have a strong effect on their own children's intellectual growth'. Stated otherwise, 'The available data suggest strongly that under existing circumstances, heredity is much more important than environment in producing individual differences in IQ'[31]. The eminent British psychologist, H J Eysenck, surveyed all the available evidence in 1979 and concluded that when intelligence is measured by IQ tests, genetic factors account for about 80 per cent of the total variance and environmental factors for the remaining 20 per cent; both figures are subject to a sizeable standard error. Richardson would doubtless have been interested in another of Eysenck's conclusions, namely that a high status family environment, compared to a low status one, can raise a person's IQ by about 15 IQ points.

Evidence of another aspect of Richardson's character, his inventiveness, was brought to light in a recent article by Professor Henry Charnock. Shortly before he left the Sunbeam Lamp Company, Richardson had filed a British patent (9423 of 1912) for an *Apparatus for warning a ship of its Approach to large objects in a fog*. Charnock conjectures that this was almost certainly inspired by the sinking of the *Titanic* after collision with an iceberg on 15 April 1912. He commends Richardson for having got his proposal to the stage of a provisional specification just five days after the disaster and suggests that he must have temporarily neglected his duties with his employers[32].

The complete specification, filed on 18 October 1912, by which time Richardson had moved to Manchester, described how a beam of sound could be sent out over the surface of the sea from a source in the focus of a parabolic reflector and how the echo from the object could be detected by a sensitive receiver. The time elapsed would give a measure of the object's distance. This is of course the principle of echo-sounding, and in itself was not new—Charnock gives information about experiments dating back to the 1830s. Richardson's contribution was to emphasise the advantage of using sound of a very short wavelength, in other words of a very high frequency, so as to obtain a well concentrated beam. Richardson pointed out, however, that the amount of sound absorbed by the air increases as the wavelength decreases and concluded that a suitable wavelength would be about 5 cm, which corresponds to a very high-pitched whistle. During the summer of 1912, while travelling around disposing of the scientific property of the Sunbeam Lamp Company, Richardson tried without success to find somebody who would finance the commercial exploitation of his ideas. In December he filed another patent (No 11125) for *Apparatus for warning a ship at sea of its nearness to large objects wholly or partly under water*. This was an adaptation of the method covered by his earlier patent to make it work under water.

Richardson's experiments with echo-sounders are well remembered by both relatives and friends. The Garnetts talk of an occasion during a holiday on the Isle of Wight when Lewis was seen on a rowing boat—with Dorothy at the oars—pointing an open umbrella towards the shore and blowing an old-fashioned tram horn. He was apparently trying to record the echoes. In a biographical memoir about his father, Stephen Richardson tells of an incident which happened many years later[33]:

> With his intense curiosity and love of inquiry he would often pounce on stray remarks. Once I was talking of the danger of collision between ships in fog (before the invention of radar). The problem intrigued him, and he suggested using high-pitched directed sounds and their echoes to spot ships. That evening we attached a high-pitched whistle to the focal point of sound of a large golf umbrella and set off looking for buildings about the size of ships. On finding such a building, we pointed the umbrella in varying directions (including that of the building), blowing the whistle and timing any echoes with a stop watch. An amused and puzzled crowd gathered—including a policeman! My father's answers to questions rapidly turned their mirth and heckling to interest and attention.

Another friend of the Richardson family remembers seeing 'a whistle attached to a rubber tube piercing an extended umbrella' during a visit to their summer cottage in Scotland at the end of the 1930s[34].

Richardson's lack of success in having his patents further developed may be partially explained by a letter sent to him in October 1914 by his former assistant scoutmaster, Maurice Wilson. He wrote: 'It is far more complicated than it appeared at first hearing and it would be difficult to get a firm to take it up—and this after all would be the only way to get a working model going'. Wilson felt that the underwater version offered the best hope and suggested that the Admiralty would welcome such an invention and carry out some experiments. 'It would' he wrote, 'be extraordinarily useful to the country just now for our fleet to be able to detect the presence of other boats and particularly submarines.... Perhaps you feel you would not like to lend your aid to the method of warfare'[35]. This last sentence was a considerable understatement of

Richardson's attitude to the military use of his ideas, and it may well have been the possibility of such applications that persuaded him not to pursue the matter further.

His ideas for echo-sounding in fact proved to be on the right lines. The main outstanding difficulty was how to generate a loud enough sound to produce a detectable echo. This problem was solved by an American, R A Fessenden, just before the First World War. In view of the obvious application of the technique to submarine detection, considerable efforts were made during the war to develop a practical system, one of the scientists involved being J J Thomson[36].

CHAPTER 4

Eskdalemuir and France 1913–19

The Richardsons never set up a home of their own in Manchester. They stayed at Dalton Hall, a student residence founded by Manchester Quakers and named after the great scientist Hugh Dalton, who had been a member of their Meeting for more than 40 years. In any case, Lewis' appointment there proved to be of very short duration. After only a few months, he applied for the post of Superintendent at Eskdalemuir Observatory, which had become vacant owing to the ill health of the incumbent George W Walker. Although the observatory was administered by the Meteorological Office, the responsibility for nominating the Superintendent rested with the Gassiot Committee of the Royal Society, the members of which included two of Richardson's former colleagues at the National Physical Laboratory, Dr Charles Chree and Sir Richard Glazebrook. There were 16 applicants for the post and Richardson was one of the 4 invited to an interview. At their meeting in February 1913 the Gassiot Committee resolved that 'Lewis Fry Richardson be nominated to the Meteorological Committee for the appointment of Superintendent of Eskdalemuir Observatory'. His initial salary was £350, nearly three times higher than what he had been earning at Manchester. In addition to this very handsome sum, the Superintendent was provided with a fine house in the grounds of the observatory. We can imagine how delighted Richardson and his relatives were when his appointment was announced.

Although he was not due at Eskdalemuir until August 1913, Richardson left Manchester before the end of the academic year. After a short period of training at Kew Observatory, he made an extensive tour of geomagnetic observatories in other parts of Britain and in Ireland, France, Germany and Poland to gain further experience and to carry out comparisons between their magnetic instruments. In addition to his travelling expenses, he received the handsome sum of £1 per day.

Eskdalemuir Observatory is located in a very remote part of the Scottish lowlands in the county of Dumfriesshire. The site was chosen very carefully in 1904 when it was decided to find a better place for the magnetic observations hitherto made at Kew Observatory, where interference from the new extension to the tramway system was causing great problems. The new site had to be as free as possible from such interference, not only for the present but also for the foreseeable future. At the same time it had to be reasonably accessible and not too inhospitable for the staff. In an article celebrating the first fifty years of the Observatory, Mr M J Blackwell (1950) wrote that the Royal Society Committee responsible for finding the best site reputedly did so 'by placing a coin on a map so as not to intersect any towns, industrial areas, or railways within a radius of some ten miles. Past and present staff of the Observatory can bear witness to the thoroughness with which they performed their task'. In an earlier article by a former Superintendent of the Observatory (J Crichton 1950), it is suggested that 'the isolation has its advantages—did not a prophet in the Old Testament find advantage in withdrawing occasionally from political life and musing under a juniper bush?'. For somebody like Richardson who loved solitude, Eskdalemuir had an attraction far beyond its freedom from man-made electrical disturbances. He wrote of this in his biographical notes[1]:

> Another temperamental peculiarity had a great influence on my career. Any individual, if left solitary, may or may not experience a distressing emotion called loneliness. I hardly know what loneliness feels like: when solitary, I am usually serene; when in a crowd, I am often embarrassed. I am now describing this trait, not praising it. As a child I played much alone. Four of my six brothers and sisters showed tendencies to solitary lives. I have always had an inner programme, often secret, often frustrated, and in retrospect sometimes silly, sometimes wise. After I had become a professional scientist and when I met obvious leaders, my usual tendency was to shy away from them. For example I met Rutherford at Manchester in 1912 to '13 at a time when many other young research workers seemed to be thinking: 'Here's a great man! How I admire him! Let me try to ingratiate myself with him, learn from him, move in his orbit, shine with the reflection of his glory'. On the contrary I thought: 'Here's a powerful man: He is a wonderfully interesting psychological exhibit. But he despises me. His researches are doubtless important; but there is not

the least risk of their being neglected. Moreover I do not share his enthusiasms. How could I attend to my inner programme in the neighbourhood of anyone so domineering? Let me go somewhere where there are fewer people buzzing around.

There were certainly fewer people buzzing around at Eskdalemuir, but the local inhabitants made up in hospitality what they lacked in numbers. Even before Richardson took up his duties, a garden party was arranged by the Acting Superintendent, J A Harker (his old NPL friend), to introduce him to his future neighbours. The party was held on 21 June 1913 during Richardson's first visit to Eskdalemuir for the magnetic instrument comparisons. In a photograph taken of this splendid ceremony—it must have been a major event in the life of the local people—we can pick out Richardson and Dorothy, and also her parents William and Rebecca Garnett, who had travelled up from London for the occasion. The Director of the Meteorological Office, Sir Napier Shaw, and his wife Sarah can also be identified.

Figure 4.1 Eskdalemuir Observatory (about 1910). The landscape is now less bare, thanks to plantations of the Forestry Commission.

Shaw has been described, with considerable justification, as the father of British meteorology. We have already noted that Richardson listed him, along with Karl Pearson, as a scientist to whom he became attached because their programmes temporarily overlapped. Shaw was educated at Emmanuel College, Cambridge, where he graduated in 1874 as a pupil of Clerk Maxwell, for whom he had the greatest admiration. He spent most of the next 25 years in Cambridge, where his appointments included that of demonstrator in the Cavendish Laboratory and university lecturer in experimental physics. He formed lifelong friendships with many of his Cambridge colleagues, including J J Thomson, R T Glazebrook and William Garnett. In 1890 he spent some months in Berlin working under Hermann von Helmholtz, the great German physicist who made some important contributions to theoretical meteorology—the man who, in Richardson's view, had eaten the meal of life in the wrong order. At that time, the Meteorological Office was administered for the government by the Meteorological Council, a body appointed by the Royal Society for the purpose. Shaw became a member of the Council in 1897 and three years later he was nominated Secretary, a post which necessitated full-time attendance at the Meteorological Office in London. He thus left Cambridge at about the same time that Richardson arrived there to commence his studies. It is not clear when they first met but it seems likely that William Garnett brought them together; there is in any case no doubt about the profound influence which Shaw had on Richardson's meteorological career and achievements.

Figure 4.2 The garden party at Eskdalemuir Observatory to welcome the new Superintendent, 21 June 1913. *Seated* (left to right): William Garnett (with beard and spectacles), Mrs Shaw, Rebecca Garnett, Napier Shaw. *Standing*: Lewis Richardson (fourth from left), Dorothy Richardson, J A Harker.

Up to the time of his appointment as Superintendent at Eskdalemuir, Richardson had not displayed any marked interest in meteorology. The first indication of a serious move in this direction was his recognition in 1911 that his methods for the approximate solution of differential equations could have applications to weather forecasting. He later wrote that this realisation first took shape as the fantasy which he was ultimately to include in his book on weather prediction by numerical process, to which we shall return in due course. During his short stay in Manchester, he may well have discussed his ideas with Sir Arthur Schuster, honorary professor of physics at the University there. Schuster was greatly interested in meteorology and for many years served as a very influential member of the Meteorological Committee. But it was not until he arrived at Eskdalemuir that Richardson gave serious attention to the problem of weather prediction; one of the main attractions of the job may well have been that he would have time for such research—with the permission and encouragement of Shaw.

Figure 4.3 Richardson's De Dion Bouton car at the entrance to Eskdalemuir Observatory, during a visit by Dorothy's relatives, the Hunt Cookes (August 1913). This was his only car.

Shaw's support for such an activity at Eskdalemuir may at first sight be surprising, for the primary purpose of the Observatory was, as we have seen, to record the Earth's magnetic field. The history of geophysical observatories shows, however, that once they have been established their programme soon expands beyond that originally planned, and Eskdalemuir was no exception. Within two years of the first magnetic observations in 1908, instruments were installed for recording earthquakes—the leading seismologist of the day, Prince Boris Galitzin of St Petersburg, came to Eskdalemuir in person to supervise the work. At about the same time, the Observatory was designated as a first-order weather reporting station and a programme of observations in atmospheric electricity was introduced.

Shaw may have been afraid that Richardson's desire to follow up his ideas on weather prediction might lead him to neglect other possibilities of research for which Eskdalemuir was better equipped. In November 1913 he gave the new superintendent some fatherly advice in a six-page personal letter[2]. He explained that although he was not in general enthusiastic about 'the continuous recording of meteorological phenomena with the routine of tabulation etc.', he felt the work of this nature at Eskdalemuir was of particular importance and should provide the basis of much co-operative research:

> Whoever is in charge has before him day by day, week by week, a continuous representation of observations of magnetism, meteorology, seismology like the pictures of a kinematograph. It is almost inconceivable that anyone with the capacity for appreciation that we all believe ourselves to possess, should have nothing to say about the panorama that is of interest and importance to his colleagues.... I cannot imagine anybody with insight and intelligence who, at the end of a year of that kind of experience, has seen nothing that arrests his attention and finds occupation for his thoughts....
>
> It is quite possible in your case that, with your want of acquaintance with the subjects, you may suppose that anything which you may notice must already have been noticed by somebody else. Do not be misled by that. Nobody has ever yet seen these things with your eyes. Even if the events have occurred before, your eyes and the brains behind them bring a new element into the situation. Of course one makes mistakes. I once spent a long time trying to find out the effect of the tides upon the barometric height at Falmouth, preparatory to solving the problem whether, if one had a ship off Falmouth going up and down with the tide, the shore barometer or the ship's barometer would show the effect of the motion. I found incidentally a marked semidiurnal variation of the barometer at Falmouth that was new to me and only found out afterwards that it is common to every barometer all over the world, that it was, in fact, common knowledge all over the Office. I did not publish that research but I learned something from it.
>
> I have written at this length because the observational subjects which are followed at Eskdalemuir are quite exceptional among all scientific subjects in being subjects of co-operative research. An astronomer with his own telescope can do many things on his own. You will find that you can do nothing in the way of research without using somebody else's observations as well as your own.... Dr. Schuster once suggested at a British Association meeting that the best thing for meteorology would be for everybody to stop observing for 5 years and set to work at making out what the observations meant.... The proper course is for the scientific staff to get a turn off from time to time in order to work out the ideas which the daily routine cannot fail to suggest, and that is the object towards which I try to steer. Meanwhile, whatever opportunities there may be for research, please do not give up thinking about what passes under your eyes and *keep a notebook*.

Richardson had little difficulty in following the above advice from

Shaw. He enjoyed making observations and was always alert for anything unusual that they might reveal. He loved using scientific instruments and discussing their merits and shortcomings with his colleagues. Furthermore, since his days at Bootham he had kept meticulous notes of his thoughts and actions. There was in fact only one aspect of the work at Eskdalemuir which did not appeal to him—the administration. He described this as a 'rather dreary' task which had to be performed with 'diligence and foresight'. 'Some scientists', he wrote, 'find their true vocation in administration. It is fortunate that there are such people. Personally, being deficient in bossiness, I get little or no pleasure from managing others'[3]. In spite of this, he maintained good relations with the rest of the staff. One of his assistants, P N Skelton, later commented on Richardson's good influence on his younger colleagues and on the way in which he gave them encouragement in their studies and hobbies[4]. They all held him in great respect and only an innocent newcomer would have dared to enter the Superintendent's office when the presence of a hat on the door handle indicated that he was at work and not to be disturbed.

Richardson's impatience with bureaucratic delays shows up in some correspondence with the Office of Works in Edinburgh. In November 1915 he suggested to them that some blinds should be placed over the library skylights as the 'Zeppelins might pay us a visit on their way to the new munition factory at Gretna' (about 20 miles south of Eskdalemuir). The Office of Works was reluctant to sanction the expenditure for blinds and asked if it would be sufficient to stipple the glass or fix cardboard shades over the lights? Richardson replied that neither would be suitable and added: 'I gather from the tone of his letter that the possibility of a raid is regarded by the Board as very remote and as the Board is responsible for the building, I may set my mind at rest on that score. Our staff will I suppose, have to look out for themselves'. Two weeks later the Office of Works agreed to supply the blinds[5].

The work at Eskdalemuir offered Richardson considerable scope for applying his inventive skills. In 1915 he presented a report to the British Association for the Advancement of Science on standardising a Milne–Shaw seismograph, one of the instruments in use at Eskdalemuir for recording earthquakes. In July of the same year he made a preliminary suggestion as to how thunderstorms might be located by observing the noises which can be heard on telephones. This idea was received with interest at headquarters, as experiments were already under way at Farnborough on the detection of thunderstorms by 'wireless' (this was the beginning of the radio direction-finding technique, still in use today, for locating distant thunderstorms). Encouraged by this favourable response, Richardson continued his observations and confirmed that some of the click-like noises were indeed due to flashes of lightning. For a time these observations were reported in code to London by the addition of two five-figure groups to the regular weather reports[6]. Within a few months, Richardson had refined his ideas and was able to make a detailed proposal on the use of telephones for detecting distant thunderstorms. A paper written jointly by Richardson and A G Shrimpton was submitted by Shaw to the Advisory Committee for Aeronautics in December 1915. It included an analysis of how the relative strength of clicks heard from two telephone lines running in different directions could provide a measure of the azimuth of a lightning flash. There was some further work on this proposal in 1916 but it was soon dropped, presumably because the techniques using wireless rather than the telephone were more promising.

To Richardson, such innovative experimental work and the administration of the observatory must have seemed like minor digressions from his main mission in life at that time, namely to develop his ideas on numerical weather prediction. To appreciate fully the significance of this work it is necessary to place it in the perspective of the meteorological situation of the period. This will be done in a later chapter. For the present account of what happened at Eskdalemuir, let it suffice to say that Richardson's basic plan was to devise a method of calculating the future weather, using the relevant mathematical equations which describe the behaviour of the atmosphere. Some of the most fundamental equations were already well known, but even these had to be transformed in such a way that they could be solved by the approximate methods which he had described in his Royal Society paper of 1910. Other equations had to be worked out from the basic physical principles and ways had to be devised of taking account of certain factors which are not routinely observed at meteorological stations. The fact that he was able to do all this in less than three years, and then only in such time as he could spare from his regular observatory duties, gives some indication of his self-discipline.

By 1916 he had completed the first draft of his book, which at that time was called *Weather Prediction by Arithmetical Finite Differences*[7]. It was communicated by Sir Napier Shaw to the Royal Society, which (in Richardson's words) 'generously voted £100 towards the cost of its publication'[8]. Before having the book published, however, Richardson wished to add a practical example to demonstrate how his method would work. We shall see how this and other factors delayed publication for almost six years.

As might have been expected, the staff at Eskdalemuir did not altogether escape involvement in the war effort. As a pacifist, Richardson was sometimes faced with a dilemma, as for example when he was asked by Chree (Superintendent of Kew Observatory) whether 'the position of a concealed gun could be determined like that of an earthquake by the tremor in the ground'[9]. Richardson replied to his director, Shaw: 'I do not know, but I suppose it is my duty to forward such information as I have'. He then suggested that 'a normal tromometer' (a device for detecting faint earth tremors) might be suitable and he named a textbook which gave more details. Shaw was clearly not very pleased with Richardson's letter. He thought Richardson should have shown more enthusiasm by carrying out some tests. 'If you can think of any simple way of demonstrating practically the use of a seismograph ... I should advise you to try it with all speed'. Richardson however was not going to allow himself to get involved if he could avoid it. He replied 'There is no simple way of demonstrating the matter here, because there are no guns firing in the neighbourhood!' This must have infuriated Shaw and he wrote sarcastically: 'I did not suppose there were any guns available for Eskdalemuir but there are other ways of producing earth tremors'. He suggested that Richardson might try dropping a heavy weight near the seismograph. He added: 'it does no good to offer an idea which may or may not work out in practice'.

Richardson still took avoiding action, and wrote to Shaw 'I gather from your letter of 9th September (1915) that Eskdalemuir Observatory may as well leave it alone. If not please send authority to spend £10 on a microscope'. By this time Shaw was losing his patience. Back went his reply: 'What you should gather from my letter ... is that such matters ... are in the hands of people whose ordinary business in life is to be professors, and it is unprofessional on the part of a professor to adopt the attitude of a learner.... If you can say that you have an apparatus that is capable of doing something, it becomes very difficult for the professor to explain to you the reasons why your suggestions will not work in practice, but otherwise it is as easy as lecturing'. He then offered to supply a microscope if needed.

This time there was no escape for Richardson. Within a few days he reported that he had modified a tromometer and would try it 'with a microscope which is my personal property'. Next day he wrote again to say that he had carried out some tests with not very encouraging results. 'The soles of one's feet are more sensitive to rapid tremors than this tromometer'.

There the matter ended for Richardson, but a year later (by which time Richardson had left Eskdalemuir), Shaw wrote to the Secretary of the Munitions Invention Department suggesting that further experiments should be made at Eskdalemuir, taking advantage of the testing of some big guns on the Solway, some 30 miles away. He remarked in his letter 'In the early days of the war I endeavoured to take the question up with the late Superintendent of the Observatory, but he had conscientious objections to being concerned with the use of scientific apparatus for locating engines of destruction'. He added that his successor 'had no objection of that character'[10]. Some further experiments were in fact made but there were no useful results.

In his biographical notes for the Royal Society, Richardson did not refer directly to this conflict between his pacifist views and his official duties at Eskdalemuir. He did however write about the persistent influence throughout his life of 'the Society of Friends (Quakers), with its solemn emphasis on public and private duty'. He mentioned that his father 'avoided, as far as he could, dealing with brewers, distillers and armament firms' and continued 'The social service of the Society of Friends pointed towards psychology, and its condemnation of war pulled me away from the many warlike applications of physics. In August 1914 I was torn between an intense curiosity to see war at close quarters, an intense objection to killing people, both mixed with ideas of public duty, and doubt as to whether I could endure danger'[11].

The conflicts and doubts facing Quakers of military age were recognised in a declaration on the war issued by the Society of Friends in November 1914. After reaffirming the Quaker peace testimony about the incompatibility of war with the teachings of Jesus, the declaration proceeded[12]:

> Today many of our fellow-countrymen are impelled to enlist by a sense of chivalry towards the weak and by devotion to high national ideals. Today again the members of our Society, especially the younger men, are entering upon a time of testing. We can well understand the appeal to noble instincts which makes men desire to risk their lives for their country. To turn from this call may seem to be a lower choice. In many cases it means braving the scorn of those who will only interpret it as cowardice. To not a few it involves the loss of employment. The highest sacrifice is to contribute our lives to the cause of love in helping our country to a more Christlike idea of service. Those who hear the call to this service, and who respond to it, will be helping their nation in the great spiritual conflicts it must wage.

There was a varied response to this declaration. Some young Friends, finding that their pacifist convictions were not strong enough to enable them to resist their patriotic urge to serve their country, joined the armed forces. Those in a second group were prepared to do humanitarian work alongside their brothers in arms, provided that they did not have to take the military oath or to engage directly in the fighting. A third group believed that even this would not be compatible with the Quaker peace testimony and refused to engage in any activity which was part of the war machine. Richardson belonged to the second group, a leading spokesman of which was J Philip Baker, who had followed a few years behind him at Bootham and Cambridge (he received the Nobel Peace Prize in 1959 and became Lord Noel-Baker in 1977). In August 1914 he wrote a letter to *The Friend* (the Quaker weekly journal) pointing out that some members of the Society of Friends wanted to 'render some service more commensurate with their powers and opportunities than is involved in the administration of war relief at home'. He believed in any case that there were plenty of well qualified people to do such relief work but that there would probably be a shortage of ambulance workers. He therefore suggested that 'young men Friends should form an Ambulance Corps to go to the scene of operations'. This letter led rapidly to the establishment of the Friends Ambulance Unit (FAU), the purposes of which were '(a) to provide through the means of an efficient Ambulance unit, a good and sound instrument, to be skilfully used in the ministry of compassion to men, of whatever nation, caught in the trammels of war or in its train of suffering and death, and (b) to provide this means in such a way as to express in some degree the principles of the Society of Friends'[13].

In the early days of the war this unofficial assistance was not welcomed by the British military authorities and most of the FAU work was with the French armies, who were only too glad to supplement their own limited medical services by a body of men who paid their own expenses. The first FAU consignment accordingly sailed for Dunkirk at the end of October 1914 to succour the wounded French soldiers who were lying in warehouses in the port while waiting to be moved on to hospitals further from the fighting. This first expedition consisted only of 43 men and 8 cars but by the end of the war there were 600 serving abroad with the FAU.

When military conscription was introduced in 1916, Quaker conscientious objectors were granted exemption from military service provided that they joined the FAU; those already serving in the FAU were automatically exempted and most of them continued to work with the Unit until after the cessation of hostilities. A few, who did not feel that it was right to escape so easily from the military tribunals, left the FAU and risked being thrown into prison. Among these was T Corder Catchpool, who was also disturbed by the thought that regular soldiers displaced from the medical service by members of the FAU were 'often drafted to the firing line and complained bitterly that I and my colleagues had sent them there'[14]. After leaving the FAU in May 1916, he spent most of the rest of the war in prison for refusing to accept conditional exemption from military service. His experiences in the FAU and in prison are graphically described in his book *On Two Fronts* (1919), based on his letters. We shall hear more about Corder Catchpool later and also about another conscientious objector with whom Richardson became very friendly, Ernest B Ludlam. At the time of Ludlam's refusal to accept military service, he was engaged in scientific research at

Cambridge. He was given the opportunity of continuing this work as an alternative to military service but refused when he found out that the results were being used for military purposes, and went to gaol[15].

As Richardson's work at Eskdalemuir was considered to be of national importance, he could doubtless have remained in the peace and quiet of the Scottish countryside until the end of the war, had he so desired. But this was not to be. As early as October 1914 he applied for leave from the Meteorological Office 'to join a Red Cross unit of the Ambulance Corps'[16], but permission was declined. After several further fruitless attempts to obtain his release, he handed in his resignation in May 1916. As a married man he was not yet subject to conscription and was able to join the FAU and to continue in its service without ever having to face a military tribunal. A few months later, having completed some initial training, he was attached as a driver to a motor ambulance convoy lent to the 14th French Army. This *Section Sanitaire Anglaise* (SSA 13), as it was officially known, worked alongside the regular French military *sections sanitaires*. Each section had 56 men, mainly drivers for the fleet of 22 ambulances. Many years later, Richardson wrote: 'I was a bad motor-driver because at times I saw my dream instead of the traffic'[17]. But there is no record of any ill effects of these distractions while he was driving ambulances in France. In fact, one of the other members of SSA 13, George Hutchinson, who used to help in the maintenance of the ambulances, wrote: 'We judged drivers by the work they brought in—particularly broken springs through careless driving into shell-holes. But L.F.R. was not one of these and we looked on him as careful and conscientious'[18].

In the French Army Medical Service, *postes de secours* (aid posts) were set up as close as possible to the firing line. The wounded soldiers were brought by regimental stretcher bearers to these posts where first field-dressings were made. They were then passed further back from the line to a *poste de triage* (sorting post) where they were sorted out according to the severity of their injuries. From the *triage*, the more serious casualties went on to the nearest *Hôpital d'Evacuation*, which corresponded to a casualty clearing station in the British Army. Here they were cared for until well enough to be passed on to hospitals in the interior for full treatment. The job of the *sections sanitaires* was to evacuate the wounded from the *postes de secours* to the *triage* and from there to the *Hôpital d'Evacuation*. On the western sector of the front, where SSA 13 was based for most of the time, an out-station was established at the *triage*. The six to ten ambulances based there were relieved every two to three weeks[19].

Little has survived of what Richardson himself wrote about his personal experiences in France. The following account is therefore based largely on the recollections, about 65 years later, of four of his SSA 13 colleagues. Richardson joined this *Section* shortly after it had left the FAU Headquarters at Malo-les-Bains on the outskirts of Dunkirk. For the next few months the *Section* was in a relatively quiet sector, as described by Robert Charnley[20], now a spritely young man in his mid-80s!

> Each day as we went on duty to the dug-out ... we stopped on our way to pick up some wood such as laths, timber frames, etc. from the ruined villages en route, to keep the fire burning. From here we went to the Front to pick up the wounded when a call came through. We slept in the back portion of the dug-outs, in full clothing so as to be ready at a moment's notice. These dug-outs were very stuffy at night and I used to waken up with a headache. Also, we had rats running around, which very often ran over us as we lay down, until someone got the bright idea of placing a paraffin oil lamp in the opening between the front and rear portions, which were the sleeping quarters, the front portion being for meals or writing letters etc. This did the trick, for the rats would not pass the light.

In January 1917

> we arrived at Chalons and were billeted in the barracks. We had a barrack room all to ourselves. This was a long room but with no beds in it, just bare walls of painted brick, with a concrete floor. We had no heating at all, so we just had to bed down as best we could and keep our clothes on, also our boots. Believe me, it was most uncomfortable for lying on one's back or side on this was sheer agony, since the concrete does not fit the contours of one's spine or hip bones. As you can imagine, we did not feel refreshed when waking up, that is if we did manage to get to sleep!

In April 1917, the French Army opened an offensive to recover the heights to the east and west of Rheims—in peacetime the centre of the Champagne district. During the ensuing battle of Champagne, Richardson's Section came under heavy shelling as they shuttled between the *triage* and the *poste de secours*. Let Robert Charnley take up the story again.

> My first incident ... was the day after we arrived. We had the use of a badly shelled farmstead for the civilian owners had left it when the war reached them. I was in the yard having a chat with another of our men when the Germans commenced to shell all round us. We could hear the guns, then the scream of the shell as it came through the air, so we knew this was not for us, but could be near. As we talked, another shell was sent off. It was peculiar for we felt a dull thud in our ears but heard no scream, although we felt some concussion, so we ran forward into a store room, or stable. Here our storekeeper was sat on a box with his back to the wall and facing the doorway. He was booking down his stock. We two then entered through the opening but we saw the stone wall bulge inward beside our storekeeper who was knocked off his box and the shell just buried itself in the earthen floor. Fortunately for all three of us, the shell did not explode for it must have been a dud.

One day one of our ambulances had to go into a wood to pick up some of the gun crew who had been hit by shell fragments. On this trip, the driver had an orderly to assist him. They got into the wood and found the battery. The wounded were in the dug-out by the side of the gun pit, so they got off the ambulance and had just got down into the shelter when a shell landed right on top of the ambulance. This just blew the vehicle all over the place. They were lucky to come out of it safe and sound.

In rather different style, the poet Laurence Binyon (1918) also describes the SSA 13 experiences during the battle of Champagne in his *For Dauntless France*, an account of British aid to the French wounded which he wrote for the Red Cross. After recalling that the French offensive was launched on that part of the front which the British ambulance sections were helping to serve, Binyon continues[21]:

For a full week the prelude of artillery had been thundering without intermission, and with a violence that seemed fantastic. The face of the landscape was pounded out of shape. Here, as in the battle of September 1915, was the same desolate country, with frequent patches of fir, the same powdery chalk soil, slimy after rain ....
Of the English sections, perhaps Section 13 had the most dangerous duty to perform. The *Poste de Secours*, from which they had to bring down the wounded, was in a wood, on the left of the line, where the attack was least successful. The *Poste* itself was continuously bombarded; so was the stretch of road down which the cars had to run; so was the village to which they brought the wounded. The stranding of some 'tanks' near the crossroads where the cars waited, attracted the enemy guns' most earnest attentions. It was not unusual in passing up and down to find sentries, who had been passed an hour earlier, lying dead at their posts, and the road in places blocked, with dead horses and some of their drivers.
The night before the attack, one of Section 13's cars was hit three times in succession by shells, absolutely smashed and burnt: almost all the cars were struck by splinters; yet no one was killed, though two men were wounded, one severely.

From these and other accounts of events in the FAU, and especially in SSA 13, there emerges a picture of Richardson's life from September 1916 until his release from the FAU in January 1919. For up to two weeks at a time he would be busily engaged in the harrowing task of transporting wounded soldiers, sometimes under shell fire. Then he would have a few days of relative respite at base camp several miles from the front. He was one of the oldest members of the section—Robert Charnley for example was thirteen years younger—and was held in considerable respect, if not awe, by his comrades. He was not by nature a very sociable person—how he must have longed for the solitude of Eskdalemuir—and spent most of his spare time on his own, meditating and writing. He was universally known as the 'Prof', allegedly an abbreviation for professor, because of his intellectual and academic prowess. But in a contemporary cartoon, drawn by a member of SSA 13, he was called the 'Prophet'. Prophet or professor? Either would have been appropriate. Less appropriate in my view is the belief, held by some members of the Richardson family, that 'Prof' was an abbreviation of 'profiteer', associated with a temporary spell of duty as steward for the Section, during which he was responsible for paying out pocket money and for issuing rations drawn from the French Army. This activity is illustrated in the lower part of the cartoon where the Prophet (incidentally a very good likeness of Richardson) is shown handing out tobacco rations. The text of the cartoon is somewhat esoteric and I have not been able to find an explanation for all the allusions. The cartoon was published, along with about twenty similar cartoons of other members of SSA 13, in *The Little Grey Book*[22], the title of which was apparently borrowed from a publication of the Pelman Institute containing a course on memory aids.

Figure 4.4 Caricature of Richardson (the Prophet) in the FAU.

One of Richardson's colleagues in the FAU, Arthur Molyneux, writes that 'he was well liked by us all and kept a weather eye on the elements'[23]. Robert Charnley recalls that 'he kept himself more or less to himself as a rule. But I think I am right in saying that, when the Germans sent over their gas shells on to our lines, he had an idea that if he could get hold of one and examine it, maybe he could learn its contents and try to sort of neutralize same'[24]. Tom Ellis, who probably knew Richardson better than most of the others, wrote that 'by contrast with our light-hearted attitude, he appeared grave and dignified. His learning had already laid a stamp upon him so that his nickname "Prof" had as much of truth as of sentiment in it. He was beloved by all. Perhaps his greatest contribution was his unconscious demonstration of the dignity of service. His simplicity and integrity transformed into acts of permanent worth the performance of the most menial of the tasks which fell to our lot in those dreadful days'[25].

Another FAU member, Herbert Morrell, who joined SSA 13 a few months after Richardson, recollects that[26]

> Although at the time he was only about 36, he always seemed to me like an old man—I was only 19. He spent a lot of his spare time setting up meteorological instruments and taking readings: we thought nothing of seeing him wandering about in the small hours checking his instruments.
>
> When he found I was studying textbooks in engineering and mathematics for a post-war career, he offered to give me lessons and help and we would sit in the back of an ambulance whilst he helped me with my problems. It was at this time that he bought himself a new slide rule to get more accurate readings and gave me his old one, which I have always kept close at hand in my desk. He had learnt to speak Esperanto as he thought it would help to bring people closer together. I remember on one occasion we were in a dressing station just behind the lines when some German prisoners were brought in for treatment; he tried to question them in Esperanto and was disappointed when they just stared blankly back at him.
>
> He used to try out new ideas to make life easier. I remember once when it was very hot he shaved off his moustache as he said he could then blow upwards and keep the flies away from his face!

George Hutchinson, in addition to paying tribute to Richardson's careful driving, wrote: 'No one ever laughed at him. He was so kindly, quiet and friendly, with perfect gentlemanly manners, a lesson to me and others. I don't remember him voicing his opinions at all. I used to wonder what he made of us and our shallow opinions. Our mouths were often too big, but he was always just friendly'. He described Richardson as 'Tall, slightly stooping. Big head, brown hair slightly curling. Big eyes and large specs. Quite properly clothed, but perhaps a puttee trailing or slightly awry, worn outside his trousers'[18].

According to Hutchinson, the only other member of SSA 13 to equal Richardson in brain power was Olaf Stapledon. His recollection that these two men often went walking together receives some support from Stapledon's own account of his experiences in the FAU in which he refers to serious talks while strolling with one of his colleagues. One can only speculate on their lively discussions about pacifism, psychology, science and other topics of common interest. Richardson certainly had a great admiration for Stapledon and it seems highly probable that this feeling was reciprocated. In one of the cartoons of Stapledon in *The Little Grey Book*, he is depicted as a small man, dressed in furs, seated by himself in a corner, concentrating on what he is writing. The caption reads:

> What he puts down I cannot guess
> But that is not my business.

Figure 4.5 Life in the *Section Sanitaire Anglaise* 13. *Top*: billet in the Champagne district; *bottom*: *Poste de Secours* at St Thomas (by Arthur Cotterell).

The cartoonist could hardly have imagined that these writings would lead ultimately to Stapledon being described as 'a genius, one of the four great writers of science fiction', as the author of[27] 'the one great grey holy book of science fiction' and as a writer whose[28] 'prophetic novels had a character as independent as those of Aldous Huxley'. It is hardly surprising that two individuals of such vision as Richardson and Stapledon, thrown together by chance, struck up a close friendship which continued after the war.

As a result of his wartime experiences and of his conversations with Olaf Stapledon and others, Richardson's thoughts turned more and more to the problems of peace and war. While still at Eskdalemuir he had already started making notes on the subject. For example, in a manuscript headed *Reflections on War*, written late in 1914, he asked himself such questions as 'What would have happened if England had remained neutral?', 'Where should we be now if it was not for the British Navy?' and 'To go to war to crush militarism, is that commonsense?' He also left notes from this period of his life entitled *Christ and War*. In this, after recording some of Jesus' sayings and actions relating to militarism and the use of force, he comments: 'When things don't fit we have three alternatives: (i) Christ didn't mean that. (ii) Christ not God. (iii) We are wrong'[29].

He wrote a more substantial ten-page manuscript in March 1915 entitled *The Conditions of a Lasting Peace in Europe*. This begins with a discussion of the four causes of wars in general and of the First World War in particular, namely: desire to extend territory or political power; friction between races; ignorance, suspicion and fear; the religion of valour. He believed that it would be 'futile to expect the status quo to endure, for history is but the record of its changes'. Hitherto, changes in political boundaries had only been achieved by war. Therefore, 'if war is to be prevented, international law must be extended so as to make it possible to shift boundaries by mutual agreement'[30]. For this purpose an international council should be formed. It should, however, not have any executive powers and there would therefore be no objection to nations having a representation proportional to their population. To reduce friction between races, one possibility would be to have more mixed marriages, subject to assurances regarding the fitness of the offspring. (Richardson returned to this question—another example of his interest in eugenics—in a paper published in 1950.) He believed that 'ignorance and suspicion' arise very largely from 'diversity of language' and he therefore advocated that the 'terms of peace should contain an agreement that each of the nations involved should teach a universal language to every child who arrives at a certain educational stage—say that of learning algebra. In fifty years this small seed would have grown into a wonderfully fruitful tree'[31]. As regards what he called 'The religion of valour', he remarked that 'Everyone admires a soldier's courage and if he is on their side they see nothing else. If he is on the other side his cruelty fills the foreground of the picture. It is as if every soldier had *courage* printed across his back and *cruelty* across his chest'. Valour should be freed from cruelty. He concluded 'A great step will have been taken when schoolboys are taught to think Livingstone a greater hero than Wellington'[32].

These kinds of questions and thoughts must have passed through the minds not only of many of Richardson's contemporaries, but also of many young men and women of subsequent generations. Richardson's unique contribution was to follow up these ideas by asking himself whether 'mathematical language can ... express the behaviour of people ... in war?' He realised that mathematical expressions were 'in general use in various parts of Sociology, for example in Economics, in Anthropometry and in all cases where statistics of large masses of mankind have to be described'[33]. Why, he wondered, have they never been used for war behaviour?

This was one of the main questions to which he addressed himself during his rest periods with the FAU in France. The result was a pioneering paper *The Mathematical Psychology of War* (1919) which he dedicated to his 'comrades of the motor ambulance convoy known as S.S. Anglaise 13, in whose company this essay was mainly written'. A large proportion of this 50-page essay is a discussion in plain language of how the attitudes of people on the opposing sides of a conflict change as the conflict develops. Most of the text would certainly have been readily understandable by his friends, but many of them must have been discouraged from reading very far by the sight of two differential equations in the introduction. Some of his colleagues were put off by the merest suggestion of using mathematical symbols. Writing in 1982, George Hutchinson (who became a school master on leaving the FAU) expressed the common attitude: 'I remember him telling me 'Let x be the will to hate—or some more obscure symbol. It beat me!'[18] If Richardson had known more about the psychology of his FAU comrades he might have avoided the use of mathematics at such an early stage in his essay.

In his psychological discussion, Richardson refers frequently to McDougall, in whose work he had in fact been interested for several years. There are, for example, numerous quotations from McDougall in some of his handwritten notes dated 1912 on 'the important elements of character according to different people'. This earlier interest in psychology was probably intensified by his wartime experiences which produced a burning desire to understand the causes of war and to contribute in whatever way he could to the maintenance of peace.

From the outset of his mathematical studies of war, as illustrated by his first published essay, Richardson was fully aware of the limitations of mathematics for this purpose. In an introductory section headed 'An apology for the use of mathematics' he wrote[34]

> To have to translate one's verbal statements into mathematical formulae compels one carefully to scrutinize the ideas therein expressed. Next the possession of formulae makes it much easier to deduce the consequences. In this way absurd implications, which might have passed unnoticed in a verbal statement, are brought clearly into view and stimulate one to amend the formula. Mathematical expressions have, however, their special tendencies to pervert thought: the definiteness may be spurious, existing in the equations but not in the phenomena to be described; and the brevity may be due to the omission of the more important things, simply because they cannot be mathematized. Against these faults we must constantly be on our guard.

In this same section he makes the first of many references to what he called a 'fundamental rule of scientific method', namely Occam's razor; frequently expressed in the form that 'Entities are not to be postulated without necessity'. Richardson elaborates this as follows[35]:

> For shaving off the superabundant growth of mathematical uncertainties and difficulties I have frequently used an analogous rule: 'Formulae are not to be complicated without necessity'. For example, if observation shows nothing except that two quantities increase and decrease together, and vanish together, then one quantity has been taken as a simple multiple of the other. Quadratic, cubic, or other more complicated terms have not been introduced without clear evidence to show that they were necessary. The formulae set down are therefore at best only rough approximations. Indeed on account of the difficulty of defining the fundamental quantities, there remains a general vagueness, which may scandalize some of those who have been trained in the exact sciences, but which, in the author's opinion, does not deprive the formulae of meaning, interest and suggestiveness.

We have quoted rather extensively from this section of the essay in view of the importance which Richardson attached to these ideas throughout his later and more mature work on the causes of war. In this first essay on the subject he need hardly have been so apologetic about the use of mathematics for most of the equations are very elementary—he had applied Occam's razor with a vengeance. At this stage however, he did not attempt to test the validity of the equations by inserting numerical values for such parameters as vigour-to-war. Later he realised that for this purpose the equations were too 'complicated and intractable' and he had to simplify some of the statements 'so as to make deductions convenient'[36]. Some of the analogies used by Richardson in this essay are very vivid. For example[37]:

Several mental dispositions or tendencies may be illustrated by comparing thought to the flow of water down a bank of soft mud. Wherever water has flowed, a channel is formed, which is analogous to a habit tending to direct any future flow. Wherever the flow ceases in any channel, the mud, owing to its softness, begins to close in, obliterating the groove and thus illustrating forgetfulness. The phenomena of distraction require us to suppose that the supply of water is limited, so that the flow down any one channel can only be made at the expense of the flows down all the other channels together. The Will may be compared to a person who can block up certain channels or scoop out others; but, to make the analogy fit, we must suppose that some of the channels are out of his reach, and that even in those which he can reach, the amount of closing or enlarging, which he can do, is very limited. On this analogy the instincts are to be represented by capacious channels in a harder mud, or in a rock even, so that they do not tend to close up when disused. Across the upper ends of many of these instinctive channels the Will has constructed a barrage of mud. But should a sudden rush of water wash away this barrage, then the instinctive channel, by its depth and breadth, draws all the water to itself, draining the others. Thus it is when a man loses his temper. His struggles to regain equanimity may be compared to pouring mud into the torrent flowing down the instinctive channel. At first the mud may perhaps be washed away as fast as he can pour it in; but as the supply of water becomes exhausted, the barrage is restored and the remaining flow diverted to other channels.

The essay also illustrates Richardson's sense of humour and successful use of parody, as for example in his discussion of the effects of boredom on those engaged in war. For men stationed in quiet parts of the front he writes that this boredom[38] 'takes the form of a yearning for a return to civil life, and with a slight admixture of fatigue, might go to the tune of 'Three Blind Mice' thus:

> I, want-to-go home,
> I, want-to-go home,
> No more bloody war,
> No more bloody war,
> I've stood two years of this blasted strife;
> I'm tired of going in fear of my life;
> I want to go home to my children and wife;
> I, want-to-go home ...

and so on, ceaselessly.' This parody could well have been based on songs which Richardson heard in France, for Robert Graves recalls in *Good-bye to all that* (1929) two defeatist ditties from the trenches each beginning *I want to go home, I want to go home*.

Richardson's principal endeavour in this essay was to determine the relation between what he called the 'vigour-to-war' of a country and its 'warlike activity' and how these factors change on the two opposing sides under the various influences to which they are subjected in the course of a war. He put forward the view that the warlike striving of either side is 'largely, though not entirely, an instinctive reaction to the stimulus of the warlike striving of the opposing side'[39]. After a detailed consideration of various influences, he reached the conclusion that[40]

> Previous to the war there existed, as causes tending to it, such purposes as: certain special business ambitions, racial jealousies, the desire for security on the part of dynastic rulers, the desire to regain Alsace-Lorraine and the love of warlike activity in itself. These desires seem insignificant in comparison with the subsequent outburst of destructive fury. But they engendered suspicion between nations, and suspicion produced armaments, and armaments in turn increased suspicion, and so on alternately, distrust and preparedness-for-war mutually increasing each other; until war broke out. From the time of the first hostile act, the intensity of the war was increased by mutual reprisals; for all warlike measures appear almost as atrocities to those on whom they are inflicted, however justifiable and necessary they may seem to the inflicting party. In other words the largest part of the total vigour-to-war was, on both sides, that due to the instinctive defensive disposition, and the second largest part, I should say, that due to vengefulness. A limit to the warlike activity was set by fatigue. This with fear, pain and boredom slowly wore down the vigour-to-war, until Germany practically confesses herself beaten. Now that fighting has ceased, the powerful distracting influence of the defensive and vengeful dispositions has abated, and Pity for the Adversary and what have been called higher influences in all countries are beginning to reassert themselves.

In view of its pioneering nature, it seems relevant to enquire into the origins of Richardson's essay. Although he usually took great pains to acknowledge the sources of his ideas, in this particular essay he did not give any clue as to what had led him to attempt a mathematical approach to the study of war. Many years later, however, he wrote that the seed from which his work on the mathematical psychology grew was *perhaps* Bertrand Russell's pamphlet *War—the Offspring of Fear*, published in 1914 by the Union of Democratic Control. Still later, in a paper which appeared just before his death, he added by way of explanation that Russell's pamphlet asserted that 'the main motives of the arms-race of 1908 to 1914 had been mutual stimulation by fear' and that his own contribution to the discussion had been 'the translation of Russell's thesis into a pair of simultaneous differential equations'[41].

I am not altogether convinced by Richardson's explanation. In his 14-page pamphlet, Russell analyses the causes of war, with special reference to the First World War, and stresses that each side in a conflict believes that it is fighting to resist unprovoked aggression. He considered that the popularly held German view of the origin of the war seemed to Englishmen an utter travesty of the facts, and vice versa. He concluded that it is 'the universal reign of fear which has caused the system of alliances, believed to be a guarantee of peace, but now proved to be the cause of world-wide disaster'[42]. Russell advocated the establishment of a 'League of Peace' which would offer to mediate between disputing parties; if one party accepted mediation and the other refused, the League should throw the whole of its armed support to the former[42]. I can readily appreciate that Richardson must have shared many of Russell's views—perhaps with some reservations about the use of force by the League of Peace—but I fail to see any direct link between Russell's philosophy and Richardson's mathematics. Nowhere does Russell's pamphlet mention an arms race or the mutual stimulation to the build up of arms which is the key to Richardson's essay.

Turning over the pages of a bound set of the pamphlets published by the Union of Democratic Control, I chanced upon an anonymous paper, issued at the same time as Russell's, with the title *The International Industry of War*. This is in the main a plea for the nationalisation of the arms industry and for the control of arms exports. The opening paragraph contains the following remarks: 'In the mind of every nation force and self-defence have been in the background of every peace arrangement, and the world has been induced to acquiesce in the most absurdly untrue of all proverbs: "The best security for peace is preparation for war". If one country alone had gone upon that assumption peace might have resulted, the peace which the unconquerable bully imposes upon his neighbours. But, when more than one nation has adopted it, a costly and burdensome armament competition has begun which could only end in war...'[43]. This seems to me to be much more in line with Richardson's ideas than Russell's pamphlet. Could it be that, with his admittedly poor memory, Richardson was wrong in attaching so much importance to the impact of *War—the Offspring of Fear*?

This is not to deny that Richardson was influenced by Bertrand Russell. He refers to Russell's pamphlet and to his book *Mysticism and Logic* in the *Mathematical Psychology of War* and sent an early draft of the essay to Russell for comment. In his reply, which Richardson quotes in the final version of the essay[44], Russell used some simple mathematical notation to demonstrate how just two aggressive individuals might be sufficient to drive their countries to war, but he did not go so far as to express his argument in the form of an equation. I understand that this is one of the very rare occasions on which he employed mathematical symbolism in the discussion of social problems.

Richardson had a strong desire to have his original work published and attributed this to a tendency, of which he had been aware from childhood, to display. Believing that such a tendency was often unsuitable, he diligently suppressed it or, at least, disciplined it; 'when disciplined', he wrote, 'a tendency to display is surely useful. Classic instances of discoverers, who were too reluctant to publish, are known in Sir Isaac Newton and the Hon. Henry Cavendish'[45]. Regarding his *Mathematical Psychology of War*, he wrote that there was no learned society to which he 'dared to offer so unconventional a work'[46]. But he did submit it to a commercial publisher, as can be seen from the following extract from a letter which Bertrand Russell wrote to George Allen and Unwin Ltd[47]:

> I have read the manuscript you sent me by Mr. Richardson on the mathematical psychology of war. I had seen an earlier stage of the same manuscript before, as he sent it to me; it is much improved on what it was then. With regard to publication, my opinion would depend upon whether Mr. Richardson has time and opportunity for further work on the subject in the near future. If he has, I should say that he ought to make his work fuller, insert more illustrations, and eliminate dogmatic statements of his metaphysical opinions where they are irrelevant.
>
> If, however, Mr. Richardson is, for any reason, unable to do further work at present on his manuscript, I consider it highly desirable, in the interests of science, that it should be published, *though I hardly think the sale would be large*. All Mr. Richardson's main ideas appear to me true, original, and important. His mathematical formulae greatly help to clarify one's ideas, and he himself is careful to point out that they give only approximations, the exact truth being probably too complex for such a mode of compression.
>
> I am perfectly willing that you should communicate what I have said to Mr. Richardson, and inform him that I should be glad to enter into correspondence with him.
>
> He has done a really remarkable piece of work, which deserves great respect.

Richardson made no reference to all this in his biographical notes. He recorded simply that he had 300 copies of his manuscript made by multigraph at a cost of about £35 and[48]

> gave them nearly all away. It was little noticed. Some of my friends thought it funny. But for me it was quite serious, and was the beginning of the investigations on the causes of wars.... It is still difficult, in 1953, to publish works on that subject. There are many anti-war societies, but they are concerned with propaganda, not research. There is a wide public interest in the subject, provided it is expressed in bold rhetoric, but not if it is a quantitative scientific study involving statistics and mathematics. There is no appropriate learned society.

This multigraphed publication, which is now something of a collector's item, was the result of one of Richardson's preoccupations during his rest periods while serving with the FAU, when as he put it, he was 'not paid to think' and so had 'abundant opportunities for meditation'[49]. But it was by no means the only subject to which he devoted his attention. It will be recalled that before leaving Eskdalemuir he had already completed the first draft of his book on numerical weather prediction, but that he had not yet worked out a practical example to show how his method would work. He took a copy of this draft with him to France, together with the best set of weather observations for Western Europe that he could lay hands on. The observations had been made at 7 AM GMT on 20 May 1910 and compiled by the International Commission for Scientific Aeronautics. The advantage of this particular data was that it was an internationally selected day for which more upper-air data were available than usual.

It took Richardson six weeks to complete his calculations of the changes which would occur in some of the weather elements at two places in the course of six hours. Although he claimed that his results formed 'a fairly correct deduction from a somewhat unnatural initial distribution'[50], the errors were so large that the forecast would have been completely useless for practical purposes. Richardson must have been very disappointed that so many weeks of concentrated effort had not yielded greater success, but characteristically he gives no sign of any disillusionment in his book. But he does relate briefly what happened to the manuscript at one stage: 'During the battle of Champagne in April 1917 the working copy was sent to the rear, where it became lost, to be rediscovered some months later under a heap of coal'[51]. Very little imagination is required to experience the anguish that he must have suffered during those months and the sublime relief at the end. He once told me that the happiness he felt after finishing a book or a scientific paper must be comparable to that of a mother after giving birth to a child. On this basis, to lose a book before it is complete would be comparable to a miscarriage—and Richardson knew only too well, from Dorothy's experiences, what this meant to a woman.

While Lewis was serving in the FAU, Dorothy was teaching mathematics at St Michael's School at Limpsfield in Surrey. This is now a boarding school for girls but in those days it was run by the Church Missionary Society for children whose parents were working overseas as missionaries. Dorothy could only see her husband during his infrequent spells of leave; they spent most of the time visiting their families. On one of her visits to the Garnetts in the Isle of Wight, Dorothy had yet another miscarriage and had to be carried off the boat on a stretcher. Each miscarriage affected her more and more seriously and on one occasion she became so distraught as to be on the verge of a complete nervous breakdown. It cannot have been easy for Lewis to say goodbye when the time came for him to return to France.

Richardson used to make notes on any serious conversations he had with his friends and relatives. One of the few that has survived recounts a walk he had in April 1918 with his FAU comrade, Roger Carter, during which they discussed the war situation. Carter remarked that life at the front had developed in him a taste for simple things. He had come to prefer the company of works foremen to that of county magnates and found himself extraordinarily angry with the general situation and the people in power. They agreed that the situation was bad and asked 'What is our personal duty?' Richardson had decided that it was to stay where they were and clean ambulances, even if that seemed very futile. He added 'Perhaps what really bears me up is the knowledge that my paper on International Voting Strength has been sent to the Foreign Office. Of course one might go home—to talk. But what would one say? One realizes that the position of the conscientious objector is rotten, unless he is personally prepared to be killed. And that sends one out here again'. Carter replied: 'Quite. If only one could get a leg knocked off', to which Richardson responded 'That would be glorious. One could shake it at them'. Carter concluded 'Then one could preach pacifism'[29].

In Richardson's paper on international voting strength, published in 1918, the leading notion was that the international assembly which would have to be created after the war 'would probably deal only with affairs arising between nations, and would be prohibited by its constitution from interfering in affairs that are purely internal to a nation; and that therefore voting strength should be a measure of internationality'. He suggested an index of internationality, having foreign trade as one of its ingredients.

He called attention to this idea again in 1926 when there was a dispute about the permanent seats on the Council of the League of Nations. Although his suggestions did not have much impact on the statesmen of the world, he was obviously convinced by the logic of his argument and was loath to let the matter drop. In a more elaborate paper, written in 1953, to which we shall return later, he noted that his 1918 proposal had been considered by a committee of the British Foreign Office but that they preferred the principle of the equality of foreign states.

Another document to have survived from Richardson's FAU days is a fragment of a play[29], probably written in 1918. It gives a picture of life as he imagined it would be in 1925. In England there would be a Labour Government and a Swiss type of conscription. Employment would be arranged through the trades unions, prices fixed by the state and railways, mines, alcohol and electricity would be nationalised. International affairs would be handled by the League of Nations. The play opens with an old colonel, who 'finds peace a bore', nostalgically visiting the area in France where he had fought during the war. He is spotted by two young men, one of whom is 'rich, brilliant, adventurous, going into a family business' and the other 'introspective, philosophic, going to the bar or to journalism'. The following extract will give the flavour of the conversation.

A	Surely that's old Colonel F over there poking about with his stick. Now I think of it, I remember hearing that his regiment fought in this part of the line.
B	What does he do now at home?
A	Oh, he lives on his estate; and writes letters to the papers to say that the Labour government is ruining the country.
B	Bored I suppose. The heroic, if murderous, endeavour in which he distinguished himself, is gone; and he finds himself a misfit in a new age.

At this point the Colonel slowly recrosses the stage and B recites:

B	Oh what can ail thee knight at arms
	Alone and palely loitering?
	The wire is rusted from the stakes
	And no bombs 'zing'."

The second scene is in a French country inn where the two young men air their French with the landlord.

Landlord	Bonjour Messieurs, vous avez fait une promenade intéressante?
A	Oui, merci monsieur; et j'ai une femme énorme.
B	(Giggling) Do you know what you've said?

A few minutes later the landlord's small daughter is called in and engages in a conversation with B in the international language Ido (here we see the influence of Richardson's brother, Gilbert).

B	Kad vu paroles la linguo internaciona?
Daughter	Yes, sioro.
B	Bone parolata!
A	But yes is English!
B	It's Ido also.

There follows a brief scene of no particular interest at a factory, and finally a conversation between a young socialist and an old aristocrat. The former wants to reform the world which he believes has been made by the aristocrat and his predecessors to their own advantage. The old aristocrat remonstrates with him: 'Well, my young friend, if you are determined to "remould the world, according to the Heart's Desire" of your Peckham Socialist club, or whatever you belong to, please first see that you grasp the "sorry scheme of it" a little more entirely than you do at present'.

Richardson's pun on *femme* and *faim* once again illustrates his schoolboy sense of humour. He used to relate a true story of an encounter he had with a French soldier while in the FAU. The soldier asked Richardson to teach him some really bad English swear words. Richardson first asked him to promise that he would never repeat them to anybody. The soldier agreed. Richardson then whispered in his ear: 'Tut, tut'!

In more serious vein, Richardson wrote some notes in France on his religious beliefs[29]:

> Ten years ago the conflict between science and religion seemed very real and poignant to me. It does not seem so now. As others here may have travelled or be travelling along a similar mental road, it may be helpful if I give a personal history.
>
> In those days people used to sympathise with one for one's 'difficulties in faith'. That was very kind of them, I thought, but very perverse, for why should anyone be pitied for being unable to believe what was manifestly untrue. I used then vigorously to *disbelieve* in lots of things, the first chapter of Genesis, many of the miracles, the virgin birth, the resurrection. I still disbelieve in most of them, but the emphasis has shifted. Disbelief does not matter so much. These particulars may be true or false, religion still lives. It is founded on an inner sense which all of us more or less possess. A sense by which we pray. A sense owing to which the lives of the saints make a direct appeal to us. These are facts just as much as the rising and setting of the sun or as chemical reactions in a glass apparatus. If a theory of the universe based on the accumulated observations of physics and chemistry should appear to show that prayer is useless, the answer is that that theory has gone too far from its facts, has overreached and is wrong. All physical theories and all theologies spread themselves beyond their observed facts. In a laboratory a hasty student takes a pair of observations, plots them on squared paper, draws a straight line through them and says that the line represents the behaviour of the material in question. Then another observer, more careful, takes some more observations away at one end of the line and finds that the line is all wrong there. Just so we may by analogy say the hasty philosopher plots his two points, one for physics and one for chemistry and draws a straight line through them and says that the line represents the Universe. But prayer and the lives of the saints do not fall on this line and therefore the line is wrong in part of its course. Ought one to believe that prayer can remove mountains or clean an automobile or bring fine weather? I think not in the direct sense.
>
> To sum up I personally believe in a spiritual universe and in a material universe which act on one another, but only through the brain.

Attached to these notes is a sheet of paper dated 1918 March 8, on which Richardson wrote: 'In so far as the ripening of your mind has been your ambition, so will its inevitable decay be your dismay'.

Another file of notes by Richardson from his FAU days, dealing with his ideas for new or improved meteorological instruments, has also survived. In the course of his research on numerical weather prediction he had realised that very few observations had been made of some of the meteorological quantities that entered into his equations—in some cases there were no observations at all. Many of his ideas were aimed at devising simple and inexpensive ways of obtaining such observations. He actually constructed some instruments from the materials at his disposal (cigar boxes etc) in France. For example he devised and tested a scheme whereby the height of the base of a well defined cloud, such as a small cumulus, could be calculated from measurements of the angle between the cloud and the Sun, the elevation of the Sun above the horizon and the distance from the observer of the shadow of the cloud. He also designed and constructed instruments for measuring radiation from the Earth at night and the percentage of the solar radiation which is reflected by the Earth (known as the albedo).

It must be remembered in particular that upper-air observations in those days were extremely scanty. The routine observations consisted almost exclusively of determinations of winds at various heights in the atmosphere by observing the motion of a freely ascending balloon (pilot balloon) and of estimates of wind speed and direction from observations by a nephoscope of the movements of clouds. At a few places, observations of the pressure, temperature and humidity at different heights were made with kites or balloons, but only spasmodically. There was an urgent need for routine measurements of these elements, and it was to this problem that most of Richardson's efforts were directed. For example, he wrote some notes on a cheap apparatus for determining the moisture in the column of air above an observer. Another of his schemes was to attach to a balloon a temperature-sensitive device which would trigger an explosive, a siren or a whistle when the air temperature surrounding the balloon reached a predetermined value. He conducted numerous experiments with various temperature sensors and with different types of explosives, organ pipes, sirens and whistles. He also had some correspondence on this subject with instrument manufacturers in England. We shall see that he was to continue experiments of this nature after the war.

As remarked by Arthur Molyneux, Richardson always 'kept a close eye on the weather'. One result was a description of a line squall which he recorded in September 1917 but did not publish until 1919. He also tried his hand at making coloured sketches of clouds for meteorological purposes, which calls for a combination of rapidity, simplicity and permanence. He discussed the best techniques for this with his FAU colleague Arthur N Cotterell who, as can be seen from the sketches reproduced in figure 4.5, was an accomplished artist. In a manuscript note on these conversations, Richardson concluded that coloured crayons on grey paper are hard to beat for rapidity and simplicity but that 'to fix them changes the colours'. On the other hand, oil paints would be quick and simple 'if a number of light tones were kept ready mixed for use'[52]. Two examples of his efforts with crayon have survived, but hardly merit reproduction. After leaving the FAU, Cotterell joined the family wallpaper business near Bristol; he continued to paint in his spare time and some of his work, including drawings made in France, has been exhibited.

In February 1918, Richardson carried out an interesting experiment on what he called a 'working model' of the atmosphere. He placed a basin, partly filled with water, on a gramophone turntable which revolved about once a second. From his measurement of the radius of curvature which the water surface then assumed, he estimated that with a paraboloidal bowl about one metre in diameter and $\frac{1}{2}$ metre deep he could have obtained a very fair representation of certain aspects of the rotation of the Earth's atmosphere. He tried the effects of heating the bowl but there were too many disturbances. In his notes on these tests[29] he made some suggestions as to how such disturbances could be avoided and how more realistic experiments could be made with two paraboloids, one inside the other. The facilities available in France were obviously inadequate for Richardson himself to follow up these ideas. Other scientists, however, have succeeded in reproducing many features of the atmospheric circulation using rotating vessels containing liquid.

In later life, Richardson rarely spoke of his FAU experiences but his two years of service in France unquestionably made some lasting impressions and had an influence on his subsequent psychological development. According to Stephen Richardson, his father's personal experiences during the war were 'highly distressing. When as a small child after the war I came upon him unexpectedly or made a sudden noise', he wrote, 'he would start fearfully and sometimes scream in terror. To comfort me, I was told it was because of shell shock during the war. These extreme reactions subsided slowly, except in nightmares'[53].

We have already mentioned that SSA 13 was involved in the Battle of Champagne in April 1917. For this they received a citation from the General Commandant of the 8th Corps of the 14th French Army. The citation referred to their service with the 16th Division for more than a year during which they 'displayed admirable bravery and devotion... going to collect the wounded only a few hundred metres from the firing line at the height of the battle. 18 of the 20 vehicles were hit by shells and 1 was completely destroyed'. The section 'never failed to give an example of discipline and devotion'[54].

The section in fact, spent most of 1917 and 1918 in or near the Champagne district some 150 km east of Paris. Richardson stayed with them until the convoy was dissolved towards the end of January 1919, when he returned to England—once again to look for a job.

CHAPTER 5

Benson 1919–20

Some ten miles south-east of Oxford, where the Thames makes a big sweep to the south, lies the village of Benson, traditionally a centre of boating and fishing, but now best known for the nearby Royal Air Force station where the Queen's Flight is based. To meteorologists who were active during and immediately after the First World War, however, it was associated with the observatory run by one of Britain's most celebrated atmospheric scientists, W H Dines. He was about 25 years older than Richardson and first achieved fame for his researches on the wind, including the invention of the Dines pressure-tube anemometer, following the Tay Bridge disaster of December 1879. A train crossing the recently completed bridge was swept with the bridge into the swirling waters below by a squall of wind. The calamity was attributed at least in part to the inadequacy of the wind data used in its design and Dines became the most active member of a committee appointed to study 'the entire question of wind-force'[1] including the relation between wind speed and the pressure exerted on buildings.

Dines was one of the early investigators of the upper atmosphere. He made his first measurements of upper-air temperatures at Oxshott in Surrey in 1902, using a meteorograph carried aloft by a kite of his own design. This technique, which involved the use of several miles of steel wire, was not without risk, especially in heavily populated areas. The French kite pioneer, Teisserenc de Bort, had been known on different occasions to hold up a railway locomotive and a river steamer with a tangle of wire, and in another incident to upset the occupants of a skiff. After the wire from his kite had fallen across a main road in Surrey, the fear that he might be involved in a more serious accident led Dines to move to Pyrton Hill in Oxfordshire in 1906 when he felt that Oxshott had become too populous. Although he had some private means, he needed financial support for his upper-air work and this was provided by an annual grant of £500 from the Meteorological Office from 1905 onwards. In his obituary notice of Dines, Napier Shaw wrote of this in 1927: 'From the modern point of view it may be regarded as an unusual arrangement. Modern administration tends to insist that anyone in charge of research shall make a precise statement of what he is going to discover before he is allowed to look for it, a limitation which W H Dines would have found irksome'[2]. At the end of 1913, Dines had to leave Pyrton Hill and, with the assent of the Meteorological Office, he moved to a neighbouring site at Benson, selected as being the best locality for continuing his upper-air observations. His work necessitated having good workshop and laboratory facilities for the construction and calibration of his meteorographs, many of which were flown from other observatories. While at Benson, he became more and more interested in the measurement of the Earth's 'radiation balance', in other words what happens to solar radiation after it arrives at the Earth's atmosphere—how much is reflected by clouds and by the Earth's surface, how much is absorbed by the atmosphere, and so on.

Richardson had a great admiration for Dines, who had visited him at Eskdalemuir. He both shared and appreciated Dines' intense scientific curiosity, his love for practical experiments and the design of instruments, and his very retiring disposition. Richardson must also have been attracted by the arrangements at Benson, where there was a minimum of red tape, in spite of funding by the Meteorological Office. To him, it surely seemed to be the ideal place to develop further his ideas for producing simple and inexpensive techniques for obtaining the additional weather observations which he felt to be necessary for the practical application of his method of numerical weather prediction.

Before leaving the FAU, Richardson enquired from Napier Shaw about the possibility of returning to work in the Meteorological Office—as he had resigned from his post at Eskdalemuir, there was no question of his being reinstated automatically. 'Now that the war seems to be approaching its long-desired conclusion', he wrote, 'I am thinking of what I shall do when it is over; and, first, I should like to bring the following proposition to your notice. It is too soon to say whether the system of weather prediction by a numerical process ... could ever become the practical system; but I think you will agree that further effort in that direction is indicated.' He then outlined two schemes, one rapid, one cautious.

> Take the rapid one first. There are 12 to 15 remaining technical difficulties, and to solve them, each might on the average require the attention of one university graduate for a year. In addition to the man in charge, whom I imagine to be myself, such a staff would require a mathematician, whose duty would be to co-ordinate, to criticise, and to build together into a

well-balanced structure the scattered portions of research. The staff would further need to be supplemented by two mechanics, a financial clerk, a correspondence clerk, and several 'orderlies'. It would need appropriate rooms, a large field for observations, the temporary use of say £3000 worth of standard apparatus (to be borrowed?) and say £1000 for special apparatus plus materials.

Now for the cautious scheme. Given the loan of a small workshop, £200 for special apparatus and materials, and the part-time services of a mechanic and of an observer, I could myself get on with the two technical problems which appear to me to be the most crucial for the success of the whole. I refer to methods of observing winds and temperatures in the upper air, with the object of making the procedure more rapid and applicable on cloudy days. I have in mind to try a projectile for wind and a special whistling balloon for temperature. If and when these methods have been worked out, whether by a large staff or by a small one, the next step should be to organise a small sample of a network of observing stations, say 5 to 13 in number, arranged at regular intervals of latitude and longitude. . . .

I am interested in the venture; but unless I can find a way of making income out of it, I shall have to seek the latter in some other sphere, where, probably, I should not have enough leisure to give any time to meteorology[3].

Shaw was too big a man to bear any lasting grudge against Richardson for the administrative difficulties he had caused by his earlier resignation. His reply was characteristically courteous, frank—and lengthy[4].

There are two sides to the question . . . . *First financial*: What you ask for . . . is not much more than was provided by the Prussian Government at Lindenberg or by the Government of the United States at Mount Weather. As further knowledge of the upper air will be of essential importance to the development of aerial navigation I see no reason to think that a scheme of that magnitude need be turned down on financial grounds . . . .

Secondly policy: Supposing a meteorological controlling authority had £15 000 a year to spend on meteorological research, would it desire to invest it in the form which you prescribe?

First consider a special point that has some general interest in this connexion. Bjerknes . . . propounded a scheme of research which has close resemblance to yours. He uses graphic methods; you propose a method of finite differences. Both require an organised system of observations that is not easily achieved on a sufficiently wide scale. After Bjerknes' scheme was propounded, the Carnegie Institution (to whom he had applied for a grant) did not take what seems a natural step for a body in possession of great resources, namely the establishment of a centre of its own for the development of its own programme. It would not cost more than the solar research institution of Mount Wilson which is theirs. . . . What happened was that the Government of Saxony set up an establishment for the purpose in the University of Leipzig and appointed Bjerknes professor....

In passing let me say that there are points of advantage in that mode of procedure. Your 12 or 15 graduates under a director are better organised under a professor in a university than as officials under an administrative director. There is more freedom and less technical responsibility and these are great advantages during the trial stages....

Next as to your particular mode of dealing with the observations. Again we come to the question if you had sufficient money for the purpose would you spend it in that particular way? On that point I suppose that different authorities would give a different answer....

For more than twenty years now I have followed a method that is different from Bjerknes's and yours and perhaps more suitable for the special circumstances of an office like this. But it is only a matter of curiosity with me whether Bjerknes's method or yours or my own will get us on the faster. I am quite sure that any of the three honestly pursued will enlarge the boundaries of our real knowledge of meteorology.

Your 'cautious scheme' is I think no more than a fully equipped observatory for the study of the upper air and more than one such observatory will be wanted in this country after the war. If you were in charge of one of them you would not find it difficult to use your opportunity for setting out the policy of finite differences ....

So, to sum up, a professorship of meteorology with a corps of advanced students who would work as in Bjerknes's establishment seems to me the ideal for your rapid scheme and that is not impossible .... And the charge of a fully equipped observatory for the study of the upper air is the literal translation of your cautious scheme. Such posts are also quite possible and indeed inevitable ....

Shaw submitted both Richardson's letter and his reply to a meeting of the Meteorological Committee in December 1918 but no decision had been reached when Richardson returned to England. In spite of this, he started working at Benson in March 1919—presumably with the tacit acquiescence of Shaw. Later that month, the Committee finally agreed to a modified version of the proposal and put aside a sum of £450 'as salary for the year 1919–20 for Mr. L. F. Richardson, now occupied in experimental work at Benson with Mr. Dines' sanction'[5]. The terms of the appointment, not formally approved until the end of May, included 'experiments with a view to forecasting by numerical process'. How happy Richardson must have been to have this opportunity of continuing his researches with official support and in the congenial atmosphere of Benson Observatory. Little did he realise that some administrative decisions which would make it impossible for him to stay in the employment of the Meteorological Office were already looming above the horizon.

The technique which Richardson had thought about in France whereby a small charge of explosive is triggered by a temperature sensor led to the design of what he called a 'cracker balloon'. As the balloon ascends, the air temperature decreases until it reaches the predetermined value at which the sensor closes an electric circuit and ignites the charge. The observer follows the balloon's ascent with a theodolite and notes the time of the explosion. As with the normal pilot balloon observations, the height of the balloon is determined either by assuming a constant rate of ascent or by measuring with the theodolite the angle subtended by a long tail attached to the balloon. A separate temperature sensor is needed for each temperature to be measured. Richardson's first cracker balloon, with only one sensor, was tested in January 1920, and his second—this time with two sensors—was launched in June. His report of the results contains the following brief account of some help which he unexpectedly received: 'While the theodolite was being set on the balloon, a group of children whose assistance had not been welcomed, began to call out that a white paper was falling'[6]. This was the signal from the first sensor which would otherwise not have been spotted.

Another of Richardson's inventions was the 'lizard' balloon for signalling the ratio of pressure to temperature. In this, a balloon carrying a long tail is covered by a piece of chiffon; as the balloon ascends, the chiffon constrains it in such a way that it can only expand in the vertical direction. Ultimately the balloon presses against a trigger which causes the tail to drop off. As Richardson put it: 'Hence the name "lizard", from the habit of some of these animals to drop their tails when disturbed'[7]. From measurements made before the ascent of the initial and final volumes of the balloon, it is possible to deduce the ratio of pressure to temperature at the level where the tail is released. From the results of four experimental flights with these balloons at Benson, Richardson concluded that they were cheaper than his cracker balloons, but not so accurate. He pointed out that they could be used at sea, having previously tested the accuracy with which balloons could be weighed on board ship—he did this in 1920 on a voyage from Newcastle to Bergen, the purpose of which we will discuss later. In his final report on this work, he noted that 'Mrs. Richardson made the chiffon cases'[8].

It will be recalled that while in the FAU, Richardson had experimented with some simple devices for measuring radiation. It is not surprising therefore that he collaborated closely in radiation experiments with Dines, who as already noted, had become very interested in the radiation balance. In his paper on *Atmospheric and Terrestrial Radiation* (1920), Dines acknowledges Richardson's help. 'For the method of calculation I am indebted to Mr. L. F. Richardson; this method is being published, but by his kind permission I am enabled to set it out here.'[9] He was in fact referring to the method developed by Richardson for his work in numerical weather prediction. By this time the first proofs of

Richardson's book on the subject were being sent to him for checking by the Cambridge University Press, and Dines was one of the select few who saw the text at this stage; in the preface Richardson thanks Dines 'for his interest in some early arithmetical experiments'. He also referred to some revisions which he made during his stay at Benson where he had 'the good fortune to be able to discuss the hypotheses with Mr. W.H. Dines'[10].

From the references in earlier chapters to Richardson's book on numerical weather prediction it will be recalled that the idea first came to him in the form of a fantasy in 1911. He followed this up while at Eskdalemuir from 1913 to 1916 and completed the first draft before going to France with the Friends Ambulance Unit. There he calculated a specimen weather forecast and shortly after his return to England in 1919 the manuscript was in the hands of the printers. Although the book did not finally appear until January 1922, more than a year after Richardson had left Benson, this seems to be the most appropriate place to discuss it in more detail.

Although Richardson had displayed some interest in the weather from his schooldays, it was not this that motivated him to direct so much of his energy towards meteorological research. The real impulse came from his realisation that his finite difference method for obtaining approximate solutions to certain types of differential equations could be applied to the problem of weather forecasting.

In these days of weather satellites, high-speed computers and other meteorological applications of advanced technology, it is not easy to picture how the weather forecasters exercised their profession at the time of the First World War, when Richardson was developing his ideas on numerical weather prediction. Their main tool was the weather map on which the observations of atmospheric pressure, air temperature, wind speed and direction and a few other meteorological elements were plotted manually. Lines joining places of equal atmospheric pressure (isobars) were drawn and from these the forecaster saw where the principal weather systems—the depressions, anticyclones etc—were located. From the observations of the way in which the atmospheric pressure was changing (the so-called barometric tendency) he inferred the future movement of these systems. Finally, guided mainly by his experience and his memory of what had happened when a similar weather pattern had occurred in the past, he produced his forecast. Richardson himself never served as a forecaster, even when he was working in the Meteorological Office, and, from his own confession that he did not have a particularly good memory, one can well imagine that he would not have been very successful in such a role.

It had long been realised that the weather was subject to the same laws as other physical phenomena. The importance of what was happening in the upper layers of the atmosphere was also recognised. The man sitting at the forecasting bench had, however, little opportunity of applying his knowledge of physics in his daily work. In the first place the only information at his disposal about conditions in the upper air was a few isolated wind measurements made by pilot balloons. The equations describing the behaviour of the atmosphere were not in a form suitable for routine use. Finally, even if he had had better observations and equations in a more convenient form, he would not have been able to carry out the enormous amount of necessary calculation sufficiently quickly for the results to be of practical use. Weather forecasting was indeed an empirical science—some might even have thought of it as an art rather than a science.

Richardson called attention to the limitations of forecasting methods in the preface of his book. As he put it: 'The forecast is based on the supposition that what the atmosphere did then, it will do again now. There is no troublesome calculation, with its possibilities of theoretical or arithmetical error. The past history of the atmosphere is used, so to speak, as a full-scale working model of its present self'[11]. He contrasted this with the methods used by astronomers. ' "The Nautical Almanac" ', he reflected, 'that marvel of accurate forecasting, is not based on the principle that astronomical history repeats itself in the aggregate. It would be safe to say that a particular disposition of stars, planets and satellites never occurs twice. Why then should we expect a present weather map to be exactly represented in a catalogue of past weather? He claimed that his book presented 'a scheme of weather prediction, which resembles the process by which the Nautical Almanac is produced, in so far as it is founded upon the differential equations, and not upon the partial recurrence of phenomena in their ensemble'[11].

Richardson was not the first scientist to tackle the problem of forecasting the weather on the basis of the mechanics and physics of the atmosphere. The famous Norwegian meteorologist Vilhelm Bjerknes had already addressed the question in 1904. His paper on the subject opens with the declaration[12]:

> If it is true, as every scientist believes, that subsequent atmospheric states develop from the preceding ones according to physical law, then it is apparent that the necessary and sufficient conditions for the rational solution of forecasting problems are the following:
> 1. A sufficiently accurate knowledge of the state of the atmosphere at the initial time.
> 2. A sufficiently accurate knowledge of the laws according to which one state of the atmosphere develops from another.

He envisaged a two-pronged attack. In the first place it would be desirable to improve the quality and quantity of the observations so as to obtain an adequate representation of the initial state of the atmosphere. From this *diagnosis* one should proceed to solve the main problem, that of *prognosis*, by integrating the fundamental hydrodynamic and thermodynamic equations. Bjerknes considered that an exact analytical solution of the equations was out of the question; such a solution would in any case not give the result in a practical form. He therefore advocated the use of a mixture of numerical and graphical methods: 'Based upon the observations that have been made, the initial state of the atmosphere is represented by a number of charts which give the distribution of the seven variables from level to level in the atmosphere. With these charts as the starting point, new charts of a similar kind are to be drawn which represent the new state from hour to hour'[12]. He felt confident that suitable methods could be worked out for doing this from the physical equations by breaking the problem down into a number of more tractable partial problems. For instance, from dynamic considerations one could determine the movements of the air masses over a short time interval and then by taking account of the thermodynamics one could assess the state in which each air mass would arrive at its new location.

At that time Bjerknes was professor of mechanics and mathematical physics at the University of Stockholm. He was not engaged in weather forecasting and his interest in the subject was theoretical rather than practical. Although he wrote of the possibility that some day his method might be used for routine weather forecasting, his primary aim was to show that the present state of the atmosphere could be derived from the preceding state by the application of strict scientific principles, in other words to demonstrate that meteorology was indeed a science.

The International Meteorological Organization, predecessor of the present World Meteorological Organization, had long been aware of the need for more upper-air observations and encouraged countries to engage in programmes involving the use of manned balloons, kites and free balloons for this purpose. The observations should preferably be made on internationally selected days so that they could be used for the construction of synoptic maps for various heights; even although the observations and the resulting maps would not be available until several weeks, or possibly years, and hence would be of no use for weather forecasting services, they would still be of great value for research workers, such as Bjerknes. The IMO scheme was organised by its Commission for Scientific Aeronautics whose enthusiastic president, Hugo Hergesell, arranged for the observations to be published.

In 1912, Bjerknes was appointed director of the new Geophysical Institute in Leipzig. There he was able to prepare and publish a series of weather maps, using the data issued by Hergesell, as part of the programme initiated by his 1904 paper. He returned to the theme of the scientific basis for weather forecasting in his inaugural lecture[13]:

> Now that complete observations from an extensive portion of the free air are being published in a regular series, a mighty problem looms before us and we can no longer disregard it. We must apply the equations of theoretical physics not to ideal cases only, but to the actual existing conditions as they are revealed by modern observations.... From the conditions revealed by the observations we must learn to compute those that will follow. The problem of accurate pre-calculation that was solved for astronomy centuries ago must now be attacked in all earnest for meteorology.

In closing, he remarked[13]:

> Our problem is, of course, that of predicting future weather. 'But,' says our critic, 'How can this be of any use? The calculations must require a preposterously long time. Under the most favourable conditions it will take the learned gentlemen perhaps three months to calculate the weather that nature will bring about in three hours. What satisfaction is there in being able to calculate tomorrow's weather if it takes a year to do it?'
>
> To this I can only reply: I hardly hope to advance even as far as this. I shall be more than happy if I can carry on the work so far that I am able to predict the weather from day to day after many years of calculation. If only the calculations shall agree with the facts, the scientific victory will be won. Meteorology would then have become an exact science, a true physics of the atmosphere. When that point has been reached, then the practical results will soon develop.
>
> It may require many years to bore a tunnel through a mountain. Many a labourer may not live to see the cut finished. Nevertheless this will not prevent later comers from riding through the tunnel at express-train speed.

Little did Bjerknes know that Richardson would start to bore the tunnel just a few months later. Still less could he have imagined that express trains would be driving through the tunnel within about forty years. In the meantime, he himself had to pass through a tragic period when his work was interrupted by the First World War. His assistants at the institute were called to the colours and five of his ten doctoral candidates were killed, one of them only about 50 miles from where Richardson's ambulance unit was working. In 1917 he left Leipzig for Bergen, where for the first time he had to pay attention to the practical aspects of weather forecasting. This was the beginning of the great Bergen School which was to have such a profound impact on synoptic meteorology in the 1920s. One of their main contributions was to develop a method of forecasting based on the concept of *fronts*, the surfaces of discontinuity between different air masses.

In the preface of his book, Richardson acknowledges his indebtedness to Bjerknes. He wrote: 'I read his volumes on *Statics and Kinematics* soon after beginning the present study, and they have exercised a considerable influence throughout it .... But whereas Prof. Bjerknes mostly employs graphs, I have thought it better to proceed by way of numerical tables. The reason for this is that a previous comparison of the two methods, in dealing with differential equations, had convinced me that the arithmetical procedure is the more exact and the more powerful in coping with otherwise awkward equations'[14]. He was of course thinking of his work on the flow of water through peat and on stresses in masonry dams.

It was Napier Shaw who told Richardson about the work of Bjerknes, as evidenced by the following extract from a letter which he wrote to Bjerknes shortly after Richardson's arrival at Eskdalemuir[15].

> The new superintendent of the Observatory at Eskdalemuir has ambitions about the solution of the meteorological future by a finite difference solution of Laplace's equation. He presented me the other day with what he called a dream of a palace at the Hague in which 500 computers were taking down the observations for all parts of the world read out from a conductor's box in the middle of a great theatre, each computer dealing with the observations for one compartment. I explained to him that you had already set out upon the programme and recommended him to digest what you had already done.

Shaw then asked if Bjerknes could help to obtain copies of his two books for Richardson as the Observatory was 60 miles from any library that was likely to have them. We shall see how this first indirect contact between Richardson and Bjerknes led ultimately to the development of a long-lasting friendship.

Chapter I of *Weather Prediction by Numerical Process* is a summary of the whole book. This is very useful, as Richardson made very little attempt elsewhere in the book to provide links between successive sections. In fact, he did not go out of his way to make it easy to read and understand, a feature which was duly criticised by several reviewers. Once exception is in Chapter II where 'Before attending to the complexities of the actual atmosphere and their treatment by this numerical method'[16], he worked out a greatly simplified case. For this he chose a set of equations which could also be solved by the usual analytical methods and then applied them to a very straightforward weather pattern. For the solution by finite differences, he divided the weather map into chequers, as on a chessboard, with red and white squares. At the centre of each red square he entered the value of the atmospheric pressure and in each white square he wrote down the momentum (which, like Bjerknes, he preferred to wind velocity). He then calculated the changes in pressure and momentum after an interval of $\frac{3}{4}$ hour at a selection of points and compared the results with those obtained by analytical methods. The agreement was satisfactory.

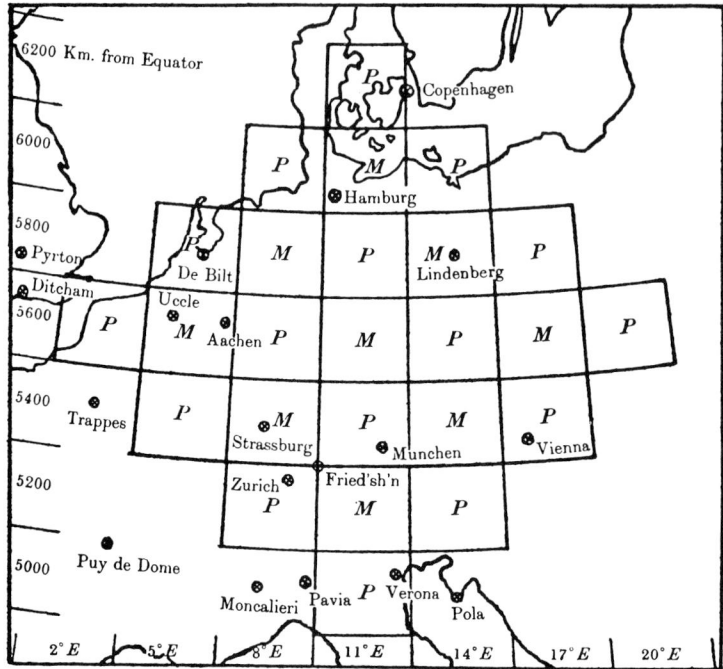

Figure 5.1 Map showing the distribution of meteorological stations from which Richardson was able to obtain upper-air observations for his attempt at numerical weather prediction. Contrast this with figure 5.2. Values of atmospheric pressure were required for the centres of the P chequers and of momentum for the M chequers.

Richardson then discussed in Chapter III the way in which the atmosphere should be divided up for a more realistic case. Ideally, the choice would have to be guided by such considerations as the dimensions of atmospheric disturbances, the accuracy required in the forecast and the cost, which increases with the number of points in time and space that have to be dealt with. He did not discuss these factors in any detail, but decided simply to let himself be guided by existing practices. He noted that the average distance between observing stations on land in the British Isles was about 130 kilometres but that there were far fewer observations on sea. Remarking that 'abundance of observations on the former does not compensate for scarcity of observations on the latter'[17], he decided that in the middle latitudes his chequers should be nearly squares with sides about 200 kilometres. For his interval of time he chose 6 hours, noting that for telegraphic purposes observations were made every 12 hours and sometimes every 6 hours. He also had to decide how best to divide the atmosphere into horizontal layers. Here he was guided by the method used in the maps published by Bjerknes—he had decided to use these as a source of data for his practical test—and chose the layers bounded by the Earth's surface and the fixed heights of 2.0, 4.2, 7.2 and 11.8 kilometres.

In the next chapter, which constitutes almost half of the book, we find a discussion of the equations to be used in numerical weather prediction. One might have expected that Richardson would have been satisfied by simply applying his finite difference method to the solution of the classical equations of Laplace and others. Instead, he considered almost every process which could conceivably affect the weather, such as the motion of water in the soil, the effects of vegetation on wind and the rates of evaporation from different types of land cover. He was not deterred by the thought that the results of many elaborate calculations might have little bearing on the outcome of his forecast. He was particularly thorough in his treatment of atmospheric turbulence, a subject in which he became especially interested and to which he made some very original contributions. It therefore merits being considered in some detail.

What exactly do we mean by turbulence? It is something which we experience every day of our lives but which is difficult to define in a rigorous scientific manner. The simplest way out is first to define the opposite! If we imagine a fluid flowing in such a way that each small volume follows precisely the same path as its predecessors, the flow will be streamlined, or laminar. Any other kind of flow is turbulent. It is rather like the difference between the orderly progress of a well disciplined company of soldiers and the wild rush of an unruly mob—the difference between order and chaos. Richardson was once seen by one of his nephews standing on a bridge pouring milk into a stream. He was using the milk to show up the turbulent eddies in the water. The term eddy is also difficult to define precisely; it may be thought of as a volume of liquid or gas which retains its identity for a limited time while moving within the surrounding fluid.

Much of the motion in the atmosphere is turbulent, especially in the lower layers, where any laminar flow of wind would soon be broken into eddies by friction with the Earth's surface and by obstructions, such as trees, buildings and mountains. But even at the levels at which jet planes now cruise we can experience on a cloudless day quite severe bumpiness due to what is called clear air turbulence. The eddies in the atmosphere range in size from several thousand kilometres down to less than a centimetre. The diameters of the cyclones or depressions which are shown up so well in satellite cloud pictures extend from about 4000 km down to some 200 km. Individual thunderclouds have diameters of a few kilometres. Then there are the tornadoes, the most violent manifestation of weather, in which the central vortex averages about 100 m across. Still smaller are the whirlwinds, just a few metres in diameter, which can raise dust or dead leaves up to heights of a few hundred metres. If we continue downwards in scale we ultimately arrive at the individual molecules of air. These are also in a continuous state of agitation, the so-called Brownian motion, but this is not quite the same thing as turbulence, at least in the way defined (or not defined!) by meteorologists.

Turbulence in the atmosphere is of great practical importance in that it is mainly responsible for mixing the air. If there were no eddies, there would be great contrasts in temperature and moisture between the air at our breathing level and at our feet. The mixing due to the Brownian motion, the molecular diffusion, is about one hundred thousand times slower than mixing by eddy diffusion and would be quite unable to distribute in a tolerable manner the heat arriving at the Earth's surface from the Sun. Turbulence likewise has much to do with the evaporation of water from the oceans, the removal of smoke and other forms of atmospheric pollution, and the scattering of seeds and pollen from plants. The formation of rain-clouds depends largely on turbulence, as does their distribution by the largest eddies, the depressions.

Weather, in all its variety, is a manifestation of the working of the atmospheric machine, in which the energy from the Sun is transformed into the energy of the wind, clouds and storms. Turbulence plays a vital role in this process and must therefore be taken into account in any scientific study of how the machine operates—hence Richardson's interest in the subject.

In opening his discussion of the effects of eddy motion, Richardson first points out that his theory must be[18]

> appropriate to the size of the element of the fluid which we treat ... in that we ignore the details of any motions taking place wholly within it. The upper limit to the size of an eddy is, like the length of a piece of string, a matter of human convenience. When an airman says that the wind is bumpy he is thinking in terms of a differential element probably comparable in size with one wing of a flying machine. For the present purpose any motion which disappears on taking the average over our coordinate intervals... has of necesity to be ignored. If it occurs in large numbers—as cumulus eddies do for example—then its general effect can be satisfactorily represented by additional terms; although unfortunately this does not help us for example to say whether it will hail or not on Mr. X's field.

He returns to this point later in the book[19]:

> We are not concerned to know all about the weather, nor even to trace the entangled detail of the path of every air-particle. A judicious selection is necessary for our peace of mind. For some such reason it is customary, at

stations which report wind by telegraph, to replace the instantaneous velocity by a mean value over about ten minutes. An extension of this process must be contemplated, for there is a good deal of evidence to show that the wind is full of 'secondary cyclones' or other whirls having the most various diameters. The arithmetical process can only take account individually of such whirls as have diameters greater than the distance between the centres of the red chequers in our co-ordinate chessboard. . . .

In telling the story of a nation, a historian cannot describe the individual lives of all the citizens. Once he has obtained a clear picture of the state of a nation, he can, however, visualise at least in broad outline the daily lives of the citizens. The state of the nation affects the lives of the citizens. In the same way, a meteorologist endeavours to assess the state of the atmosphere in terms of the behaviour of the large weather patterns; having based his general forecast on this assessment, he can then try to determine how the smaller-scale phenomena will be affected—even if he cannot quite get down to the scale of Mr. X's field!

Richardson epitomised much of what he was trying to say in his famous—and often misquoted—ditty[20]:

> Big whirls have little whirls that feed on their velocity,
> And little whirls have lesser whirls and so on to viscosity—
>   in the molecular sense.

These lines, which appear in Richardson's book as prose, are usually described as a parody of some lines from a poem by Jonathan Swift, published in 1733 in his book *On Poetry a Rapsody*. The context can be seen from the following extract from this diatribe against false poets[21]:

> The Vermin only teaze and pinch
> Their foes superior by an Inch.
> So, Nat'ralists observe, a Flea
> Hath smaller Fleas that on him prey
> And these have smaller Fleas to bite 'em
> And so proceed *ad infinitum*.
> Thus every Poet in his kind
> Is bit by him that comes behind;
> Who, tho' too little to be seen,
> Can teaze, and gall, and give the Spleen.

Others have more correctly attributed the source of Richardson's inspiration to the following lines by Augustus De Morgan, a nineteenth century mathematician of some note[22]:

> Great fleas have little fleas upon their backs to bite 'em,
> And little fleas have lesser fleas, and so *ad infinitum*.

De Morgan may well have had Swift's poem in mind when he wrote his ditty, but he himself did not make any acknowledgment to his distinguished predecessor. The ditty first appeared in a review of a book by Nicholas Odgers entitled *The mystery of being; or are ultimate atoms inhabited worlds?* Together with many of De Morgan's other contributions to the *Athenaeum*, it was re-published posthumously in *A Budget of Paradoxes* (1872). The association with Swift was not mentioned until this book was re-issued in 1915 when the editor added an appropriate footnote. De Morgan's ditty continued:

> And the great fleas themselves, in turn, have greater fleas to go on;
> While these have greater still, and greater still, and so on.

Perhaps Richardson was not aware of this, for he might well have continued his parody:

> And the big whirls of bigger ones partake in the rotation,
> Until at last we reach the gen'ral circulation—in the global sense.

In relation to the total length of the book and the rather small impact of the results on the final calculations, the section on turbulence seems somewhat out of balance. But it could have been much worse, for Richardson refers in the preface to 'various excrescences' having been removed for separate publication and these almost certainly include three papers on atmospheric turbulence which were published in 1919 and 1920; in all they would have added an extra 50 pages. The 30 pages we are left with consist in essence of a series of short essays on the various ways in which eddy motion enters into the subject, with explanations of how Richardson decided to treat each aspect in his calculations. The details are of great interest to specialists but need not detain us here.

Another topic which Richardson discusses very thoroughly in Chapter IV is the way in which the effects of radiation can best be taken into account in his equations. He begins with the absorption by various atmospheric gases of the radiation emitted by the Earth. In some cases he found that the existing experimental data were inadequate, but rather than waiting for the situation to be remedied he remarked that 'In the meantime meteorologists must carry on business on premises which are, so to speak, in the hands of the builders'[23]. He introduced the concept of a 'parcel' of air, which he defined as a 'geometrical figure formed by two closed curves not in one plane'[24]. This term later entered into the everyday language of meteorologists, but mainly when talking of convection, which may be considered as the upward motion of air parcels. The rest of this section deals with the absorption and scattering by the atmosphere of solar radiation. Here, as elsewhere, he often has to introduce simplifications in order to reduce the labour involved in his calculations: for example, the effects of dust in the atmosphere, which in fact vary greatly both with place and time, are handled statistically.

The concluding section of Chapter IV is entitled *Beneath the earth's surface*. One of the most important factors here is the sea temperature, but as it varies more slowly than the air temperature, Richardson makes out a case for using climatological values. He lists the items which would have to be taken into account if the sea surface temperature had to be predicted and adds: 'It may come to that, but let us hope that something simpler will suffice'[25]. Once again he was using Occam's razor. In discussing soil and vegetation he refers again to one of his own experiments, namely an attempt to measure the porosity of peat dust. Earlier in the book he had given the results of observations he had made in France of the dispersal of smoke and at Benson of the reduction of wind by a field of ripe wheat. Later he provides evidence about the local variability of rainfall provided by two raingauges he had kept in France. He was indeed a rare combination of experimenter and theoretician.

In what can only be a rather superficial account of this highly complex book, we will pass quickly over the next three chapters; they are of a very technical nature and in any case rather short. The first describes how Richardson dealt with a quantity which entered into several of his equations but which is not observed regularly and for which very few measurements of any kind were available, namely the vertical velocity of the air—the speed at which it is ascending or descending. The second relates to the uppermost atmospheric layer in Richardson's model, the stratosphere. This differs from the other layers in having practically no variation of temperature with height and in not having a definite upper limit. For these reasons, it requires special treatment. The third of these chapters deals with what Richardson called the 'arrangement of points and instants', in other words the timing of the observations and their distribution between the red and white chequers.

We now arrive at Chapter VIII, which discusses the practical sequence of operations in carrying out a numerical weather prediction. Whereas in writing Chapter IV Richardson's ideal was to 'obtain a description of atmospheric phenomena which should be in the first place correct, and which, secondly, might be used in prediction'[26], in this chapter he reversed the order of emphasis[26]:

> The ideal now is to make a scheme first workable and secondly as exact as circumstances permit. After a new machine has emerged from the experimental stage, its workability is tested by the cost and the value of its product, by its satisfaction of human needs. But the present scheme has

not yet emerged. The questions still are: does it conform to the nature of the external world? will the wheels go round at all? So the essence of workability is here taken to be that, when we have made a step forward in time, we should find ourselves provided with the data for making the next step. The initial data are arranged in a pattern which, by borrowing a term from crystallography, we may call a 'space-lattice'. Wherever in the lattice a pressure was given, there the numerical process must yield a pressure. And so for all the other meteorological elements.

He mentioned briefly, in the proper sequence, the operations which had already been adequately described and then discussed in more detail the remaining problems. He stated unapologetically: 'The procedure to be described is certainly very complicated, but so are the atmospheric changes'[26]. Many years later he made a similar remark about his mathematical treatment of wars.

For the actual calculations, Richardson prepared a set of 23 computing forms, copies of which were placed on sale separately from the book 'to assist anyone who wishes to make partial experimental forecasts from such observational data as are now available'[27]. He himself subsidised the printing of these blank forms from his own pocket—I wonder how many times they were ever used?

In Chapter IX, which has been described as the quintessence of the book, Richardson gives details of his worked example of a weather forecast—the first attempt ever made to do this by numerical methods. Strictly speaking, it could be argued that it was not a prediction, in that the result referred to something that had happened several years earlier. The calculations were based on the state of the atmosphere over middle Europe at 0700 h GMT on 20 May 1910. To the extent possible, Richardson used the actual observations as reproduced in Hergesell's publication, but for some elements and locations he had to supplement these by reading off values from Bjerknes' maps. Starting from this initially observed state, Richardson applied the methods described in the preceding chapters of his book to calculate changes, in the next six hours, of pressure, wind, temperature etc at the centre of two of his lattices. Even this limited forecast occupied him for six weeks, in the intervals, it will be recalled, of transporting wounded soldiers in France. The results were little short of disastrous, the most spectacular error being a calculated change of pressure of more than 145 mbar at a place where the actual change was only about 1 mbar. The difference between the highest and lowest pressures ever recorded over the British Isles is less than 130 mbar! Richardson considered that the errors could have arisen from 'the errors of observations with balloons, or from the finite differences being too large, or thirdly from the process by which the winds at points, arranged in a rectangular pattern, are interpolated between the observing stations'[28]. It is now generally recognised that the principal cause lay in the initial wind data, in which case Richardson may have had some justification for his claim that the forecast changes formed 'a fairly correct deduction from a somewhat unnatural initial distribution'[29].

Chapter X deals once more with the problems arising from the turbulence of the atmosphere, but this time from the point of view of its effects on the observations. As a result of the turbulent eddies, the wind, temperature and other weather elements all fluctuate quite substantially; at one moment the wind speed may be 15 knots and a few seconds later it may be 20 knots. How can the observations best be smoothed so as to obtain the most representative value? Richardson considers five different ways in which this might be done but they all have serious disadvantages and he does not come out strongly in favour of any particular solution. This is not surprising, for even today the non-representativeness of the observations is a very serious problem.

The last substantive chapter is entitled *Some remaining problems*, the two most important being the incompleteness of the observing systems and the lack of adequate computing facilities for carrying out the calculations with sufficient speed for operational purposes. This topic is introduced as follows[30]:

> The scheme of numerical forecasting has developed so far that it is reasonable to expect that when the smoothing ... has been arranged, it may give forecasts agreeing with the actual smoothed weather. When that stage has been attained, the other difficulties will tend to group themselves with questions of the desirability of weather forecasts and their cost. We need here an estimate of the economic value of a forecast reliable for n days ahead, given as a function of n. As with improved methods n is likely to increase, so forecasts will become of more value to agriculturalists. Now the annual value of the world's food crops is at least £1000,000,000, so that a very tiny fractional saving would correspond to a large sum.

Since Richardson's day, many studies have been made of the costs and benefits of meteorological services. The costs of providing the services can readily be determined but it is more difficult to make a reasonable estimate of the benefits, some of which are almost impossible to quantify. In 1966 The Director-General of the Meteorological Office, B J Mason (later Sir John), made a bold attempt to assess the value of meteorology to all branches of the national economy, including agriculture[31]:

> The annual value of the United Kingdom agricultural production (including horticulture and forestry) is about £2,000 m. If all weather forecasts and meteorological advice were accurate and properly utilized by all farmers there is no doubt that the productivity would be much increased, surely by at least 5 per cent or £100 m per annum. Although the

present meteorological services are neither perfect nor fully utilized, their present contribution can hardly be less than £20 m per annum or 1 per cent of the annual gross production.

Mason concluded that the probable overall benefit/cost ratio of the civil national weather service was about 20 to 1, a figure which has been widely quoted.

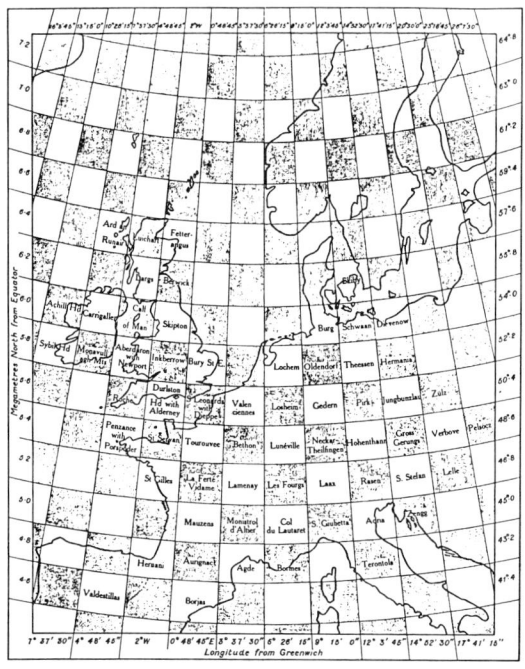

Figure 5.2 Richardson's idealised network of meteorological stations, designed to fit with the mechanical properties of the atmosphere.

On the problem of obtaining better initial observations, corresponding to the first part of Bjerknes' 1904 programme, Richardson first complained about the irregular distribution of the observing stations. This is, in fact, a recurring theme in the book. In the preface he wrote[32]:

The present distribution of meteorological stations on the map has been governed by various considerations: the stations have been outgrowths of existing astronomical or magnetic observatories; they have adjoined the residence of some independent enthusiast, or of the only skilled observer available in the district; they have been set out upon the confines of the British Isles so as to include between them as much weather as possible;

or they have been connected with aerodromes in order to exchange information with airmen. On the map the dots representing the positions of the stations look as if they had fallen from a pepperpot. The nature of the atmosphere, as summarized in its chief differential equations, appears to have been without influence upon the distribution.

A map showing what he felt would have been a rational distribution was reproduced in the frontispiece. In spite of the efforts made under the World Meteorological Organization in its World Weather Watch programme, we are still far from attaining this ideal.

Richardson then considered how it might be possible to obtain more and better upper-air observations. Here it must be remembered that at that time radio was still in its early stages of development; it was not until 1928 that the first successful ascent was made with a radiosonde, a device in which sensors to measure pressure, temperature and humidity are sent aloft attached to a balloon and the measurements are immediately transmitted to the receiving station on the ground by radio. We have already described some of Richardson's ingenious methods for making such measurements, and more will be said later.

In the next section, on the speed and organisation of computing, we at last come to Richardson's fantasy, which had been responsible for triggering his research on weather forecasting. It surely merits being quoted *in extenso*[33].

> It took me the best part of six weeks to draw up the computing forms and to work out the new distribution in two vertical columns for the first time. My office was a heap of hay in a cold rest billet. With practice the work of an average computer might go perhaps ten times faster. If the time-step were 3 hours, then 32 individuals could just compute two points so as to keep pace with the weather, if we allow nothing for the very great gain in speed which is invariably noticed when a complicated operation is divided up into simpler parts, upon which individuals specialize. If the co-ordinate chequer were 200 km square in plan, there would be 3200 columns on the complete map of the globe. In the tropics the weather is often foreknown, so that we may say 2000 active columns. So that 32 × 2000 = 64,000 computers would be needed to race the weather for the whole globe. That is a staggering figure. Perhaps in some years' time it may be possible to report a simplification of the process. But in any case, the organization indicated is a central forecast factory for the whole globe, or for portions extending to boundaries where the weather is steady, with individual computers specializing on the separate equations. Let us hope for their sakes that they are moved on from time to time to new operations.
>
> After so much hard reasoning, may one play with a fantasy? Imagine a large hall like a theatre, except that the circles and galleries go right round through the space usually occupied by the stage. The walls of this chamber are painted to form a map of the globe. The ceiling represents the north

polar regions, England is in the gallery, the tropics in the upper circle, Australia on the dress circle and the antarctic in the pit. A myriad computers are at work upon the weather of the part of the map where each sits, but each computer attends only to one equation or part of an equation. The work of each region is coordinated by an official of higher rank. Numerous little "night signs" display the instantaneous values so that neighbouring computers can read them. Each number is thus displayed in three adjacent zones so as to maintain communication to the North and South on the map. From the floor of the pit a tall pillar rises to half the height of the hall. It carries a large pulpit on its top. In this sits the man in charge of the whole theatre; he is surrounded by several assistants and messengers. One of his duties is to maintain a uniform speed of progress in all parts of the globe. In this respect he is like the conductor of an orchestra in which the instruments are slide-rules and calculating machines. But instead of waving a baton he turns a beam of rosy light upon any region that is running ahead of the rest, and a beam of blue light upon those who are behindhand.

Four senior clerks in the central pulpit are collecting the future weather as fast as it is being computed and despatching it by pneumatic carrier to a quiet room. There it will be coded and telephoned to the radio transmitting station.

Messengers carry piles of used computing forms down to a storehouse in the cellar.

In a neighbouring building there is a research department, where they invent improvements. But there is much experimenting on a small scale before any change is made in the complex routine of the computing theatre. In a basement an enthusiast is observing eddies in the liquid lining of a huge spinning bowl, but so far the arithmetic proves the better way. In another building are all the usual financial, correspondence and administrative offices. Outside are playing fields, houses, mountains and lakes, for it was thought that those who compute the weather should breathe of it freely.

The 'staggering' figure of 64 000 computers is considerably larger than the 500 which Shaw had mentioned in his letter to Bjerknes. Even so, it has since been pointed out that the revised estimate was based on an uncharacteristic error by Richardson, the correct figure being no less than 256 000[34]. Richardson's light-hearted mood spilt over into the next section which begins[35]:

It is conceivable that by a change of variables the equations could be much shortened. But as we are always required in the end to arrive at quantities of direct interest to the public, namely wind, rain, temperature and radiation, so it may be that analytical simplicity does not simplify the arithmetic. There is a tale of a philosopher who succeeded in reducing the whole of physics to a single equation $H = 0$, but the explanation of the meaning of H occupied twelve fat volumes.

The final chapter summarises the units and symbols used in the book.

Richardson had included so many different quantities in his equations that the English and Greek alphabets were insufficient to allow him to assign a separate letter to each parameter. He therefore introduced a number of Coptic letters and even invented a few rather picturesque symbols of his own, such as a small leaf to represent the rate of evaporation from foliage. Recognising that a mathematician may be able to 'grasp the purport of a book if he knows the meaning of the symbols', Richardson would have liked to have explained his symbols not only in English but also in 'the second language for mankind'[36]. In the absence of any international agreement on the subject, he selected Ido, partly as this enabled him to consult his brother Gilbert—an enthusiastic advocate of Ido, it will be recalled—on the correct translation of the more obscure expressions. The English terms include two new words, 'scatterivity' and 'turbulivity', neither of which have found their way into regular usage—not even by meteorologists. The only new word which he proposed for Ido was *entropio* for entropy, which seems fairly non-controversial.

Мечта Ричардсона.

Figure 5.3 A Russian artist's concept of Richardson's fantasy weather forecast factory (Gandin 1965).

In the last paragraph of the book, Richardson expresses some hesitation about the use by Napier Shaw of the expression 'lapse rate' for the change of temperature with height (unlike Richardson's two new words, this term is now in daily use); he would have preferred 'up-gradient' or 'down-gradient' as the case may be. The book concludes with the sentence: 'In the foregoing pages "grade" is sometimes used for "gradient" but that was perhaps a mistake'[37]. What an anticlimax to such a monumental work! It would surely have been better to finish with something on the lines of what he said in the preface[38]:

> The scheme is complicated because the atmosphere is complicated. But it

has been reduced to a set of computing forms.... Perhaps some day in the dim future it will be possible to advance the computations faster than the weather advances and at a cost less than the saving to mankind due to the information gained. But that is a dream.

It will be recalled that the Royal Society had in 1916 made a grant of £100 towards the cost of publishing the book. During the 5-year interval between this and the submission of the final version of the text to the printers, Richardson had revised the manuscript very considerably with the result that the cost of publication turned out to be far higher than the original estimate. In June 1921, the Cambridge University Press asked Richardson if he could obtain a further subsidy of the same amount. He was not optimistic about the possibilities and became even less hopeful two months later when he heard that the Royal Society were not willing to vote any additional funds. He informed the publishers of this decision and added 'I am in the painful predicament of having to draw on my personal resources. I send herewith a cheque for £50'[39]. As he still needed £50, he called on the Director of the Meteorological Office (by that time Dr C G Simpson) who took a sympathetic interest but said that a formal application would have to be forwarded to the Treasury as he himself had no direct control of money. In reporting this to the publishers, Richardson asked rhetorically 'How much does it cost the nation to make a grant of £50?'[40]

In the end the grant was authorised and the book duly appeared at the beginning of 1922 at a price of 30 shillings—by no means cheap for those days. The publishers described the sales as being 'certainly not spectacular'[41] and although only 750 copies were printed they were still able to supply a copy nearly 30 years later. In spite of the subsidies, the Cambridge University Press were out of pocket, a fact that Richardson would remember with regret in 1938 when he had occasion to seek a publisher for another book.

We now know that Richardson's dream (not his fantasy) came true. But few of the book's reviewers held out hopes of any such expectation. The first review appeared just a few days after the publication date in the *Methodist Recorder*, the main interest of the writer being that by that time Richardson was teaching at the Methodist training college. One of the purposes of a review—not always honoured—is to assist the reader in deciding whether to buy the book, to borrow it from a library or simply to ignore it. In this case the advice was clear:

> Let none of our readers be deceived by the title, and think that they have here some easy process whereby they will be enabled to discover suitable weather for Sunday School excursions. A stiffer work in mathematics would be hard to find, and to the lay reader, as he turns over the pages, there is the additional irritation of coming across every now and then

paragraphs which give him hope ... only to find himself once more plunged in a page of mathematics.

The reviewer nevertheless congratulated the author on his 'remarkable volume' and also Westminster College 'that he should be one of its staff'[42]. He then stated that Richardson had translated the English technical terms into Ido because the Meteorological Office was distributing the book to all the national observatories in the world. I have not been able to find any evidence to support this statement and suspect that the only observatories to receive free copies where those in Britain; Richardson's use of Ido was simply an expression of his international outlook.

The only other review in a non-scientific publication (it is strange that the book was not even mentioned in *The Friend*) was by an anonymous 'meteorological correspondent' in the *Manchester Guardian*—the style of his writing makes me wonder if he could possibly have been Sir Napier Shaw? He began with a lengthy extract from Richardson's account of his fantasy and commented: 'At first blush, it must be confessed, this sounds like the rhapsodies of an irresponsible visionary. Actually, it is an attempt to picture a "forecast-factory" of the future in which the weather of the whole globe will be predicted on highly rational and scientific lines'. The reviewer then noted that the rate of improvement in weather forecasting was lagging behind the advance of knowledge in other branches of meteorology. He felt that Richardson's book showed a method whereby forecasts could be greatly improved but that there were several practical reasons why it could not be brought into operation at once. He concluded: '... it will be for future generations to consider the desirability of reorganising their forecasting systems in accordance with the new methods. One has little hope of seeing it done unless we can breed a race of millionaires to find the money and of Richardsons to organise the work'[43].

In the course of the next few months reviews were published in various scientific journals on both sides of the Atlantic. One of the most enthusiastic was in the American *Monthly Weather Review*; the enthusiasm may have been in part attributable to the youthfulness of the reviewer, Edgar W Woolard, who was only 23. After an excellent summary of the historical background and of Richardson's method, he concluded[44]:

> The book is an admirable study of an eminently important problem; being a first attempt in this extraordinarily difficult and complex field, it necessarily possesses, self-confessedly, many imperfections; and is by no means, of course, the final word; however, it indicates a line of attack on the problem, and invites further study with a view to improvement and extension .... Perhaps the most serious handicap to studies in this field

is the lack of adequate observational material. So far as the purely mathematical difficulties which the complexity of the subject introduces are concerned, they are surmountable. For the Philosphy of Mathematics teaches us that the power of pure mathematics is unlimited—its development can not be stayed—it meets difficulties by a creative act which leaps over them; whence the nature of the peculiar role it plays in the physical sciences. It is sincerely to be hoped that the author will continue his excellent work along these lines, and that other investigators will be attracted in the field which he has opened up. The results can not fail to be of direct practical importance as well as of immense scientific value.

Contrast this with the words of a fellow American, Alexander McAdie, professor of meteorology at Harvard University (Woolard was serving in the US Weather Bureau). Writing for the *Geographical Review*, he surmised that Richardson's book 'can have but a limited number of readers and will probably be quickly placed upon a library shelf and allowed to rest undisturbed by most of those who purchase a copy'. He nonetheless recognised that the book was 'perhaps the most remarkable of all meteorological treatises.... It is a strikingly bold attempt to wake up weather forecasters; indeed, to make forecasting a science rather than an art'[45]. He also admitted that Richardson did not lack a sense of humour.

In Britain, the reviews were by some of the most distinguished meteorologists of the day. Let us give pride of place to their doyen, the man who had continued to support Richardson in spite of some trials and tribulations and had in fact given him the opportunity to carry out his meteorological researches—Sir Napier Shaw. In a 3-page review in *Nature*, he hailed the book as a *magnum opus* on weather prediction and proceeded to analyse the contents in his own amusing but somewhat discursive style—not exactly 'Shavian'! For example, he invokes Lewis Carroll in the following cautionary words inspired by what he felt was a slight inconsistency by Richardson: 'Mathematicians in dealing with the elusive atmosphere are not infrequently inspired by Jabberwocky,

> One two, one two, and through and through,
> The vorpal blade goes snickersnack;

but they ought to make sure that they get the right Jabberwock by the neck before "Galumphing back" with his head'[46]. He refers to Richardson's fantasy as an orchestra of computers and remarks that the 'trial specimen is not such a good example of the art of forecasting that it tempts the reader forthwith to become one of the great orchestra'. He continues[47]:

> .. the wildest guess .. at the change in this particular element would not have been wider of the mark than the laborious calculation of six weeks.

Nor is that all. Many of the chapters end in parenthetic expressions of regret or of suggestions for improvement. There are also many supplementary paragraphs which indicate that when the author comes to make another edition, as he or somebody else undoubtedly will, he will write somewhat differently. And the reader will not be sorry, for in many ways the book makes hard reading. It is full of mathematical reasoning, a good deal of which is conducted 'by reference'. The reader who wishes to follow it must have a very handsome library and a few stepladders which Mr. Richardson does not provide. A reviewer with less than the ordinary sufferance of his tribe might easily murmur: forecasting by numerical process seems so arduous and so disappointing in the first attempts that the result is a sense of warning rather than attraction. He might also wonder for whom the author is writing, and regard the book as a soliloquy on the scientific stage. The scenes are too mathematical for the ordinary meteorologist to take part in and too meteorological for the ordinary mathematician. But such complaint would be as misleading as the computed forecast. On the road to forecasting by numerical process nearly every physical and dynamical process of the atmosphere has to be scrutinised and evaluated; the loss of view into the future from the first summit is compensated many times by the insight which one gets into the working of Nature on the way.

A very different review appeared in the *Mathematical Gazette*. It was by a much younger man, N K Johnson, who was later to follow in Napier Shaw's footsteps as Director of the Meteorological Office. At the time he was head of a team of meteorologists carrying out some brilliant research at the chemical warfare station at Porton on Salisbury Plain on atmospheric turbulence (in view of its effects on the distribution of poison gases). He devoted half of the review to a discussion of this and other applications of mathematics in meteorology. He wrote[48]:

> The consummation of the mathematical method is to be found in this latest work of Mr. Richardson. The author compares his book with the *Nautical Almanac*, but the task is in reality a far more complex one .... The originality of the whole conception is such as one has learnt to associate with Mr. Richardson's name. This process of forecasting is necessarily extremely laborious, and for this reason alone has little chance of competing with the present empirical method, for the present at least. But the main value of the book lies in the fact that it presents a co-ordinated dynamic treatment of meteorological processes which has not hitherto been attempted.

In a rather short review in the *Philosophical Magazine*, H Jeffreys (later Sir Harold) expressed the view that Richardson's method[49]

> is one that appeals strongly to the mathematical physicist. It is necessarily laborious in its present form, and probably could not be worked with sufficient speed to make it a practical method of forecasting; but when forecasters have acquired experience in its use, they will probably find a sufficient number of the quantities allowed for are comparatively small to

make it possible to expedite the calculation considerably without great sacrifice of accuracy. The value of the work is not confined to the application to forecasting, though the possibility of predicting the disturbing occasions when cyclones cause merriment in the daily press by moving in the wrong direction makes this the feature of most general interest.

Jeffreys gave no indication of the errors in Richardson's trial forecast.

Sidney Chapman, who later wrote an introduction to the paperback edition of the book, reviewed it for the *Quarterly Journal of the Royal Meteorological Society*. He began: 'The enterprise contemplated in this book is of almost quixotic boldness'[50]. After comparing the respective merits of Richardson's and Bjerknes' methods, he describes the results of the computed forecast as being 'rather discouraging'. He felt that the absence of data from uninhabited regions would always seriously limit the systematic application of numerical forecasting and concluded[51]:

> Perhaps the best immediate outcome that can be hoped for at present is that Mr. Richardson's discussion may help towards the solution of atmospheric problems by assisting direct research (after the event) into specially simple or peculiar states of weather. It may also help by calling attention to the kind of observations most needed for theoretical studies. The book is in any case one of great interest and individuality, and full of evidence of a mind widely versed and keenly interested in every phase of meteorological activity.

A former colleague of Richardson, F J W Whipple, reviewed the book for the *Meteorological Magazine*. As scientists with a mathematical bent combined with experimental skill, the two men had much in common and the review is duly appreciative[52].

> The mathematician when confronted with a physical problem usually simplifies the conditions so as to make it amenable to his analytical machinery. The sledge-hammer method developed by Mr. Richardson requires no such preliminary paring down of awkward corners; it demands, however, conscientious attention to detail and no shirking of laborious arithmetic.

Later on he remarks that even if Richardson's scheme were perfected[52],

> the really interesting weather would not be forecasted, for thunderstorms and tornadoes as well as the secondary features of the cyclones of the temperate zone would be 'smoothed out'.... In view of this criticism, which, it is fair to say, is put forward by Mr. Richardson himself, it hardly matters that the one forecast six hours ahead at one place, which he has computed, is sadly in error.

This review of reviews would hardly be complete without mentioning the opinion of at least one of the leading European meteorologists, Felix

M Exner, professor of geophysics at the University of Vienna. He had achieved a considerable reputation, not least as a result of his textbook on dynamic meteorology, the first edition of which appeared in 1917. In the *Meteorologische Zeitschrift* he wrote[53]:

> One will succeed only gradually in extracting new ideas from this work, not only because the reading is difficult, but also because it contains too much heterogeneous material to be satisfactory. It is the opinion of the reviewer that the author would have rendered a better service to meteorology if he had set forth his basic studies of the separate processes independently of each other. A very valuable theoretical work would have been the result, whereas there will be only a few readers who will work through the book in its present form, for it carries too much of the stamp of personal intent. Whoever is convinced at the outset—and this will be the case with the majority of meteorologists—that the way to weather forecasting mapped out by Richardson is wrong or at least is very premature, will hardly muster the patience to study this work. It would be very advisable, therefore, for the author either to make known in separate papers what is new in his calculations, or—still better—to publish a book on theoretical meteorology, which is free of the design of and immediate application to prediction.

Exner struck a somewhat more personal note in a letter to Richardson: 'I should really be glad if your book would be made easier to read....For us Germans it is really very difficult to study the book and I should want so necessary [sic] to know all about it'[54]. He later sent Richardson a presentation copy of the second edition (1925) of his *Dynamische Meteorologie*, which is now in my possession. As with many books from Richardson's personal library, there are copious marginal notes on some pages from which it can readily be seen which parts he found to be of special interest. He rated the book very highly and wrote to the Cambridge University Press in support of a proposal by Exner that it should be translated into English—but without success.

Summing up, it can be said that *Weather Prediction by Numerical Process* was widely reviewed and was generally assessed as a remarkable achievement, combining great originality with laborious perseverance. Most of the criticisms were very fair and some extremely pertinent. I especially like Shaw's reference to a 'soliloquy on the scientific stage' and Chapman's suggestion of 'quixotic boldness'. I can also sympathise with Exner's difficulties, for the book makes difficult reading even to meteorologists of English mother tongue.

In view of the difficulties in implementing Richardson's scheme for numerical weather prediction, and bearing in mind the poor results of his attempts to demonstrate the method, it is hardly surprising that his book's immediate impact on the routine work of the meteorological services was negligible. Reminiscing in 1982 about his impressions on joining the British Meteorological Office in 1927, R C Sutcliffe wrote[55]:

> ...that was more than half a century ago when British weather forecasting unsupported by any systematic research seemed to the newcomer to be in the doldrums expanding organizationally but going nowhere scientifically. L.F. Richardson's dynamics had passed over with not a ripple in the weather service and even the much talked of frontal and airmass concepts fresh, or fairly fresh, from Norway had made little impression on London practice. Weather charts were not 'analysed' and there was no recognizable technique or methodology of prediction.

The above comments appeared in Sutcliffe's obituary of C K M Douglas, whom he described as having been for a long time the most distinguished synoptic meteorologist and practical weather forecaster in Britain. Douglas' success was partly due to his prodigious memory of past weather situations but he was also a very competent theoretician. In surveying the weather forecasting methods in vogue in 1931, he wrote[56]:

> Mention should be made of Richardson's gallant effort to work out a scheme designed to place weather forecasting on an exact basis. Unfortunately the attempt must be classed as a splendid failure, owing to the complexity of the whole problem and particularly the difficulty of dealing with small irregularities.

In retrospect, it is easy to understand why Richardson's book had no immediate practical impact. What is more surprising is that, in view of its contribution to meteorological theory, it received little recognition in meteorological textbooks for many years. For example, in two of the most important books on dynamic meteorology in German (by Exner in 1925 and by Koschmieder in 1933), there is but the briefest reference to numerical methods. Richardson's ideas are not even mentioned in the leading American book of that era, *Physics of the Air* by W J Humphreys. The well known Russian book by S P Chromov, published originally in 1934 and in a German translation in 1940, dismisses the subject in one sentence: 'An exact integration of the atmospheric equations is beyond the present means of mathematical analysis; even an approximate solution by numerical integration (Richardson) involved such a complicated and ponderous operation and led to such unfortunate results that it can be of no practical value'[57]. Even in his classic *Physical and Dynamic Meteorology*, Sir David Brunt, who had known Richardson since his Aberystwyth days, only referred to the book in an appended list of supplementary reading. That was in the first edition of 1934; in the second edition of 1939 it is not cited anywhere!

The most positive treatment of Richardson's method in those days was in the second edition of Napier Shaw's *Forecasting Weather*, published in 1923. Although this is a more elementary text than those already mentioned, it includes a separate chapter on weather prediction by numerical process devoted entirely to Richardson. Shaw admitted that his method was unlikely to be adopted by official forecasters but felt that he had clarified many of the processes of weather and had presented a large amount of useful information about the atmosphere. He concluded[58]:

> If visitors from Mars, unfamiliar with engines and, at the same time, familiar with Cartesian co-ordinates, happened upon the engine room of some big ship, they might endeavour to unravel the mystery of that great development of power by dividing the space occupied by the machinery into points arranged according to co-ordinate axes. If they did so it would take a vast amount of computation to arrive at an idea of the constituent parts of the machinery which would certainly be disclosed to them in time if they examined enough points; but as working knowledge what they knew would be less useful.
>
> It is possible to carry the process of minute analysis too far. There is a story of a conversation between Maxwell and Tyndall which may or may not be in print. Talking of the ultimate molecular constitution of matter Tyndall pointed out to Maxwell how curious it was that the pulling force depended on the mutual attraction of separate molecules, and that the traces by which a horse drew a cart were not really continuous, but were held together by attraction at a distance. 'Yes,' said Maxwell, 'but when you are in the cart it is very comforting to know that the traces are there'. In like manner, the grouping of the phenomena of the atmospheres into entities, which are sufficiently persistent to be dealt with as working units in selected situations and which can be recognised in maps, remains as a possible alternative to a generalised system which is equally applicable to every situation and at all times. The application of Mr. Richardson's method to such selected types should prove very illuminating.

Richardson would have been especially interested in the immediate reaction to his book by V Bjerknes and his collaborators in Bergen. Unfortunately, Bjerknes apparently did not write anything on the subject until 1932 when, together with H Solberg, T Bergeron and his son J Bjerknes, he finished his monumental 3-volume work on physical hydrodynamics. In the annotated bibliography at the end of this book, written personally by V Bjerknes, he describes how some of his papers fit into the framework of his 1904 programme. For example, he states that the Bergen system of frontal and airmass analysis was a step towards the realisation of the prognostic part of this programme, but qualifies this by adding 'with the sole difference that in the routine of a meteorological service it is necessary to replace complicated calculations by rapid estimates'[59] He still believed that further progress could be made and his statement continues: 'There is no reason why in due course these rough estimates should not give way to more accurate calculations, especially in view of the publication in the meantime of

L.F.Richardson's remarkable work'[59]. Bjerknes sent a complimentary copy of the French version of this book, from which the above quotations are translated, to Richardson. In acknowledging the gift, Richardson wrote[60]: 'It is a special pleasure to me that I have met all four authors of this important book, which I expect will be a standard work of reference for the next generation'.

As regards Richardson's own book, we have seen how it was received by his contemporaries with a mixture of admiration for his persistence and of ridicule for his failure. His trial forecast was so far removed from reality that it may have served to discourage any further attempts at using numerical methods for weather forecasting. Even so, might it not have been expected that his sheer audacity would have fired the imagination of some of the more enthusiastic and knowledgeable amateur meteorologists? Apparently not, for in 1925 his former science master, James Edmund Clark, wrote a popular article for *The Friend* on modern weather forecasting; there was not even a whisper of Richardson.

There we must leave for the time being the story of Richardson's book on numerical weather prediction and return to the story of his life in Benson. As so much of his time there was spent in revising the manuscript of his book, his social life must have been rather limited. He did, however, revive the local troop of Boy Scouts with Dorothy serving as Cubmistress. One of the assistant scoutmasters was Basil Lewis who also worked at the Observatory. Another was Russell E Munday who became an Exhibitioner at Jesus College, Oxford, where he studied mathematics and then transferred to modern history. His name will go down in the history of meteorology as the joint author with Richardson (1926) of a meteorological memoir entitled *The single-layer problem in the atmosphere and the height integral of pressure*. He remained a good friend of the Richardson family and, probably through Lewis, became interested in eugenics. During a spell in Buenos Aires as a teacher he carried out some tests of the intelligence of children with grandparents from the same nation and compared them with children having grandparents from more than one nation (1932). Richardson had occasion to use the results in an article on war and eugenics, to which we shall return later. Russell Munday also taught for a time in Jersey where he died in 1933 at a very early age; the news came as a great shock to the Richardsons and the Garnetts, whom he had met during holidays on the Isle of Wight.

Some of these holidays had been marred for the Richardsons by Dorothy's miscarriages. In spite of these she still hoped against hope that some day she would have a child of her own (even at the age of 45 she had the courage to tell a niece that, after another disappointment, she would not shirk from 'trying again'[61]). In the meantime, with encouragement from the Garnetts, she and Lewis decided to start adopting children. The main reasons were doubtless their love of youngsters and their desire to help children without a home or parents. An additional motive may well have been the thought that Dorothy's health would benefit from having a child to bring up. It has also been suggested that Lewis was keen to learn about the reaction of children to a caring and relatively affluent home environment; if by good chance he and Dorothy ever had children, he might even be able to contribute to the nature–nurture study which he had proposed to the Eugenics Society.

The first child to be adopted was a 3-year old boy who came to live with them in Benson in February 1920. They named him Olaf Kenneth Morley Richardson, the Olaf in tribute to Lewis' FAU colleague Olaf Stapledon, Kenneth after Dorothy's favourite brother and Morley after Lord Morley, for whom both Lewis and Dorothy had a great admiration. This was not so much because he had begun his political career as member of Parliament for Lewis' home town, Newcastle upon Tyne, but that he had taken a strong stand against both the Boer War and the First World War; his unpopular views led to his resignation from the Cabinet in 1914. He was of course also well known as editor of the *Fortnightly* and as biographer of Gladstone.

The second child to be adopted was another boy, just a few weeks old; they called him Stephen, after Dorothy's favourite saint, and Alexander, after a cousin who, like Kenneth Garnett, had been killed in the war. The family was completed several years later in 1927 by the adoption of a daughter, Elaine Dorothy; the name Elaine came from the legend of King Arthur, probably through the poetry of Tennyson, one of Dorothy's best-loved writers.

The responsibility of bringing up a family does not appear to have affected in any way Richardson's devotion to his scientific work. From this point of view, Benson had for him the big advantage over Eskdalemuir that it provided better opportunities to discuss his ideas with other scientists. He could talk things over daily with Dines and was within easy reach of London to attend scientific meetings at the Meteorological Office and at the Royal Meteorological Society, of which he became a Fellow in April 1919. In November of that year he met, for the first time, Vilhelm Bjerknes, who had been invited to address the Society on the structure of the atmosphere when rain is falling[62]. The President of the Society, Sir Napier Shaw, asked Richardson to open the ensuing discussion as he was the only person, apart from Bjerknes, who was attempting to integrate the physical equations which govern the behaviour of the atmosphere. Richardson remarked that he was greatly honoured to have his name associated with that of Bjerknes and proceeded to amuse the audience by pointing out that the picture which

Bjerknes had shown of a typical cyclone was rather suggestive of a dog chasing its tail—the dog being the cold air, and its tail the warm air.

Bjerknes wrote to his wife that evening[63]:

> ...and so on Monday evening I shall go out into the country to Dines—a special invitation was delivered to me ... by Mr. Robertson [sic], who is at present living there and working with Dines. This R. is a remarkable young man, a gifted theoretician who independently takes the same path that we take. I was therefore not in the least doubt that it was my duty to accept the invitation.

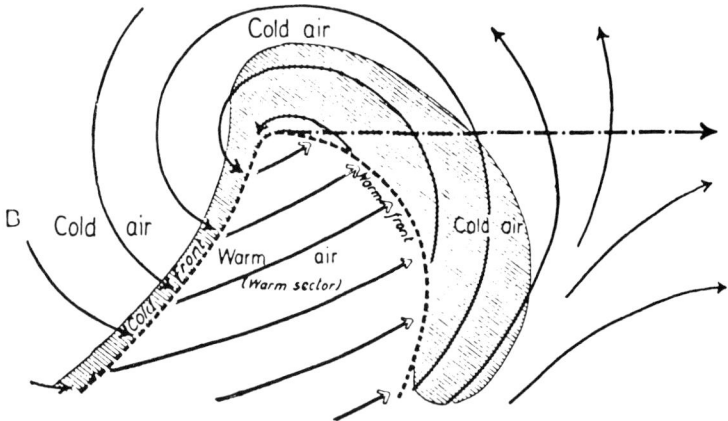

Figure 5.4 Horizontal section of a typical cyclone, according to the Bergen School, which Richardson likened to a dog chasing its tail.

Four days later he wrote to her again:

> Arrived here yesterday where Dines has his observatory and am staying at Mr. Richardson's who is a very valuable theoretician exactly in our direction. I wish Jack could come and stay here a while, he would be most welcome....

Bjerknes had many of the same qualities as Richardson. He was very human but shy, dignified yet with a keen sense of humour, and always ready to lend a helping hand. During his visit to Benson, for instance, he was told by Richardson of some difficulties in obtaining copies of the publications of the Geophysical Institute in Leipzig. On his return to Bergen, he immediately wrote to his successor at Leipzig, Dr Robert Karl Wenger, to seek his assistance. He described Richardson as 'a first-class theoretician, full of ideas, and moreover a tireless worker'. He continued[64]:

> This can best be demonstrated by the fact that, using the Leipzig synoptic charts, he has computed the weather situation 6 hours ahead by means of

the complete system of equations. The result was meaningless, but he has nonetheless made the first attempt, developed the method and proved that it is practical.... This man, who by the way is an antimilitarist Quaker, wants to obtain a full set of Leipzig publications.... We agreed that the interpolar discontinuity was most probably one of the main causes of his failure and he is afraid of attacking the big problem again.

After some further correspondence with Wenger, Richardson received the publications he needed.

By this time Richardson began to realise that the almost ideal working conditions at Benson could not last much longer. He mentioned his worries in a letter to Bjerknes early in 1920[65]:

On Jan. 2 we had another successful flight with a 'cracker balloon'.... So the two researches, for which I came to Benson 11 months ago, are emerging from the experimental into the practical stage. The next thing was to have been a network of stations in rectangular order. But what will happen now I do not know. For personally I am sorry that our good old Meteorological Office has been swallowed up by the Air Ministry, as I do not like preparations for wars. I should very much like to come and work with you at Bergen, if you would permit it; but indeed that is only an idle wish, as I do not see how I could finance it.

To appreciate the significance of this change, it may be helpful to recall that, prior to the First World War, the Director of the Meteorological Office had enjoyed a remarkable degree of freedom from ministerial control, being responsible only to Parliament through the Treasury for the expenditure of the public funds granted for running the Office. Important policy questions were referred to the unpaid Meteorological Committee, which had replaced the Meteorological Council in 1905. To meet the growing demands of the armed forces during the war, three other meteorological services had been set up by the Admiralty, the War Office and the Air Ministry. Although these bodies received support from the Meteorological Office, there was inevitably some duplication and confusion which led to much discussion about the best way of organising things after the war. In December 1918, the Meteorological Committee was informed of a decision by the Committee of the War Cabinet for Home Affairs that the Office was going to be attached to the Government through the Committee of the Privy Council which controlled the grant for Scientific and Industrial Research, which meant, of course, that it would lose its semi-independent status. Although this decision was not apparently announced officially, it was believed by members of the staff that they would in due course come under the Department of Scientific and Industrial Research. This would have been quite acceptable to Richardson as the conditions would have been similar to those in the National Physical Laboratory where he had twice been employed.

The real blow fell on 8 May 1919 when the Committee for Home Affairs changed its mind. The actual wording of their decision was rather nebulous and it was not until a meeting on 9 July that the situation was finally clarified: all the other meteorological services were to be taken over by the Office, which would henceforth come under the Air Ministry[66]. The process of assimilation was of course gradual and was not in fact completed until the middle of 1920. Sir Napier Shaw, who had originally planned to retire in 1915, stayed on to ensure a smooth transition and then handed over the reins to G C Simpson (later Sir George). He himself was appointed on a part-time basis as the first professor of meteorology at Imperial College in London.

We can well imagine the dilemma which faced Richardson when he heard about the future arrangements for the Meteorological Office (an announcement about the impending changes appeared in *The Times* on 31 July 1919). On the one hand he had a job for which he was ideally suited and which suited him ideally, while on the other he was strongly opposed to working directly for the armed services. Looking back at this period of his life just a few months before his death, he wrote: 'To me doubt is mostly pleasant. Once, in 1920, I did experience a doubt so intense as to be distressing: it was as to whether to leave the Meteorological Office'[67]. By the summer of 1920 he had made his decision: he would have to follow the dictates of his conscience by severing his connection with the Meteorological Office. A future Director of the Office, Sir Nelson Johnson, later described this decision as 'one of the tragedies in the history of meteorology'[68]. The late Sir Charles Normand, who was Director of the India Meteorological Department at the time, told me of the shock wave in the meteorological world when the news of Richardson's resignation was received[69].

Richardson was not alone in having misgivings about the transfer of the Office to the Air Ministry. In February 1920, C J P Cave (a well known meteorologist of private means and a close friend of Shaw) presented a paper to the Royal Meteorological Society expressing concern that the funds available for the Meteorological Office would in future be liable to drastic reductions whenever the Government decided to cut defence spending. The Society took the unusual step of adopting a resolution calling for a public inquiry. The members present who worked for the Office and were thus civil servants, including Shaw, were careful not to express opinions nor to take part in the vote for the resolution. Cave again expressed his dissatisfaction in his presidential address to the Society in 1925: 'There was no inquiry into the desirability of this step and no reasons were given ... the public have some right to know why such a drastic change was made'[70]. But Cave and his supporters were unsuccessful and the Meteorological Office remains in the Air Ministry or its equivalent to this day; their fears about inadequate financial support have proved to be largely unfounded and the Office has, in fact, prospered.

Even if Richardson had elected to remain in the Meteorological Office, there are two good reasons for supposing that the favourable conditions for his research would probably not have lasted much longer. The first is that under the new administration basic research was no longer encouraged to the same extent. One result of this was the resignation of one of the top scientists, H Jeffreys (later Sir Harold); he described what happened as follows: 'I had hoped that most of my work would be research in meteorology and that I should do it in the Meteorological Office. However, after Shaw's retirement I found that it would have to be in administration and I left in February 1922 and returned to Cambridge'[71]. The second reason is that the days of Benson Observatory were already numbered. W H Dines resigned in June 1922, leaving one of his two meteorological sons in charge. This proved to be only a temporary arrangement, for all the official work of the Observatory was transferred from Benson to Kew in September 1923. It can well be imagined that the rather informal relationship between the Office and W H Dines was not particularly attractive to Sir Napier Shaw's successor, Sir George Simpson, whose previous experience as Director-General of Observatories in India probably led him to prefer more conventional arrangements. This perhaps helps to explain the lack of enthusiasm in the official announcement of the closure of Benson in the annual report of the Meteorological Office; the text reads 'Administration has been facilitated by closing the somewhat inaccessible station at Benson'[72]. Some of the observatory buildings were dismantled and re-erected elsewhere in Benson for other purposes. The site was taken over in the mid-1930s as part of a new airfield for the Royal Air Force; the only relic in present-day Benson of the great Dines era is a road named 'Observatory Close'. Fortunately for everybody interested in the history of meteorology, the Royal Meteorological Society published as 'an inspiring monument to the memory of a great meteorologist' the *Collected scientific papers of William Henry Dines*[73]. In his copy of this fascinating book, which I have before me as I write, Richardson wrote: 'I should like to record Mr. W.H. Dines' quiet courtesy, the patience with which he listened to argument and the gentleness with which he put forward his carefully thought out opinions'.

CHAPTER 6

Westminster Training College 1920–29

Richardson's resignation from the Meteorological Office took effect on 1 September 1920, but he actually left Benson in July. One reason for his early departure was to visit Bergen in Norway as a member of a small delegation at the invitation of V Bjerknes. As soon as the First World War was over, Bjerknes was keen to bring together meteorologists from both sides to discuss with them the exciting discoveries made by the Bergen School and the means of obtaining better weather observations for the continuation of his research. He had originally hoped that it would be possible to have them all at the same time but in the end he decided that it might be more diplomatic to convene two separate meetings. In his letter[1] to Sir Napier Shaw inviting him to send four or more British meteorologists, he mentioned Richardson's name and expressed the hope that Shaw himself would be able to participate. As Richardson was about to leave the office and as Shaw was on the point of retiring, there would have been good reason for omitting them from the delegation; Shaw was, however, a very broad-minded administrator and he readily accepted Bjerknes' suggestions. In informing a Scandinavian colleague of Shaw's decision, Bjerknes once again referred to Richardson as an extremely talented theoretician and mentioned his Quaker pacifist convictions, in view of which it seemed to be 'not improbable that he may be moved to remain for the meeting with the Germans too'[2].

During the meeting[3], which lasted from 19 – 30 July, there was a series of lectures by participants from both groups. At the opening, V Bjerknes described how a relatively dense network of weather observing stations had been set up during the war, in the hope that this would to some extent compensate for the lack of information from other countries. It had in fact led to the discovery that within a cyclone there are usually two lines of discontinuity, associated with abrupt changes in air temperature, wind speed and direction and other elements. These are of course what we now know as the warm and cold fronts of a depression; in Bergen they were called the 'steering line' and the 'squall line' respectively. Shaw suggested the unwieldy names 'anaphalanx' and 'kataphalanx', but fortunately received little support. Bjerknes then gave a detailed account of his concept of the structure of a cyclone and explained that his theory was not entirely new; somewhat similar ideas can be found in the work of Helmholtz and others—including Shaw. On the following day, Jack Bjerknes spoke about the application of the new theoretical ideas to routine weather forecasting. The visitors had ample opportunity of seeing how this worked out in practice for there was a discussion each day on the current weather situation, based on charts analysed in accordance with the Bergen system. Richardson joined wholeheartedly in these discussions, as did one of his young colleagues, C K M Douglas, later to be described as the most distinguished synoptic meteorologist and practical weather forecaster in Britain. He played a key role in the famous D-day forecast for 6 June 1944 on the accuracy of which the success of the Allied invasion of Normandy depended so heavily; a glance at the published weather maps for that crucial period shows how closely the analyses were based on the concepts of the Bergen School.

The British meteorologists also lectured on recent scientific developments. Richardson gave no fewer than four talks, mainly on his new observing techniques and on numerical weather prediction. One of the few comments on the latter was by Shaw to the effect that when contemplating Richardson's efforts he had been reminded of the lines from Juvenal 'it is pleasant to stand on the heights and watch a ship toiling in the waves'[4].

In his invitation to Shaw, Bjerknes had written: '... the conference will have a perfectly unofficial and informal character. No receptions or feasts will be given, nor will evening dress be required'[1]. This was the kind of atmosphere which appealed to Richardson and he would doubtless have been happy to stay on for the meeting with the German meteorologists but there were too many things he wanted to do back in England before taking up his new job. He did, however, talk to Bjerknes about the possibility of returning the following summer to work in Bergen during his summer vacation—of which more later.

Richardson's main concern on leaving Benson—apart from the need to have an income adequate for looking after his family—was to find a position where he could still carry on with his meteorological research.

The offer of a university chair would have been most welcome at this stage, but this was not to be forthcoming until 20 years later. In early 1920 there was in fact not a single professor of meteorology in the United Kingdom—Shaw's appointment at Imperial College dated only from September that year. Furthermore, although Richardson's brilliance in meteorological research was beginning to be recognised by a few leading meteorologists, he had not yet demonstrated to the academic world at large his suitability for a professorial appointment. He still had only a bachelor's degree, his record as a lecturer at Aberystwyth and Manchester could hardly be described as outstanding—and his book on numerical weather prediction had not yet been published. Finally, it must be remembered that, at least in the immediate post-war years, there was considerable prejudice in university circles against those who had taken a pacifist stand.

Under these circumstances Richardson must have welcomed the announcement of a vacancy for a lecturer in physics and mathematics at Westminster Training College. Although he had no formal qualification in mathematics, he was considered unanimously by the Governing Body to be the most suitable of the three candidates interviewed on 18 May 1920. They could not take a final decision, however, as Richardson had intimated that 'he had another appointment in view and desired a few days grace before giving his decision in regard to the offer'[5]. We do not know anything about this other appointment (was Richardson still hoping for something to come of his approach to Bjerknes?), but in any case he accepted the offer at Westminster and started work there at the beginning of September.

This was a period of rapid development for the College, under its dynamic principal, the Reverend H B Workman, who wanted to raise its academic level to that of a university college. Westminster Training College had been founded in 1851 to provide training for teachers at Wesleyan day schools, of which there were more than 350 in the country at the time. The site for the college was deliberately chosen in the densely populated area of Horseferry Road, London, as the aim of Wesleyan schools was to provide education for the 'lower classes'. Students attending the College came from elementary schools and had completed a five-year apprenticeship as pupil-teachers. After successfully completing the two-year course they were awarded a teaching certificate and most of them were then immediately appointed to the headship of a school. Over the years, the educational standards of the College were raised, and in 1894 two students took an external BA degree from London University. On his appointment in 1902, Dr Workman set out with the deliberate policy of attracting well qualified lecturers for the new science department. One of his early appointments was Dr T Martin Lowry, who later was elected FRS and became professor of physical chemistry at Cambridge after a distinguished war career in explosives. He was succeeded in 1913 by Dr A S Russell, who was expected to return after the war but instead became a reader in physics at Christ Church, Oxford. This created the vacancy which Richardson filled when the College returned to Horseferry Road in September 1920—the buildings had been used as military headquarters for the Australians during the war.

Richardson's starting salary at the College was £450 per annum, well above the average for this type of post at the time. One of the other carrots which Workman dangled before the eyes of prospective candidates was the prospect of having a reasonable amount of time for research work. This was doubtless one of the main attractions for Richardson who is described in the official history of the College as 'perhaps the most brilliant man who ever held a position on the staff', and as 'a fiery man in many ways, a law unto himself in all ways and once he had arrived the compilation of the time-table became no easy task'[6]. Elsewhere it is recorded that, when Richardson took up his appointment, the laboratory was in a 'state of chaos', but that he 'soon restored it to a condition more congenial to his own orderly mind'[7].

The trend towards College students taking London external degrees continued and by 1924 all had to stay at the college for three years and take a degree. Richardson had, of course, graduated at Cambridge and, again according to the official College history[8],

> was not conversant with the London University Intermediate examination, so, partly to satisfy his curiosity, he sat the examination himself. One of the examiners later told how Richardson had spent the entire time allotted to one of the mathematics papers in pointing out the various solutions which could be given to the various interpretations of the question!

However this may be, he passed his Intermediate examination in July 1922 and, not satisfied with that, went on to take a pass BSc in psychology with pure and applied mathematics in 1925. Richardson later wrote that this was the first mathematical degree that he had tried for: 'I was placed in the first division. Of course I have read, understood and enjoyed mathematics widely beyond that standard, but I cannot memorize even the formula connecting $\cos^2 \theta$ with $\cos 2\theta$'[9]. Some corroboration for this is provided by the story that he was observed by a student one day working out the square root of 9 on a slide rule!

When I read that Richardson had taken the London examination 'partly to satisfy his curiosity', I myself became curious. What were his other motives for this unusual action? I found a possible clue in the following extract from the obituary of W H Dines, written by Sir Napier Shaw; '...he held resolutely to the opinion that a self-respecting graduate could not apply for the higher degree of M.A. so long as it could be acquired simply by payment'[10]. Without an MA, Dines was precluded from taking a doctorate at Cambridge and he remained a simple Bachelor of Arts all his days. Could it have been, I wondered, that Richardson shared Dines' views on this, as in so many other matters? If so, might his motivation not have been a desire to obtain a doctorate from London University, thereby avoiding the need to purchase an MA from Cambridge? Confirmation that this had indeed been the case ultimately came from somebody who was very close to the Richardson family in the 1920s, the Reverend Donal Browne—about whom more will be said shortly. He remembers quite clearly that Richardson considered on principle that higher degrees should be earned rather than bought. This had in fact made such an impression on him as a youngster that he himself had refused to buy an MA from Oxford University.

The regulations at London University differ from those at Cambridge in that candidates for a Master's degree have to sit a written examination. The regulations are not however completely rigid, as can be seen from a minute of a meeting of the University Senate in October 1925[11]:

> Mr. L.F.Richardson, B.Sc.(Lond.), B.A.(Cantab.) and a fellow of the Royal Institute of Physics [sic] requests for permission to become a candidate for the D.Sc. without passing the M.Sc. examination. In support of his application the candidate submits a list of 25 works published in book form or in journals of various learned societies. The candidate acted as special external examiner in M.Sc. mathematics. His request was approved.

Richardson did not waste any time. Within a few months he applied for a DSc in physics on the strength of his published work. The 18 papers on mathematics, meteorology and physics which he submitted are reported to have caused some consternation to the examiners—it would have taken them several months to examine all this material. The historian of Westminster College was told that each of the papers was 'worthy of the degree'. In any case, Richardson was awarded his DSc on 24 March 1926. Later that year he was elected Fellow of the Royal Society, the highest scientific honour bestowable in Great Britain. In congratulating him, the editor of the Westminster College magazine wrote[12]: 'We are pleased he thought it advisable to add these distinguished letters to his name'.

Most scientists cease to seek further academic qualifications once they have been awarded a DSc or equivalent. Not so with Richardson. The inclusion of psychology in his pass degree in 1925 was indicative of his increasing interest in this subject and he now decided to take an advanced course by attending evening classes at University College,

London—we shall see later how he challenged the scientific findings of one of his lecturers, Dr S J F Philpott. In 1929—at the age of 48—he passed the London external special BSc in psychology with second class honours.

In following Richardson's academic attainments, we have jumped ahead of the rest of the story. Although he was always held in high regard by his colleagues and students at Westminster, this was thanks to his intellectual stature rather than his success as a teacher. He was generally felt to be 'a man of outstanding ability ... absorbed with problems of a scientific nature which lifted him above ordinary mundane affairs'[13]. One of his students described how he used to write everything on the blackboard in his slow methodical way, with the result that too little was achieved by the end of the lesson. Nobody dared to pull his leg, for he had the reputation of being very fierce if roused from his usual calm. The arts and crafts teacher, N C Patten, told me that some students once borrowed a few accumulators from the physics laboratory for lighting some lamps needed for a play. When Richardson found out about this he was absolutely furious, because they had not asked for his permission. This was by no means the only occasion on which he lost his temper. There is another story which, even if not entirely true, illustrates how his colleagues found him odd. Richardson, who had a sweet tooth, invariably arrived late for lunch in the staff dining room and to make sure of having some dessert he would start his meal with this and work his way backwards to the first course if any was still left.

At Westminster Richardson soon found opportunities to apply his ingenuity in the design of simple apparatus for teaching purposes. In 1923, he gave a practical demonstration of one such device at a meeting of the Physical Society of London. He mounted two bicycle wheels side by side and wound a piece of stranded picture wire round the rims and over the pulley of a small electric motor in such a way that the wheels would be driven in opposite directions. When the motor was switched on an electro-motive force was generated by the revolution of the wheels in the Earth's magnetic field. This force could easily be calculated and could also be measured by means of a galvanometer connected to the hubs of the wheels; agreement to within about 4 % was attainable. This, claimed Richardson, was a means of introducing students at an earlier stage than usual to a quantitative conception of electromagnetic induction[14].

A few years later he collaborated with seven of his students in writing a paper for the Physical Society on a simple means for the absolute measurement of an electric current. He pointed out that although such absolute measurements were included in the syllabus for the BSc examinations in physics, the theory of the most important instrument for measuring electric current—the current balance at the National Physical Laboratory—went beyond the level of mathematics called for in the syllabus. He had therefore designed a current balance which removed this inconsistency. The students taking part must have been thrilled to be involved in original work—and to see their contribution acknowledged in a scientific journal[15].

Students taking the practical examinations in physics were required to make two experiments in the allotted time, but as the experiments were not the same for all the candidates it was difficult for the examiner to place them in order of merit. Richardson had accordingly introduced a new method whereby the experiments were identical for everybody: he also increased the number of experiments so as to cover more of the syllabus. In consultation with a classical colleague, he called the new type of examination *pantopiric*, derived from Greek words meaning attempted by all. A report on the results obtained from a trial of this new system at Westminster College was presented at a discussion on practical examinations organised by the Physical Society[16].

At this time, the Richardsons were living in Bridge Lane, Golders Green. Their choice of this residential area may have been influenced by its proximity to Hampstead Heath where Lewis could walk and find relative solitude in the wide open spaces. Another factor was doubtless that Dorothy liked to be near the Garnett home. Her brother Maxwell's children were among the frequent visitors at Bridge Lane, for, as always, the Richardson home was a mecca for youngsters. It was a typical semi-detached house with barely sufficient accommodation for the growing family, but one room on the ground floor was converted into a laboratory/study, where Lewis could read, make scientific experiments and write his papers. As the room was rather dark, he erected a large mirror in the garden outside to reflect in as much daylight as possible. He also placed a notice on the door 'No wogging outside the lab.', wogging being the family slang for making too much noise. At the rear of the house he constructed a coal-shed of his own design. The special feature was the roof, made of canvas mounted on bamboo to give it rigidity: as the coal stock became depleted, the roof could be rolled back as necessary, thereby avoiding the need to duck down.

After her earlier miscarriages, Dorothy had been nursed back to health by Ellen Swan, who had been nanny with the Garnett family since 1895. She came to live permanently with the Richardsons after the adoption of the two boys, and by all accounts she was in many ways closer to the children than Dorothy who was more interested in their education and spiritual well-being than in their material needs. In her late 60s, when confronted with a grandchild in dire distress, Dorothy had to confess that she had never changed a nappy in her life. Having Ellen made it possible for her to engage in good works outside the family, as for example her weekly visit to a youth club in Clerkenwell, a somewhat poorer district than Golders Green—rumour has it that Charlie Chaplin used to frequent this same club as a boy.

There are several stories about manifestations of Richardson's sense of humour during this period of his life. In his lectures on magnetism at Westminster, for example, he liked to tell the students that Nelson upset ships' compasses—because he had an iron constitution. Stephen remembers an occasion when his father had discovered that by slitting a piece of card in a certain way it could be made to produce a loud noise when banged against an object. One day some visitors at Bridge Lane were taken aback when Lewis told them that a serious matter had come up and that he would have to punish his children. 'I want you to know that I do this', he added. Thereupon he used the card to spank Olaf and Stephen, with whom the whole joke had been carefully rehearsed. It did not hurt in the least but made an ear-splitting sound which greatly alarmed the visitors. The two boys rather spoilt the effect by bursting into laughter in the middle.

Sometimes his humour verged on what the Germans call *Schadenfreude*. During a camping holiday at Wheelbirks, Lewis demonstrated to his family the correct way to cut wood. Dorothy proceeded to cut some wood by some other method and cut her hand in the process. Instead of commiserating with her, Lewis found the incident very entertaining. This story illustrates another aspect of Lewis' character, his appreciation of manual skills. Stephen recalls his father pointing out how a professional labourer develops a steady rhythm—in digging, for example—which he can keep up all day, whereas a novice would rush at the job and soon become tired.

Not content with having adopted Olaf and Stephen, the Richardsons also opened their home to three children of an Anglican missionary while he was working for the Church Missionary Society in China—the two families had come into contact through Dorothy's association with St Michael's School. From about 1920 until 1925, William, Aileen and Donal Browne spent all their school holidays with the Richardson family. There was little rivalry between the two groups of children; Olaf and Stephen were several years younger and tended to be left in the care of their nanny, Ellen, while Dorothy herself looked after the Brownes. Some 60 years later Donal still remembers with gratitude the kindness and generosity of his guardians. He writes[17]:

> I was utterly uncouth after seven years up country in China. I learnt a little of how to behave in society and of how to appreciate art and the theatre. We saw the whole of Shakespeare in the Pit at the Old Vic. At this time I had no conversation but I listened and learnt something of really intelligent talk.

Donal Browne recalls being taken behind the stage during the interval in a performance of Eden Philpott's *The Farmer's Wife* to meet Ralph Richardson and his first wife, Muriel Hewitt, who were playing the roles of a rather simple young farmer and an attractive but flirtatious farmer's daughter. Ralph Richardson remarked that it was pleasant to be acting with his wife, to whom he had only been married a few months, in parts which involved them in some light-hearted conversation. The young couple lived for a time in Hampstead, not far from Lewis and Dorothy. Ralph was very fond of his Uncle Lewis and told me that he was[18]

> a warm and delightful man. I used to go and see him and talk to him about his work. He said I gave him an idea—he acknowledged it in one of his published papers—amazing! He did not quite know what I was doing in my job—don't think he went to any theatre at all.

Ralph was of course still at an early stage in his theatrical career and only became established as a leading actor in the West End in the 1930s; he was knighted in 1947.

Donal Browne confirms that Lewis never accompanied them when Dorothy took them to the theatre; he was always too busy with his scientific work when he was not teaching. He did, however, take his responsibilities as a father seriously, and whenever possible he would take the boys to sporting events. Sometimes he would join the whole family on a camping holiday during the summer holidays, but at Easter Dorothy usually went with all the children to a house on the Isle of Wight near the Garnett home. There she loved to organise games of hockey on the beach, boating trips and, no matter how cold the water, lots of swimming.

Donal and his elder brother William both became Anglican ministers while Aileen had a successful career in teaching. She was apparently not so happy at the Richardsons' as her brothers for she felt that Dorothy had inherited the Garnett trait of giving preference to the male members of the family.

The fact that Richardson acknowledged his nephew's suggestion, which many scientists—and apparently Sir Ralph himself—would have regarded as being too trivial to merit being mentioned, illustrates how scrupulous he was throughout his life to give full credit to other people's original ideas. He naturally expected his fellow scientists to reciprocate and was disappointed when he felt that somebody had used one of his ideas without an appropriate acknowledgment. A case in point was when R V Southwell published a series of papers in the mid-1930s on a new mathematical technique, which he called 'relaxation', for the approximate solution of certain types of equations[19]. Richardson considered this to be similar to his own finite difference method and suggested to Southwell that he should have acknowledged this earlier work. Southwell did not agree; to him the two methods were quite distinct. This difference of opinion was no secret, for Richardson referred to it—in polite terms—in two of his later publications and in some correspondence. It was even mentioned in the biographical memoir of Southwell written for the Royal Society by D G Christopherson[20]. After pointing out that Southwell's basic idea was not new, Christopherson stressed that the relaxation method had several novel features, its chief merit being that it enabled the operator to shorten the calculations by applying his physical insight.

In his later writings, and especially in his *Relaxation Methods in Theoretical Physics* (the second volume of which did not, however, appear until after Richardson's death), Southwell did pay tribute to Richardson's 'notable examples' and to his 'ingenious device whereby approximate solutions may be improved'[21]. Southwell was a rare example of an engineer with outstanding mathematical abilities. He received a knighthood for his brilliant academic career at the universities of Cambridge, Oxford and London. Unlike Richardson, he was an excellent lecturer and public speaker, sociable and popular. After a street accident in Glasgow at the age of 71, his memory became unreliable and he spent his last years in seclusion; he died some ten years later in 1970.

A frequent visitor to the Richardson home at Bridge Lane was Hubert H Lamb, grandson of the mathematical physicist Sir Horace Lamb. He was at the same school as Olaf for several years and was often invited to tea. 'The great man used to appear and was affable to us kids', he writes, 'but of course it was Mrs. Richardson that presided at the tea table'. Through Olaf and Stephen, Lamb became interested in the Boy Scouts; a few years later at Oundle School he himself became a Scout as an alternative to being in the Officers Training Corps, one of his first acts of independence from his 'very dominating father'. As a youngster, Lamb was already interested in the weather, but he did not discuss the topic at all with Richardson, possibly because of his antipathy to mathematics, again a reaction to his father who 'was always goading' him to do pure mathematics, contrary to all his natural inclinations. He subsequently had a successful career in the Meteorological Office but resigned in 1972 to become professor of climatology at the University of East Anglia. Regarding the Richardsons, Lamb writes[22]:

> My impressions were then and always of a kindly and interesting family living simply and even frugally. In fact, it struck me that they took the demands of the Christian faith so seriously in terms of abstinence from all sorts of pleasures that my impression of them acted for many years as a brake upon my own interest in possibly joining the Society of Friends. In all honesty, I didn't think I had it in me to impose on myself the austerity that they practised. And it was only after long associations with other Quakers of other outlooks that I finally did join the Society.

Much of the austerity and frugality which Lamb noticed resulted from Dorothy's enthusiastic adoption of her husband's Quaker standards. Although they were by no means wealthy, they could well have afforded some of the small luxuries in which even some of their Quaker friends indulged. Dorothy was very generous but prided herself on not wasting money on food or clothes. Her manner of dress was not only sober; it was also economical. Each year she ordered just one new costume from a tailor she had found in Eskdalemuir, always the same pattern and always in a plain colour. The meals she offered to family and visitors alike were wholesome but notably lacking in frills and variety. The children knew what menu to expect according to the day of the week; their greatest treat was the Sunday roast.

The staff at the preparatory school where Hubert Lamb met Olaf included a student-teacher, Arthur Hunt Cooke, whose father was a half-brother of Rebecca Garnett; he remembers Olaf as a real terror in the classroom, but a very nice boy outside. The Hunt Cookes also frequented the Richardson home and were enthusiastic participants in the holidays with the Garnetts on the Isle of Wight. Arthur described his 'cousin' Lewis as the finest man he had ever met, a rare example of somebody who had learnt to reconcile science and religion and who acted out his religious beliefs in everyday life. He and his three brothers were handed down an old suit of Lewis', consisting of a heavy Norfolk jacket and a pair of knickerbockers, equally appropriate for Lewis' favourite activities of walking and cycling. To the youngsters these were highly desirable articles of clothing, mainly because of their warmth, and the privilege of wearing them was reserved for whoever had got up first. Arthur visited the Richardsons one cold day and was about to take off his coat when he was greeted by Lewis, warmly clad in a thick dressing gown and carpet slippers. 'Keep your coat on, my boy', said Lewis. 'The house is very chilly, for my wife would still be warm in a blizzard.'

Arthur told me of another member of the Isle of Wight 'clan', Cecil Rowling. Like the Browne boys, he was the son of missionaries and a pupil of Dorothy at St Michael's School. During a month's stay with the Richardsons at Benson, while still a schoolboy, he helped Lewis with some of his lizard balloon experiments, his assistance was duly acknowledged in Richardson's report on the subject. After graduating from Cambridge, Rowling entered the colonial service and spent many years in Nigeria. He was keen on yachting, a sport in which he had doubtless developed his skills at the Garnett home. In the mid-50s his boat became waterlogged in rough weather in the Solent; Rowling was drowned while attempting to swim to the shore to get help.

Arthur Hunt Cooke, who is now living in retirement on a farm in a remote corner of Kent, also had a career in the colonial service, mainly in Nigeria but with spells in Mauritius and Malawi. He continued to visit the Richardsons when on leave, and was almost the first outside the immediate family to call on Dorothy after Lewis' death.

Lewis and Dorothy both gave service to the Friends Meeting in Golders Green, as can be seen by glancing through the Minutes of the Preparative Meeting (the meeting at which Quakers conduct their local business). In November 1921, for instance, Dorothy presented a report on *Local activities with regard to the present state of affairs in Ireland*, while in September 1923 there was a report in Lewis' unmistakeable handwriting about a Quaker scheme for providing allotments to the unemployed.

Richardson was also busily engaged in the activities of various scientific societies. He was elected a Founder Fellow of The Institute of Physics in 1920. Two years later he became a Fellow of the Physical Society. But it was above all to the Royal Meteorological Society that he gave so much of his very limited spare time. He accepted an invitation to serve as one of the three honorary secretaries at the end of 1920, when the Society was about to complete the purchase of its new premises in Cromwell Road, South Kensington—a convenient location for members working at the headquarters of the Meteorological Office or studying with Shaw at Imperial College. He soon found himself on the Editing Committee, the New Premises Committee and the Finance Committee, and acting as referee for many of the papers submitted for publication in the Society's *Quarterly Journal*. He also helped the Society by reviewing books for the *Journal*, including one by J Bjerknes.

From 1919 to 1926, Richardson was a regular attender at the monthly scientific meetings of the Society and one of the most frequent contributors to the discussions of the papers presented. No matter the subject, he seemed to have something to say. He suggested improvements in observing techniques, offered explanations of errors in weather forecasting and advocated a closer relationsip between observations and theory. In the reports of these meetings there are several references to his efforts to introduce the name 'empyrean' for a warm layer of the atmosphere now regarded as part of the stratosphere. In 1921 the phenological report for the previous year was presented by James Edmund Clark; in asking some questions, Richardson referred to him as a former schoolmaster who had first roused his interest in phenology.

In December 1922 C J P Cave and R A Watson Watt (later to become famous for his contributions to the development of radar) presented a paper on the relation between meteorology and atmospherics—the noises heard on the radio resulting from electric discharges in thunderstorms. Richardson recalled the 'swishing' sounds he had heard on the telephone at Eskdalemuir (see p 48) and asked Watson Watt for an explanation. He described how the pitch of the sound decreased with time, just like a shell passing high overhead. Richardson said that he had the idea, perhaps quite mad, that the swish on the telephone might be produced by a meteorite. Watson Watt's reply is not on record, but it now seems highly probable that what Richardson had observed was a 'whistler', a phenomenon which was not explained until about 1950. It is caused by an electromagnetic disturbance from a lightning discharge which, unlike most of the atmospherics heard on a radio receiver, has penetrated the ionosphere and been guided back to earth by the magnetic field in the upper layers of the atmosphere. There are very few earlier reports of whistlers and Richardson might therefore have become a pioneer in the subject if he had pursued his Eskdalemuir observations along the right lines[23].

Richardson ceased his service as Secretary of the Royal Meteorological Society in 1924 and was then elected as one of the four vice-presidents, along with his old friend W H Dines. This was to be the highest honour bestowed on him by the Society during his lifetime.

At this time, when meteorology meant so much to Richardson, it was by no means his only scientific interest. He loved to attend lectures by eminent scientists covering a wide range of subjects and would make summaries of what they said, sometimes illustrated by thumb-nail sketches of the speaker. Among his papers are notes on a lecture by Niels Bohr on atomic physics, by Sir William Bragg on crystalline structure and by Sir James Jeans on astronomy. He also left copies of letters to well known scientists, pointing out what he thought were mistakes in their books. The following extract from a letter to Jeans shows how tactfully he did this[24]: 'The more one admires a statue, the more one regrets a chip off its little toe. Because I have so often admired the clearness of exposition in your 'Theory of Electricity and Magnetism' I now draw your attention to a suggested amendment for the next edition ...'.

Richardson himself was by no means infallible. In 1927 he wrote to *Nature* to call attention to what he felt was a mistake in a paper by the German physicist E Schrödinger on the wave theory of the atom. The editor passed the letter on to Schrödinger who failed to see any error. Richardson subsequently realised that it was he who had been wrong and wrote to Schrödinger admitting his mistake and apologising for having troubled him[25]. In the following year he attended a lecture by Schrödinger but history does not record whether he spoke to him personally.

Richardson took full advantage of the encouragement given at Westminster College for members of the staff to engage in scientific research. In the early 1920s he devoted his talents almost exclusively to meteorology, which in his own words 'had developed by 1926 into something like a passion'[26]. With but a few exceptions, all his important meteorological papers were published between 1919 and 1930. Towering above them all was his *Weather Prediction by Numerical Process* which we have already described. But before its final publication in 1922 the 'various excrescences' which he had removed from the book had already been published by the Royal Society in four separate papers, three on atmospheric turbulence and one on the measurement of water in clouds. The last of these[27] need not detain us long; it is a fairly straightforward survey of the problem of how to obtain the data on the water content of clouds needed for numerical weather prediction—one of the 'remaining problems' which he had referred to briefly in the penultimate chapter of his book.

Richardson was not the first British scientist to investigate atmospheric turbulence. He was preceded notably by G I Taylor, who together with the Austrian Wilhelm Schmidt is generally recognised as having originated the systematic study of the subject. In 1913, Taylor had the good fortune of being selected as meteorologist aboard the sailing ship *Scotia*, which had been commissioned by the British Government to observe the distribution of icebergs in the North Atlantic following the loss of the *Titanic* the previous year. He spent most of his seven months at sea studying the properties of the lower layers of the atmosphere with kites and made some valuable discoveries about turbulence. He developed an equation for the rate at which such factors as heat and moisture can be transferred by eddies from one layer of the atmosphere to another, depending on a coefficient known as the eddy-conductivity. He also introduced the concept of *mixing length*, the vertical distance over which an element of fluid would move before assuming the physical properties of the surrounding fluid. Taylor was only concerned with atmospheric layers which are close together and was therefore able to assume, for simplicity, that the density did not vary with height.

As Richardson's work on numerical weather prediction, which had led him to study atmospheric turbulence, covered a height range which involved large variations of density, Taylor's equation was not really applicable. He accordingly introduced a new stirring coefficient, which he called turbulivity, to replace the eddy-conductivity. From estimates of precipitation and of the vertical gradient of moisture over the whole globe, he calculated mean values of the coefficient at heights of 0.5 and 8.5 kilometres. He also estimated the value near the Earth's surface from some of his own observations. From all this he concluded that the turbulivity varied remarkably with height, the maximum value being at about 0.5 km. This work was discussed in his paper on *Atmospheric stirring measured by precipitation*, published in 1919. In his second paper, entitled *Some measurements of atmospheric turbulence* (1920) he elaborated his theoretical ideas and discussed the results of a variety of estimates of eddy-conductivity and turbulivity which he had derived from published data and from experiments he had made in France and elsewhere with the help of such people as Olaf Stapledon and George Hutchinson. He paid tribute to the work of Taylor but expressed some reservations about the concept of mixing length which, in his view, did not 'lend itself easily to measurement, except in the case of cumulus eddies'[28]. He went so far as to state that his own theory had been contrived specially in order to avoid this parameter. As pointed out by Taylor[29], this paper was unfortunately vitiated by mistakes—this once again illustrates Richardson's fallibility.

The third of this group of papers, also published in 1920, was destined to have the biggest impact and is still frequently cited by scientists. Nearly 40 years earlier, Osborne Reynolds had announced the results of his famous experiments on the flow of liquids in long straight pipes from which he had deduced a criterion, now known as the Reynolds number, to determine the conditions under which a laminar flow will become turbulent. Richardson set out to extend Reynolds' theory to make it applicable to the case of eddies in the atmosphere. His approach is well described by the title of the paper *The supply of energy from and to atmospheric eddies*. He related this energy exchange to the change of temperature and wind velocity with height and derived a criterion which he later described as a criterion for 'just-no-turbulence'.

Not satisfied with a purely theoretical result, Richardson tried to confirm his findings by field experiments at Benson. He measured the air temperature and the wind velocity at two different heights and used the range of gusts at the higher level as an indication of turbulence. In a paper presented at the British Association meeting in Toronto in 1924 (*Turbulence and vertical temperature difference near trees*), he claimed that his criterion was in rough agreement with the observed facts. The anemometers available to Richardson were somewhat crude for the purpose and he had to introduce some approximations in his criterion when testing it against this particular set of observations. It was to this shortened and, in Richardson's view, inaccurate form that the name *Richardson number* was originally given[30]. In 1942 Richardson received for comment a paper by O G Sutton entitled *Note on the use of the Richardson number in meteorological problems*. In his reply, he remarked[31]:

> I deeply appreciate the honour of having a number called after me. Yet it makes what should be an objective study, for me embarrassingly personal. From the point of view of rationing, G I Taylor deserves to have lots of numbers called after him, say Ta_1, Ta_2, Ta_3....

The Richardson number is of course now universally recognised, along with the Reynolds number, as one of the fundamental parameters in the turbulent motion of fluids.

One of the practical applications of an understanding of atmospheric turbulence is the study of atmospheric pollution, which has become of great importance in relation to the distribution by the atmosphere of harmful gaseous products of industry. Turbulence helps to prevent the dangerous accumulation of such products by the process of atmospheric diffusion, which can be readily observed by watching a smoke plume from a factory chimney. Richardson frequently made use of observations of diffusion as a measure of turbulence. His son Olaf remembers being asked as a young boy to make puffs of tobacco smoke simultaneously from two cigarettes so that his father could observe how rapidly they became separated. Also at a very early age his second son Stephen took part in experiments to measure the relative dispersion of pairs of toy balloons. On several occasions, Richardson observed the rate of scattering of the downy parachutes of dandelion seeds. He sought the advice of the Botanical Department of the British Museum and was delighted to receive some strophantus seeds which have large white parachutes and long stalks by which they can be held conveniently. In a paper published in 1925 *Diffusion over distances ranging from 3 km to 86 km*, he made use of the results of a competition organised by the League of Nations Union in which large numbers of toy balloons were released at Regent's Park in London. Each balloon carried a postcard which the finder was asked to return to the organisers, stating the time and place at which it had been found. The Secretary of the Union, Maxwell Garnett, doubtlessly helped his brother-in-law in the arrangements.

All these observations confirmed Richardson's earlier conclusion that the so-called constant K in the classical equation for diffusion in fact varies enormously, depending on the scale of the phenomenon under consideration. This led him to attempt 'to comprehend all this range of diffusivity in one coherent scheme', to quote from the opening paragraph of his most celebrated paper on the subject *Atmospheric diffusion shown on a distance-neighbour graph*, published in 1926. The paper continues: 'Lest the method which I shall adopt should strike the reader as queer and roundabout, I wish to justify it by showing first why some known methods are in difficulties'. The next section begins with the provocative question: 'Does the wind possess a velocity?' followed by the equally provocative comment: 'This question, at first sight foolish, improves on acquaintance'—hardly the language one would expect to find in the *Proceedings of the Royal Society*.

To appreciate the relevance of Richardson's question, we must recall that instantaneous velocity is the limit of the ratio of the distance travelled to the interval of time as this interval becomes infinitely small. This concept is easy to comprehend when we are thinking of a solid body, such as a train; but in the case of air motion the meaning is not so clear, for as we reduce the scale we have to think in terms of smaller and smaller eddies until we arrive ultimately at the motion of individual molecules. In mathematical terms, Richardson was suggesting that we may have to describe the position of an air particle by a function which does not have a derivative. In the case of the wind, it is legitimate to talk of a one-minute velocity or a six-hours velocity but we should think twice before speaking in terms of an instantaneous velocity. At the time that it was written, this section of Richardson's paper was fairly revolutionary.

We now come to the method which Richardson was afraid might be considered *queer* or *roundabout*. In the classical approach to diffusion, distances are measured from a fixed point; with this system of coordinates the mean rate of separation of a cluster of molecules increases without limit as the volume under consideration increases. Richardson's idea was to measure the distance l of each of a pair of particles from its neighbour, rather than from a fixed origin. With this system he found that the mean rate of separation has a specific relationship with the separation distance l, so that particles far apart (or large smoke clouds) separate more rapidly than particles near together (or small smoke clouds). From all the available observations he deduced empirically that K was proportional to $l^{4/3}$, which we shall henceforth refer to as the $\frac{4}{3}$ law. When this same law was derived theoretically many years later by the Russian scientists A N Kolmogoroff (1941) and A M Obukhov (1941), Richardson commented that it would be desirable 'to notice to what extent $\frac{4}{3}$ is a coincidence or a confirmation'. He continued: 'The atmospheric observations could have been fitted passably by any index between 1.2 and 1.5. The $\frac{4}{3}$ was chosen partly as a rough mean, and partly because it simplified some integrals'[32].

Although Richardson elected for simplicity in this particular case, he had no illusions about the behaviour of the atmosphere being simple. He once wrote[33]:

> Einstein has somewhere remarked that he was guided towards his discoveries by the notion that the important laws of physics were really simple. R.H. Fowler has been heard to remark that, of two formulae, the more elegant is the more likely to be true. Dirac sought an explanation alternative to that of spin in the electron because he felt that Nature could not have arranged it in so complicated a way. These mathematicians have been brilliantly successful in dealing with mass-points and point-charges. If they would condescend to attend to meteorology the subject might be greatly enriched. But I suspect that they would have to abandon the idea that truth is really simple.

At an international symposium on atmospheric diffusion and air pollution held in Oxford in 1958, Sir Geoffrey Taylor (who by then had become the 'grand old man' in this field) expressed the view that Richardson's paper on the concept of the distance–neighbour graph had 'initiated the modern approach to the subject' but added that it was 'perhaps rather surprising that he did not take the step' which would have led him to provide a theoretical justification for the $\frac{4}{3}$ law. He also made the pertinent comment that Richardson was 'a very interesting and original character who seldom thought on the same lines as his contemporaries and often was not understood by them'[34].

As Taylor was one of the 'contemporaries', it is of some interest to speculate on whether he himself had sometimes failed to understand Richardson and indeed on the relationship between these two pioneers in the study of atmospheric turbulence. Unfortunately, the available evidence is scant. Shortly after his arrival at Benson, Richardson was invited by Taylor to take part in a meteorological expedition at sea during the summer of 1919. Richardson was not enthusiastic: 'I do not know what ties I may be bound by at that season. Later on perhaps I could give a definite answer'[35]. While agreeing that the sea was the best place for such observations (presumably to measure the frictional force exerted by the wind on the Earth's surface), he wondered (being a 'miserable sailor') whether they could not be made from a lighthouse or a coral reef rather than from a boat? There is no further mention of the proposed expedition in Taylor's papers. His next publication on turbulence (1921) was in fact a mathematical analysis of *Diffusion by continuous movements*, which was later described as being well ahead of its time. Richardson had been shown the paper prior to its publication and had remarked that he had already proved one of Taylor's theorems; at Taylor's request Richardson's proof was appended to the published version of the paper. The only other piece of evidence I have found is a letter dating from 1933 in which Richardson wrote that he had been pleased to read that Taylor had been awarded a Royal Medal (by the Royal Society). He added: 'I have read most of your meteorological works with admiration for your combination of scientific insight with an engaging style of presentation' and concluded 'Don't trouble to reply'[36]. My own feeling is that for some reason the relationship was formal rather than intimate.

Throughout his stay at Westminster Training College, Richardson encouraged the students to take part in his scientific work. He always acknowledged such assistance and whenever somebody made a significant contribution he was invited to be a joint author of the resulting scientific paper. In general, the students—or members of Richardson's family or friends—would help in making observations or in calculating the results, while Richardson provided the ideas and direction. There was however one co-author who made some highly original contributions to Richardson's theoretical work, namely J Arthur Gaunt. He was the son of missionaries serving in China and had been one of Dorothy's most brilliant pupils at St Michael's School during the

First World War. He then went on to Rugby and Trinity College, Cambridge, where he gained a distinction in the Mathematical Tripos in 1926. During the vacations, he frequently stayed with the Richardsons or the Garnetts on the Isle of Wight and thoroughly enjoyed the boating, swimming, hockey, charades etc with the members of the extended families, including the Brownes and Russell Munday. While still an undergraduate, he wrote to Richardson in March 1926[37]:

> I am rarely so happy as when meddling with other people's mathematics. Consequently, some remarks you made last term started me playing with your Centered Differences, when I ought to have been attending to the Tripos. Having hit on a new (to me) quite unexpected result, I started writing the theory down in a state fit for your perusal.

The result was a two-part paper published by the Royal Society in 1927 under the daunting title *The deferred approach to the limit*—it was recently described to me as having been very advanced for its time. This was in effect an important elaboration of Richardson's 1910 paper on finite differences. Writing in 1954, Gold expressed the view that 'it was a great achievement to introduce this procedure for dealing with practical problems of physics and engineering and securing a degree of accuracy far surpassing that previously obtainable'[38].

After graduating, Gaunt became a research student at Trinity. During the next three years he wrote a number of important mathematical papers, mainly relating to quantum mechanics, a subject which was then in a rapid state of development. In 1929, he was elected a research fellow at Trinity, but prior to the announcement he had already offered his services to the Church Missionary Society. He spent the rest of his life as a teacher at St Stephen's College—a school for Chinese boys—in Hong Kong. He was captured by the Japanese in 1941 and died in a prisoner-of-war camp two years later at the age of 39. His last scientific paper, another joint effort with Richardson, was published in 1930 under the title *Diffusion as a compensation for smoothing*. This was a sequel to an earlier paper of Richardson's in which he had come to the conclusion that 'no differential equation in which position and time are the independent variables, and mass of diffusing substance per length is the dependent variable, can describe horizontal atmospheric diffusion'[39]. It had since been pointed out that this conclusion seemed contrary to most of the established results on vertical diffusion. Richardson and Gaunt resolved this dilemma by showing that diffusion, which implies averaging over time or space, follows different laws depending entirely on the arbitrarily chosen method of averaging. Richardson's earlier conclusion, which he now described as 'embarrassing', was only valid for a certain kind of spatial averaging as applied to a spreading cluster.

Richardson had been very disappointed when he heard that Gaunt, whom he rightly considered to be a brilliant young mathematician, had decided to give up his research work to become a missionary; he felt that this was a great loss to science. He was, of course, deeply upset when news came in about Gaunt's tragic death. Dorothy wrote to his father[40]:

> It's difficult to write of Arthur; he was so brilliant, and so modest, a very charming young man. We wanted him not to leave this country ... a distinguished career in mathematics was obviously opening before him, had he chosen to follow it. But Arthur said, 'There was a man named Levi who abandoned his talent for mental arithmetic; and he wrote a Gospel.' ... We well remember how on his last leave he took a teacher's training course in London: and when my husband asked him 'Why not at Cambridge, and pick up your friendships and the Cambridge life again?' Arthur said merely 'That would bring it all back again'; by which doubtless he meant 'I have given that up. I won't go back, the wrench would have to be taken all over again.' It was the only time we heard him refer to his great sacrifice.

Richardson's association with other scientists studying atmospheric turbulence is also of interest. In a tribute to her husband after his death in 1953, Dorothy wrote: 'then came a time of heart-break when those most interested in his "upper air" researches proved to be the "poison gas" experts. Lewis stopped his meteorological researches, destroying such as had not been published. What this cost him none will ever know!'[41] As Richardson himself had never referred to this rather important episode, I had often wondered about the accuracy of Dorothy's recollections of something which must have happened more than twenty years earlier; then in 1968 some corroboration came from Sir Graham Sutton, who by that time had retired from his position as Director-General of the Meteorological Office. He wrote[42]: 'I feel sure that the "poison gas experts" referred to were primarily the late Sir Nelson Johnson and myself'. He recalled that in the 1920s the Meteorological Office had hardly any programme of organised research. The exception was at the Chemical Weapons Experimental Station at Porton where a team of scientists, seconded by the Office, 'devised a thorough-going attack on the basic problems of the lower atmosphere with particular reference to the eddy diffusion of matter'. When Sutton went to Porton in 1929 the data collected had made it evident that theories of atmospheric diffusion were wide of the truth. Johnson, who was in charge, was well aware of the work of Richardson and would have liked to have enlisted his aid but nothing came of it. Sutton continued:

> About 1930 I began to make headway with my own statistical theory approach and ... I tried to get Richardson interested. I wrote to him from

my private address without mentioning chemical warfare to save him (as I thought) embarrassment, for I knew his Quaker views. He sent me a short courteous reply (I well remember his rather schoolboy handwriting) congratulating me on the progress made, but clearly he did not want to continue the correspondence, and I respected his wishes....

I am sure that no arguments would have persuaded him to take part in work that, however desirable it may have been in the interests of the science to which he was devoted, had connections with warfare. This gentle man was unshakable in his decision to conduct his life in accordance with the principles of his creed. It may be, of course, that he would not have fitted easily into any team, for it is clear that he was very much the individual worker. But what an inspiration he would have been as a consultant! Richardson, in my view, ignored the fact that it is never possible to ensure that any original work will not be used by others in ways that the originator would condemn. But that is a risk that must be accepted, and one should have regard for the potential good as well as the possible evil. However, I don't think that he would have accepted this argument.

We have seen that some of the papers on turbulence which Richardson published in the early 1920s were based largely on work carried out at Benson. This applies also to four papers[43] which appeared in 1923 and 1924 on his technique for estimating upper winds, and even temperatures, from observations of small spheres fired almost vertically from a gun. On a perfectly calm day, a ball fired vertically will of course land on the same spot from which it was fired; any displacement will give an indication of the strength of the wind between the Earth's surface and the highest point of the trajectory. Richardson's idea was to balance the displacement due to the wind by tilting the gun sufficiently from the vertical so that the ball lands near enough to the observer (standing under a steel shelter by the gun) for him to be able to hear it hitting the ground and hence estimate its distance from the gun. The wind can then be calculated from this measurement and from the tilt of the gun. By increasing the quantity of the explosive charge, it is possible to fire the ball up to greater heights and thus to obtain the average wind for successively higher layers of the atmosphere.

The chief advantage of the technique over the usual pilot balloon method is that it can be used to measure the wind above fog or low cloud. But there is also a big snag, of which Richardson was fully aware, the risk of somebody being hit by a stray shot. He made an analogy to driving a car[44]:

> A motor car would be an exceptionally dangerous machine were it not for the fact that selected persons can be trusted to move the controls so as to avoid accidents. The same applies to the meteorological gun. It can hardly be made fool-proof; but there is an abundance of discreet persons who can be trusted to handle it safely.... In order to make sure that there

is no one hidden from view in the mist within the danger zone, it is important to call out before firing 'Hullo! Danger! Bullets will fall! Keep off!' or something to that effect. A fog horn might be more effective.

These and other recommendations were based on 'the experience gained in observing the fall of 280 spheres of the size of a lentil, 330 of the size of a pea, and 48 as big as a cherry'.

Figure 6.1 Benson, 1920. W H Dines (with straw hat) and B C Lewis testing Richardson's method of observing the wind by shooting spheres upward.

Richardson was convinced that his technique was both useful and practical, at least for wind measurement. His proposal to extend it to the measurement of temperature (the trajectory of a projectile depends on this as well as on the wind) was not received very enthusiastically when he presented it at a meeting of the Royal Meteorological Society. He must have been dismayed when one of the speakers suggested that his work had obvious applications in correcting the shooting of field guns in foggy weather. More to his liking would have been the comment of the anonymous reporter of the meeting: 'L.F. Richardson is furthering the biblical prophecy; if he has not literally turned swords into ploughshares, he has made guns serve the purposes of the peaceful meteorologist'[45].

Among the distinguished visitors to the Richardson home in the mid-1920s was Jack Bjerknes, who made a lasting impression on Olaf and Stephen, as a tall friendly man. He had been invited to spend six months in London, giving lectures and working with the forecasters in the Meteorological Office in an effort to convince them that the methods of frontal analysis were scientifically sound and of great practical value. We have already seen that, in spite of this and the earlier efforts of his father, Vilhelm, these methods were still not in regular use in 1927, but there can be no doubt that the visit helped to pave the way for their eventual adoption. Jack Bjerknes later had a brilliant acadamic career as professor of meteorology at Bergen and then at the University of California, fully maintaining the high standards of scientific achievement set by his father and grandfather.

It was shortly after the visit of Jack Bjerknes that Richardson decided to make a deliberate break with meteorology 'in order to attend to rival ideas'. In reminiscing about this decision many years later, he recollected that his first serious work in meteorology had begun in 1913 as an official duty. From then on he had been carried away by the process which Gordon Allport called 'the functional autonomy of motives'[46]. According to this, each motive in the life of any individual has a definite point of origin. As the individual matures, the bond with the initial motivation may be broken but the motive itself may develop until it takes possession of the individual and becomes the alpha and omega of his life. This was what Richardson felt had happened to him by 1926 when meteorology had become 'something like a passion'.

It has sometimes been assumed that the 'rival ideas' to which Richardson wished to give attention related to the causes of war and that the reason for his taking a degree in psychology was to equip himself better for research in this field. This is not the case. His true desire was to apply his knowledge of mathematics and physics to pure psychology; he did not revert to his peace studies until the mid-1930s.

Richardson himself stated that the original stimulus for the research on which he embarked in 1926 was an idea which had occurred to him in 1918 that some of McDougall's theories, as propounded in his *Physiological Psychology*[47], might be expressed in the form of mathematical equations. This led him to try to estimate the intensity of the mental imagery which he experienced when endeavouring to think of a particular object. In a typical experiment, he looked at a word in a dictionary, selected at random, and then tried to think of a word with the opposite meaning. Having succeeded, he repeated the process with some other words and then came back, after an interval of 10 seconds or so, to the original word—not surprisingly, at the second trial, he found that he could think of the opposite much more quickly. Throughout each experiment he noted roughly the intensity of the imagery and plotted a graph showing how it varied with time. The shapes of the resulting graphs had several features in common and Richardson accordingly tried to find a mathematical equation which would produce similar properties. Once again he used Occam's razor, expressing it this time in the form that 'formulae should not be complicated without good reason'. He claimed that most of the features of his curves could be reproduced by a fairly simple equation. The results were published in a paper (1929) with the accurate but somewhat esoteric title *Imagery, conation and cerebral conductance*. The language he used was extremely terse; only once did he allow himself the luxury of a simile. In describing how the auditory image of a word came to an end after about a second and then often repeated itself he wrote: 'The conglomerate is perhaps what some writers call "thought". If so, the thought is like jam containing in it certain recognizable berries which are like the images'[48].

Two eminent psychologists who read the paper in draft were incredulous and found it difficult to believe that Richardson's estimates of imagery could have any significance. The prevailing view at the time was that sensations could not in fact be measured, although the whole question had been the subject of much controversy among scientists for many years. Believing that psychology would never be accepted as an exact science unless psychic intensities can be measured, Richardson set out to explore the possibility of estimating the magnitude of different kinds of sensation, such as colour, loudness and pain. His most famous paper on the subject, *Loudness and telephone current*, appeared in 1930. He wrote it in collaboration with J S Ross, who was then a lecturer in education at Westminster College, but later became Principal.

The paper describes a number of experiments in which an observer was asked to estimate the loudness of the signal in a pair of headphones; the electric current producing the signal was independently measured by a second observer. There is, of course, no absolute scale of loudness. What the observer had to do was to estimate the ratio of the loudness of the signal to that of a standard signal, the standard having been chosen at a level considered to be pleasant. In the first tests Richardson made the estimate of loudness while Ross measured the current and the results showed an approximately linear relationship between the logarithms of the two quantities, at least for loud and for moderate sounds. The tests were then repeated with ten different observers, including the College organist. All of them were willing to assign numbers to their estimates of loudness but most were unable to describe how they arrived at their figures; the organist stated that he did it by thinking of how many organ pipes would produce the sound. A more curvilinear relationship was found for other observers. The slope of the best-fitting line differed from one to the other.

The researches described in his paper on loudness formed but a small part of an extensive series of experiments which Richardson carried out immediately after his decision to give up meteorology, all of them being aimed at showing that quantitative estimates of sensation could be meaningful. One of his papers related to the sense of touch. It describes some tests which he made with his wife and, once again, with his colleague J S Ross, in which they were asked to estimate the separation of the two points of a pair of dividers pressed lightly on the forearm. Both observers found it difficult to do this when the points were less than about 20 mm apart but were more successful for greater distances; the most frequent result was an underestimate. Richardson concluded from these experiments that, rather than being clearcut, the threshold for double touch was 'confused by a vagueness, uncertainty or scatter'[49]. In the final sentence of the paper, which was entitled *Thresholds when sensation is regarded as quantitative*, he modestly remarked that 'The new facts in this paper are the two hundred opinions kindly given me by Mr. J.S. Ross and by my wife'.

Richardson's main interest seems to have been in the estimation of colour. He had been provoked into making experiments in this field by a remark of William James in his famous *Principles of Psychology* (1901) that 'To introspection, our feeling of pink is surely not a portion of our feeling of scarlet'. For some of his tests he used a circular disc with a white outer rim, a scarlet central core and an intermediate zone $\frac{1}{3}$ white and $\frac{2}{3}$ scarlet. When the disc was rotated at high speed, the intermediate zone looked pink and the observer was asked to assign a number to the pink on the scale from 0 (white) to 100 (scarlet). Following a suggestion of his nephew Ralph, Richardson simplified the observers' task by giving them a piece of paper on which he had ruled a straight line 100 mm long. One end of the line was marked 'white' and the other 'scarlet' and the observer had to put a mark on the line to show where he felt that the pink came on this scale of redness. He conducted the experiments with various groups of people, including a meeting of Fellows of the Royal Society, students at Westminster College, some art students and a number of Quakers at a social gathering. Of the 164 persons invited to take part, only five said that the task was impossible. The mean value assigned to the pink varied between 39 and 59 for the different groups, with women tending to give a higher value than men. The most internally consistent results came from the science students while the greatest diversity was shown by female art students—and the Quakers[50].

These experiments were continued by another Westminster teacher, R S Maxwell (1930), who used five discs with different ratios of white and scarlet in the intermediate zone. The observers had little difficulty in assigning numbers to each of the resulting shades of pink and there was a fairly close relationship between the average of their estimates and the ratio of scarlet to white on the disc. As in Richardson's experiments, there was a marked tendency to underestimate the amount of scarlet.

The published results of the tests by Richardson and Maxwell were severely criticised by some of their contemporaries. For example, T Smith (with whom Richardson had worked during his second spell at the National Physical Laboratory) suggested[51] that the observers might have unintentionally been estimating brightness rather than colour; in any case he was not convinced that the experiments had in any way discredited the opinion of William James. Richardson and Maxwell accepted some of the points made by Smith but continued to think that 'by attacking James' dictum about pink and scarlet, we have cleared the way towards a more scientific study of quantitativeness in sensations of various kinds. The rare mistakes of a genius like James are apt to enslave the rest of us'[52]. This controversy continued for many years and we shall return to it later.

Although Richardson described his break with meteorology in 1926 as 'deliberate', it was not sudden. Over the years he had been an enthusiastic supporter of the British Association for the Advancement of Science, usually known as the BA. As early as 1907 he had presented a report to the annual BA meeting (in Leicester) and he reported to subsequent meetings, including one in Toronto in 1924, on various meteorological topics. On the transatlantic voyage, on board the SS *Caronia*, between Liverpool and the mouth of the St Lawrence, he and some fellow meteorologists prepared daily weather maps and made some observations of the variation of temperature with height using an electrical thermometer of his own construction. Sir Harold Jeffreys recalls that on the first day at sea 'Richardson did not turn up for breakfast and I went round to see him. He was lying prostrate with just enough strength to work the plugs in a resistance box'[53]. One of the other members of the party, M A Giblett, who had earlier prepared the index for *Weather Prediction by Numerical Process*, met a tragic death a few years later in the crash of the R101 airship.

During the 1920s, the BA Committee for Investigation of the Upper Atmosphere, under the chairmanship of Sir Napier Shaw and with C J P Cave as secretary, had supported a limited amount of meteorological research, for example by providing radiation equipment. In June 1927 Shaw invited Richardson to an informal lunch meeting to discuss the annual report of the Committee. In accepting the invitation, Richardson raised some questions and made a few suggestions. The draft report deplored the decline of interest in meteorology on the part of universities and schools. Richardson asked 'Is not the following cause potent:— Formerly the question "what is the temperature up aloft?" was interesting, because so little was known. But nowadays the main facts are known, and curiosity is active only in those who care for details'[54]. (In parenthesis, one might comment that in many fields more people are interested in the details than in the generalities.) Richardson continued: 'In general I hope that the Committee will set out to answer interesting questions and not merely to collect observations in the hope that they may be of use to somebody sometime'. In putting forward his suggestions for additional projects he added 'But very likely it will be felt that the Committee has enough on hand as it is. I know that I have; and I dread further entanglements'. One of his suggestions was for a balloon 'as big as a house to be constructed to float at a height of say 3 km for as long as possible'; this would help to answer the question 'Where does the air go to?'

The following year, Shaw wrote to Cave suggesting that they had not been very effective on the Committee and wondering how they 'could have their mantle fall on Richardson and Jeffreys'[55]. Cave duly reported to the BA that although the Committee had not been very active it would be a pity if it came to an end. He added that he had written to Richardson and Jeffreys to see if they could discuss the future of the Committee at the BA meeting in Glasgow that year. In spite of his decision to break with meteorology, Richardson agreed to serve on the Committee, but it was in any case disbanded in 1929.

Richardson also continued his activities with the two international bodies which had specific responsibilities in meteorology, namely the International Meteorological Organization (IMO) and the International Union for Geodesy and Geophysics (IUGG). In August 1927, for instance, he attended a meeting in Leipzig on the IMO International Commission for the Investigation of the Upper Air, of which the president was V Bjerknes. It was in fact Bjerknes who, remembering Richardson's readiness to return to Bergen, had first invited him to attend a session of the Commission as long ago as the summer of 1921. In his letter of invitation[56], Bjerknes inquired if Richardson could stay longer in Bergen—where the meeting was to be held—and mentioned the possibility of providing some financial support from his Carnegie funds. By this time, Richardson had planned to spend his summer vacation 'writing out for publication the numerous experiments in projectiles made at Benson'[57] but he agreed to pay a short visit to Bergen. Initially, he accepted the offer of financial support, but shortly after his arrival in Bergen he wrote again to Bjerknes: 'I have good news for you .... Quite recently the College at which I teach has given me a substantial increase of salary, so that now, while thanking you for the offer, I think that I ought and indeed should prefer, to pay my own expenses'[58]. He evidently preferred to be independent.

Richardson took a very active part in the Bergen discussions and pressed on the Commission some of his own ideas about upper-air work, including the ideal distribution of observing stations to meet the needs of numerical weather prediction, and the way in which the observations should be smoothed. Being no longer on the staff of the Meteorological Office, he had of course no reason to be inhibited by the presence at the meeting of the rather formidable Assistant Director, Ernest Gold, who as usual tended to dominate his international colleagues.

Figure 6.2 Bergen, July 1921. International Commission for the Investigation of the Upper Air. *Centre row*: V Bjerknes (fourth from left); *front row*: E Calwagen (extreme left), Napier Shaw (centre), E Gold (third from right), C J P Cave (extreme right); *back row*: Richardson (fourth from left), J Bjerknes (third from right), G I Taylor (extreme right).

Bjerknes told the meeting about the new forecasting methods which he had developed in Bergen and about the observations he required from other countries. As at the gatherings of the previous year, he arranged for some practical demonstrations of frontal analysis by his collaborators, one of whom was Dr E Calwagen of Sweden. He was a pioneer in the use of aircraft for weather observations and was killed in 1925 when his research plane crashed. Richardson was impressed by the accuracy of Calwagen's forecasts which he compared with the notes he kept of the changes in the weather from day to day. In these notes he described Bergen as a 'combination of North Shields [not far from his birthplace] with Ryde [on the Isle of Wight, near the Garnett summer home], Skiddaw [in the English Lake District] being seen in the background and the climate being as rainy as that of the English Lakes'[59]. He also listed the prices of various items of food, clothing, imported articles and lodgings—he paid only 85 kroner (equivalent to about £3 at that time) for his room for just over two weeks.

The report of the Bergen session contains a 12-page appendix entitled *Memorandum on the upper-air work* and attributed to 'V. Bjerknes in collaboration with L.F. Richardson'[60]. Admirers of these two outstanding scientists must regret that their only joint publication dealt with the work of an international commission rather than with the results of their own researches.

Many years later, another famous Norwegian meteorologist, Dr Sverre Petterssen, wrote about this meeting[61]:

> I came to Bergen early in 1923, that is, less than two years after Richardson's visit.... From what I heard when I joined the Bergen group I understood that Taylor [who had also been at the meeting] and Gold were highly thought of, while Richardson was generally regarded as a strange sort of character. Richardson was seen to play with his little gun, with the aid of which he could measure the shear along the vertical of the wind, and many considered this just a toy and the man as an overgrown boy scout; no one seemed to have become impressed with the philosophy underlying Richardson's Numerical Process.

Petterssen also related the following anecdote:

> Richardson used to draw isobars which, as seen by Bergen-school eyes, seemed somewhat unorthodox. The philosophy of smooth fields was dominant while Richardson's isobars represented rather the opposite extreme. On one occasion an analyst invited Richardson's attention to the absence of smoothness, but Richardson was quite undisturbed and answered, 'It doesn't matter what they look like as long as we know the values at grid points.' This in the year 1921!

Petterssen's reference to Richardson's 'little gun' links up with an account by Richardson of experiments he made at the bottom and top of mountains near Bergen to determine the effect of atmospheric pressure on the behaviour of his gun. He also mentions having purchased a new gun in Bergen. At the next session of the Commission, in London in 1925, which he attended as a full member, he tried once again to stimulate interest in this technique by proposing an additional item for the agenda: 'What, if anything, is being done to observe wind above clouds?'[62] In presenting the item, he recognised the limitations in shooting spheres into the atmosphere in heavily populated areas but suggested that the method could still be useful in places such as Greenland where there are few people and where the presence of low clouds or fog frequently precludes the use of pilot balloons. There was no enthusiasm for his idea and I am not aware of any further efforts to develop his technique for routine use. But Richardson himself never forgot it. Reminiscing in 1948 about his meteorological publications[63], he still maintained that his spheres had certain advantages over radiosondes, which by then were in worldwide use for making upper-air measurements, in that their relative cheapness would permit more frequent observations. Had he forgotten what he said in 1925 about the risk of injury from falling spheres?

At the Leipzig meeting of the Commission, in 1927, which he attended in spite of his resolve to give up meteorology, Richardson again proposed an additional item for the agenda in the form of a draft resolution on absolute temperature, but it was refused on the grounds that it had been submitted too late. This was his last appearance at an IMO meeting, but he remained a member of the Commission until it was replaced by the Aerological Commission at Warsaw in 1935.

Figure 6.3 Leipzig, July 1927. International Commission for the Investigation of the Upper Air. Group photograph signed by most of the participants.

One of the two IUGG bodies for which Richardson continued to work after 1926 was its meteorology section, now known as the International Association of Meteorology and Atmospheric Physics. Unlike the IMO, which was essentially a club for the Directors of National Meteorological Services and hence primarily concerned with the international exchange of weather information needed for operational purposes, the IUGG Meteorology Section was of a strictly scientific nature; its main function was to serve as an international forum at which individual meteorologists could report the results of their latest researches. Richardson's principal interest was in the Section's research programme on the reflectivity of the Earth's surface. In his *Weather Prediction by Numerical Process* he had called attention to the importance of knowing how much of the solar radiation reaching the Earth's surface is absorbed and how much is reflected. One of the simple instruments which he constructed while in the FAU was a photometer for measuring the reflectivity of the Earth's surface (the albedo), and he presented in one of his Royal Society papers[64] the results of some measurements he had made with it in Eastern Champagne, the Isle of Wight and other places. The need for more albedo measurements was recognised in 1924 at the Madrid meeting of the IUGG Meteorology Section, when a grant was given for the purchase of four of Richardson's photometers. At the next meeting in Prague in 1927, where Richardson again met V Bjerknes, he submitted a report on the action taken, including instructions on how to use the instruments and an account of how he had calibrated them at Westminster Training College. The photometers were then allocated to France, Great Britain, Italy and the United States. Richardson himself continued his measurements of the albedo in 1927 and 1928, mainly from an aircraft flying over woodland, fields and gardens in Hertfordshire. For this he received a grant from the Government and assistance from the De Havilland Aircraft Company in mounting his instrument on a DH9 and a Moth aeroplane. In presenting his results to the Royal Meteorological Society[65], Richardson stressed that they were but a small contribution to the goal of having maps of the world showing the distribution of reflectivity month by month throughout the year, for studies of climatic change. They were nevertheless some of the first measurements of the reflectivity of the Earth to be made from an aircraft. Nowadays, of course, such measurements are obtained from much more sophisticated radiation instruments mounted on artificial Earth satellites.

The other IUGG body on which Richardson continued to serve was the British National Committee, where he represented the Royal Meteorological Society from 1923 to 1933. From the official minutes of the meetings of the Committee, it seems that he contributed little to their deliberations. Perhaps he was discouraged by the rejection, in spite of support from Napier Shaw, of a proposal he made in 1927 to the effect that all scientific papers should have an abstract in an international language, preferably Esperanto or Ido.

Throughout Richardson's stay at Westminster Training College, progress was being made towards the goal of attaining university status. Dr Workman would have liked it to become a college of London University, but the funds were inadequate for equipping the science laboratories to a sufficiently high standard. It was therefore decided that from 1930 there would be a special arrangement whereby Westminster students would attend three-year courses at King's College, University College or the London School of Economics and then spend their fourth year entirely at Westminster. Under the circumstances there would no longer be a need for science teachers at the College. Richardson had been keeping his eyes open for other appointments for some time. He was, in fact, offered a position in a university in New Zealand in 1926 or 1927, but turned it down after much deliberation and consultation with the family; he felt that it would be too far away from the centre of the scientific world. By 1929, it was clear that he would have to move and he applied for the post of Principal at Paisley Technical College; Paisley was rather distant from his beloved Cambridge and London, but at least it was more accessible than New Zealand. According to a rumour which persists to this day in Paisley, the Board of Governors were so overwhelmed at the thought of having a Fellow of the Royal Society as Principal that they never even considered the other 38 candidates. This is not, however, borne out by the official records[66], according to which it was first decided to concentrate on 9 candidates, from which a shortlist of four was drawn up for interview. Richardson was finally selected and was offered the appointment from 1 October with a salary of £1000 per annum and the right to occupy an official residence in Castlehead. This represented a substantial increase over the £650 which he was then receiving from Westminster and he had little hesitation in accepting the offer. One of the unsuccessful candidates was another former member of FAU, Dr Ernest Ludlam, who became a good friend of Richardson. He was a lecturer in Chemistry at Edinburgh University and a member of the Edinburgh Quaker Meeting; they must have met at the Scotland General Meeting which is attended by Friends from all over the country.

Richardson left Westminster with some regrets. He had taught there for nine years, an unprecedented length of time for him to stay in one job. In the early years he had benefited from the stimulation of meeting the leading British meteorologists at the Royal Meteorological Society. The high level of his scientific achievements had been recognised by his DSc from London University and his election as a Fellow of the Royal Society. When he decided in 1926 to abandon meteorology for psychology, the necessary academic facilities had been readily available. In his work at the College he had been relatively free from administrative responsibilities and had been encouraged to pursue his own research activities.

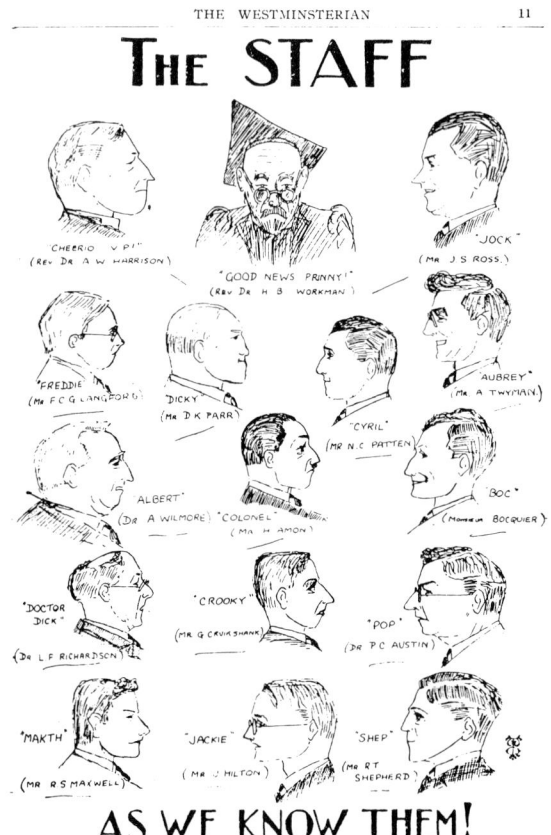

Figure 6.4 Staff at Westminster Training College, 1928. Ross and Maxwell collaborated in Richardson's early psychological work. Austin joined Richardson at Paisley in 1930. Patten was the source of several Richardson anecdotes.

Although more than 50 years have passed since he left Westminster College, his name is still remembered there with respect, if not awe. A photograph of Richardson adorns the walls of the College archives. Unfortunately, none of the laboratory equipment which he constructed, often with the aid of the students, has survived—most of the College records were destroyed by fire during the Second World War. The College itself is now located in a very pleasant campus on the outskirts of Oxford where it is officially associated with the University.

On hearing of Richardson's appointment at Paisley, Sir Napier Shaw sent him a letter of congratulation, or rather condolence. 'I am sorry you have to go so far away', he wrote, 'and I wish the Royal Society and the

Universities could be induced to realise that research in the dynamics and physics of the atmosphere ought not to be left entirely to a government service'[67]. Richardson doubtless shared this wish but refrained from comment. In his reply he referred instead to the high proportion of evening students at Paisley. From his own experience in studying for a degree in psychology, he had concluded that: 'Working for examinations in evenings is rather like taking off for a high-jump from ground covered by brambles'. He went on to discuss a technical meteorological problem which Shaw had raised and came up with a different line of approach. 'Is this any good?' he asked. 'I see no prospect of ever finding time to work it out in detail. And am much more attracted by a prospective Swiss holiday than by calculations'[68].

In this letter Richardson also made no comment on Shaw's remark about Paisley being 'so far away'. But this must have been how he too felt about moving to Scotland. He knew how much he would miss the facilities, such as the excellent libraries, which he had enjoyed in London. Perhaps he even foresaw, as later proved to be the case, that one day he would need to have ready access to the libraries of Cambridge University, for which it would be advantageous to have an MA. Whatever the reason may have been, his resolve not to purchase a degree finally broke down and he took his MA by proxy shortly after his arrival in Paisley.

CHAPTER 7

Paisley Technical College 1929–40

Richardson's predecessor as Principal of Paisley Technical College was Angus MacLean, a very hardworking and conscientious individual. When the Board of Governors was considering his replacement, he informed them that, in addition to discharging his administrative duties, he lectured in mathematics, dynamics and physics for an average of 38 hours per week! Nowadays a Principal would expect to have one or two full-time assistant administrators and would rarely, if ever, give lectures himself. When Richardson accepted the post, he must have realised that he would have to assume a heavy teaching load on top of the administration; any serious research work would have to be confined to his spare time at home.

The official residence for the Principal was, and still is, a large substantially-built house in Castlehead, a pleasant residential area not too far from the Technical College. Richardson was able to convert a large room on the first floor into a combined study/laboratory where he happily spent his limited hours of leisure, surrounded by workshop equipment (including a lathe), physical apparatus, books and all kinds of gadgets. A smaller adjoining room was set up as a well equipped chemical and photographic laboratory. Richardson kept his precious papers, such as unpublished manuscripts, in a safe as a precaution against fire rather than theft. To discourage burglars from attempting to blow the safe open, he painted a conspicuous notice: THIS SAFE IS NOT LOCKED.

Although I must have already seen Richardson several times at Glasgow Quaker Meeting, my first definite memory of meeting him is an occasion when, at the age of about 15, I was invited to his home and, as a special privilege, into his study. I remember him as tall and broad-shouldered, but slightly bent. His gold-rimmed spectacles and somewhat quizzical expression gave him an owl-like appearance—a combination of wisdom and an intense curiosity in what was going on around him. Sometimes during a conversation he would suddenly become withdrawn and I could not be sure whether he was still listening to me; he might even fail to respond immediately to a direct question. But I knew he was not being impolite; it was simply that his mind had been switched to a new idea which had temporarily taken full control of his attention. For the moment, it was as if I were no longer there.

It may have been on this or a subsequent visit that the 'Doctor' showed me an experiment in which a pair of small electric lamps were flickering on and off. They were Osglim lamps in which the light is produced by an electric discharge through neon gas. As I watched, the lamps would glow one at a time, never together, in a completely random manner. Richardson explained to me how the behaviour of these lamps was in some way analogous to the working of the human brain and in particular to the psychological phenomena of concentration and change of attention. The experiment was in fact a sequel to his earlier work on imagery, conation and cerebral conductance, the object being to explore the analogy, first pointed out by McDougall (1901), between the way thought occurs and the behaviour of an electric spark.

Richardson's attention had been drawn again to this analogy by a well known experiment in which the voltage applied to a commercial neon lamp, such as an Osglim, is set just below the value at which it will light up immediately. In this condition the lamp will eventually light up but the time required for this to happen is very variable. In a paper entitled *The analogy between mental images and sparks* (1930), Richardson remarked that this 'fickle behaviour, so unusual in physical things, reminds one of the fickle behaviour of one's mind when the same problem has to be solved on the blackboard once a year for many years in succession'[1]. More specifically, he compared 'the strain of expectation, which we feel while we wait for the solution of a problem' with 'the electric intensity in a spark-gap before the spark passes'[2]. Pursuing this similarity, he listed nine facts about the neon lamp which had a mental analogue. Encouraged by this success, he tried to reverse the process by listing some well known mental phenomena and predicting the corresponding behaviour of a neon lamp. For example, on the basis of the common experience that a thought is readily extinguished by its successor, he predicted that it should be possible to connect two lamps in such a way that the lighting of one would extinguish the other. Of six such predictions, four were subsequently fulfilled by very simple experiments of the kind which he had shown me.

His further research on neon lamps over the next few years led to a series of papers mainly on the physical aspects of their behaviour. As usual, he acknowledged in each paper any assistance he had received. In the first paper he went so far as to acknowledge the benefaction of Mr William Bow, who had 'gifted for the use of the Principal of Paisley Technical College a dwelling house large enough to contain a laboratory'[3]. A lecturer at the College, James Carson, collaborated with Richardson in writing a paper on the mathematical aspects of certain biological phenomena which behave like waves. In trying to explain some of his Osglim lamp experiments, Richardson had used a mathematical equation which he considered applicable to a wide range of oscillations that occur in nature, such as breathing or the rhythmic contractions of the muscles in animals when treated by certain chemicals. The paper contained a series of numerical tables, laboriously calculated by one of Richardson's finite difference methods, listing the values of the variables in the equation for a few selected parameters. The authors justified this in the introduction with the comment that[4]

> Those mathematicians who hold the doctrine, suitable for examinations, that a differential equation is solved when it has been reduced to an integral, may see little of interest in the following pages. Other mathematicians, including Lord Kelvin and Karl Pearson, have held that a differential equation is not usefully solved until it has been reduced to numerical tables.

The paper was submitted to *Biometrika* (the journal for which Richardson had prepared an index 30 years earlier) but it was rejected and never saw the light of day. In informing Carson of this Richardson wrote[5]:

> I am disappointed, not only on my own account, but because I feel guilty of having persuaded you to do much which (apparently) will come to nothing. What move to make now I don't quite know. I might send the paper to Woods Hole ... where they put all (!) my publications in their library; at least so their librarian wished to do. The best plan would probably be to rewrite and extend the paper.... But I see no prospect of my doing that during the war.

Although Richardson had apparently never visited Woods Hole, a small town at the south-western tip of Cape Cod in Massachusetts, he referred to it in a short note *A holiday resort for geophysicists*, published in 1924. By that time, Woods Hole was already well known as the location of a most successful Marine Biological Laboratory. Richardson attributed much of its success to the attractiveness of the site for scientists and their families and advocated the establishment of a similar institute for geophysicists so that they could carry on with their research during their holidays accompanied by their families: he suggested the

Isle of Man or Holland as possible locations. This wish was to some extent fulfilled in 1931 when the Oceanographic Institution, with a substantial programme in geophysics, was set up at Woods Hole.

The paper on biological waves never reached Woods Hole; in fact, I have what is marked the 'Best Copy'. Carson did not stay long in Paisley. He served in the Meteorological Office during the Second World War and later became director of education in Dundee.

One of the characteristics of neon lamps is that they light up more quickly in the presence of external radiation, such as x-rays, Richardson allowed himself to speculate about a mental analogy to this. 'Do cosmic rays' he asked, 'start the release of stored brain-energy and so occasion sudden thoughts?'[6] It may have been an attempt to answer this question that led him to carry out some experiments on cosmic rays—the very penetrating radiation which falls upon the Earth from outer space. In order to determine the extent to which these rays are absorbed on their passage through the Earth, he measured the variations in their intensity while travelling by train through the tunnel between Greenock and Gourock on the Firth of Clyde. He was accompanied by his son, Olaf, who told me that the cosmic ray detector was mounted in a lead-lined deed box which was placed on the carriage floor. They travelled to and fro between the two stations and had made several successful series of measurements when the train suddenly filled up with shipyard workers returning home after their day's work. They were duly impressed by the sight of the 'Doctor' counting the discharges of an Osglim lamp. After exchanging a few good-humoured remarks, one of them tapped Richardson on the shoulder and enquired innocently: 'Have ye no found the rabbit yet, mister?' Thereafter the experiments were continued in the undisturbed luxury of a first class compartment!

Olaf also related what happened when he and his father were travelling by tram to the railway station on one of these occasions. Richardson paid the fare for them both but did not feel inclined to pay also for his cosmic ray detector which was occupying a seat. The resulting dispute with the tram conductor had not been settled before the tram arrived at the station. Father and son had to rise and dismount in a hurry without having paid for the extra seat. Richardson was dumbfounded when he realised what he had done. He hastily abandoned his apparatus, ran after the tram and deposited the additional fare in the little red box which was provided for uncollected fares.

Richardson's interest in cosmic phenomena extended to the aurora borealis or northern lights, whose streamers are produced by collisions between particles from space, mainly of solar origin, and the rarefied gases of the Earth's upper atmosphere. He mentioned this to me by letter: 'Did I tell you about my speculative theory concerning the greatest possible degree of sharpness of the edges of auroral streamers? They are never quite sharply defined. I believe'[7]. His idea was that measurements of this sharpness might provide some insight into the philosophical interpretation of Planck's constant, which is related to the quantum theory and the uncertainty principle. He sent a note† about this, ready for publication, to the Meteorological Office in 1936, but so far as I know the suggestion has not been followed up. Much more is now known about the aurora and it seems likely that Richardson's intriguing proposal would be invalidated by the effects on auroral particles of phenomena which were unheard of in his day.

Figure 7.1 Richardson in 1931.

Richardson's researches on the behaviour of neon lamps illustrate very well his interest in the grey areas where physics and psychology overlap. Some of his results were published in journals devoted to physics while others appeared in psychological periodicals, but few psychologists were able to understand his physical and mathematical

†entitled *Quanta and diffusion*

explanations and few physicists were impressed by his psychological arguments.

Not long after Richardson's arrival at Paisley Technical College he proposed to inaugurate a course in psychology. This proposal was not received with great enthusiasm by the Board of Governors but they eventually agreed to start a class on an experimental basis. The 14-week course of five hours per week was divided equally between general psychology and industrial psychology. Richardson himself lectured on the former while Mr C A Oakley of Glasgow University was appointed as part-time lecturer on the latter. He described Richardson as 'a kindly, well-intentioned man, but better suited to the academic calm of Oxford than to the rough-and-tumble of Paisley'[8]. On one occasion Richardson warned another member of the staff not to be alarmed if he heard a shot—it would only be because he was studying the reactions of the students to an unexpected noise! He wrote an article for *Education for Commerce* in 1938 entitled *Psychology Class at an Evening Institute*. This contains useful advice for anybody planning to organise a psychology course, based on his experience at Paisley. The concluding paragraph reads[9]:

> How does this class satisfy known needs? It does so because there is a persistent demand for education in the selection, training, and management of personnel. There is, from another quarter, a persistent demand for education in citizenship, for understanding of other people, for ability to recognize unfair propaganda when it is encountered. Other influential people stress the importance of the 'Humanities' by which they usually rather restrictedly mean ancient literature. The classwork described above is in the wider sense a 'Humanity', supplementary to literature; it is experimental and is in touch with a rapidly growing body of truth.

In 1932 Richardson took the opportunity of a lengthy stay in a nursing home to continue his experimental work on the measurement of sensation, in this case the sensation of pain—he was suffering from an acute attack of cellulitis of the right thumb caused by a bee sting. In his paper to the Psychological Society entitled *A quantitative assessment of pain*, he described his attempts to estimate the intensity of pain by its effects in distracting from thought. He suggested a five-degree scale increasing from '(A) a pain so slight that it was perceived only when carefully attended to' up to '(E) a pain so intense that ... the pain arrested the movement of thought, leaving a stationary mental state consisting of just oneself and pain'. Fortunately he was able to state that he only experienced intensity (E) 'a few seconds at a time, during surgical operations'[10]. Assigning a figure of 100 to the intensity of pain which just prevented him from going to sleep, he estimated (A) as having a value of about 3 and (E) of about 500. He observed that a pain of intensity 300[11]

> had after six hours continuance begun to split the personality into a querulous child making uncontrolled movements versus an ashamed but helpless spectator, who had an awful sense of further disruption impending. But further disruption did not occur, for at this stage the physician gave me morphia.

He also found that a pain of intensity 70 could be reduced to 30, either by having a meal or by paying close attention to a story he was reading.

This paper did nothing to resolve the controversy which was still raging about the meaningfulness of subjective measurements of sensations of any kind. In June 1932 the Physical and Optical Societies held a joint discussion on vision at which R S Maxwell read a paper by Richardson on *The measurability of sensations of hue, brightness or saturation*. In it he claimed that he had shown that it is possible to make direct intuitive mental estimates of the intensities of various visual sensations, in other words, that sensations are vaguely quantitative. 'Accordingly', he argued[12],

> the measure of a stimulus, however precise, is for psychological purposes no substitute for the intuitive estimate of a sensation, vague though it be. The reliability of instruments, when we wish to measure sensations, is like the reliability of a paranoiac, who, when asked a particular question, invariably gives the same wrong answer. For example, a spectrograph records a spectrum extending continuously past the wave-length at which sensation ceases. Regarded as a psychologist the spectrograph is evidently suffering from a delusion.

Richardson's claim did not go unchallenged. During the discussion of his paper, Mr J Guild admitted that sensation has a quantitative aspect but he could not agree that subjective estimation was a form of measurement; in his view a stimulus and its resulting sensation were non-comparable phenomena. The point was made even more forcibly by Dr N R Campbell in a paper read to the Physical Society in 1933. He contrasted an accurate physical measurement of the temperature of a bowl of water with the guess which can be made by feeling the water. He saw no reason to believe that Richardson could measure sensation. If he was measuring anything it could well be the stimulus and this could be measured much better by other means. Once again there was a lively debate. Richardson was defended both by Maxwell and by Dr R A Houstoun of Glasgow University. The former pointed out that the sensation of warmth is surely a different thing from temperature, while the latter suggested that the relation between visual estimates and physical measurements of the magnitude of stars indicated that sensations could indeed be measured. As he was not present at the meeting, Richardson had to reply to Campbell in writing. He considered that the serious question was not whether sensory events are quantitative, but how accurately they are quantitative. 'That can be found out by psychophysical experiment,' he wrote 'but never by argument'. To illustrate the difference between sensation and stimulus he produced the following parody of Hamlet's epigram from his letter to Ophelia:

> Doubt how electrons behave;
> Doubt if anything really is there;
> Doubt whether sound is a wave;
> But never doubt you hear,
> Unless of course you really *are* quite certain.

By way of explanation he added: 'This comic anticlimax supports the thesis, for "not-being-quite-certain" is a personal mental state, very different from a stimulus'[13].

The controversy spread to the British Association at its meeting in York in 1932, when Houstoun and Richardson made a joint presentation on *Quantitative mental estimates of saturation with colour*. They invited the audience to take part in an experiment involving individual estimations of an intermediate shade of colour on a rotating disc. The results were similar to those reported earlier by Richardson and Maxwell. The Association then appointed a committee, consisting of physicists and psychologists, to try to resolve the problem. Six years later, the committee submitted a lengthy interim report which stated both sides of the case but with no conclusion. It was then reappointed to consider whether the different views were reconcilable. Some progress was made, but in the final report, dated 1940, the committee had to admit that on certain important points there was still disagreement and that 'no practicable amount of discussion would enable them to express an agreed opinion'[14]. Members of the committee, including Richardson and Campbell, expressed their individual views in a series of appendices to the report. This was the end of the matter for Richardson, who by this time had turned his attention to what to him were more important questions. But before we embark on these we must first catch up with his life in Paisley.

From the outset he had been unhappy at the very large number of classes which he was expected to take in person. In July 1930 he reported to the Governors that in addition to working all day he had to give over 16 hours per week of lectures at evening sessions. He indicated that this was 'unusually much, personally disagreeable, and not for the good of the college'[15]. The Governors accordingly relieved him from teaching the most elementary classes. Even so, his routine throughout his stay in Paisley was to go to the College after breakfast, stay there all day, come home for a meal and a brief nap in the early evening, and then return to the College to teach mathematics and physics up to the standard required for the London external degree. He would be too tired to do anything serious when he got home in the late evening, and his research was strictly confined to weekends and holidays.

On the administrative side, Richardson's main contribution to the College was to complete the planning and construction of a major extension of the buildings. He probably found this more stimulating than the routine administration, which, as at Eskdalemuir, he considered to be 'rather dreary'[16]. His tasks included the preparation of the annual budget (amounting only to about £12 000!) for approval by the Board. One of the important sources of income for the college was a grant from the Scottish Education Department, which was calculated on the basis of 2d per class hour for students following a regular course and $\frac{1}{2}$d per hour for casual students. For budgetary purposes it was thus necessary to forecast the number of each category of students for the following year. For this Richardson devised a formula based on a weighted average of the number of students for each of the preceding four or five years. The Governors were greatly impressed by the accuracy of his forecasts. In order to claim the grant at the end of the academic year, it was necessary to submit a statement showing how many hours of classes each student had actually attended. It was therefore important to maintain accurate registers of attendance. The story goes that on one occasion when Richardson forgot to take the register during a class, he was asked by the Governors to complete it some time later—otherwise the College would lose some of the grant. Richardson refused, on conscientious grounds, to fiddle the register in this way and even went so far as to threaten to resign if the Governors insisted.

As at Westminster Training College, Richardson did not leave behind a reputation of being a brilliant teacher. In spite of all his efforts to understand the psychology of teaching, he never learnt how to rouse and maintain the interest of his students or how to lecture to them at a level which they could understand. He was extremely kind, conscientious and sympathetic, but could not communicate very well with the average student. Perhaps this goes to show that teaching is more of an art than a science and that only those who are born teachers will ever be really successful in the profession. While I was studying at Glasgow University in the mid-1930s, I attended Richardson's evening classes in physics for two years. In the diary that I used to keep in those days (how I wish now that it had been more detailed) a typical entry would have been 'Went to Paisley in the evening. L F R gave rather dull lecture on electromagnetic induction'. By far the best part of his course was the practical work. The experiments were always interesting and much of the apparatus, some of which was home-made, was very ingenious. Sometimes the students (there were only about six of us taking physics) would help to make a gadget needed for one of the experiments.

Looking through my file of personal correspondence with Richardson, I find a letter which illustrates his caring attitude towards his students. 'As you weren't present yesterday evening, I write to correct a mistake in my lecture of 23rd ... '[17]. He enclosed two pages from his old lecture notes to illustrate the point. He may not have been among the world's best lecturers, but he was certainly one of the most conscientious.

In the practical work Richardson always called attention to the limitations imposed by the accuracy attainable in difficult kinds of measurement. As he put it in one of his books, 'We must be on our guard against ... fictitious accuracy'[18]. For example, if a thermometer can only be read to the nearest whole degree, it would be misleading to report the mean of two readings to the nearest hundredth of a degree. Some unfortunate student gave the result of calculations based on rather crude measurements to no less than 10 significant figures. Richardson allegedly commented: 'This is the first time that anybody has achieved such an accurate result. We must certainly report it to the Royal Society'.

Another student of the 1930s, Robert Barbour, recorded in some notes for use in preparing his memoirs that Richardson conveyed to his classes a respect for ethics and morality by insisting that all experimental data should be written down as measured and never altered. If the result of the experiment turns out to be unacceptably in error, the source of the error must be found and an appropriate note added to the report. Morality, he maintained, comes first.

Richardson also tried to pass on to the more advanced students some of his ideas about the relations between the physical, biological and social sciences. This is illustrated by the following extract from a lecture which he gave to a class which was about to proceed to degree standard after having successfully completed the intermediate course[19].

> You are on your way to become specialists in the mathematical–mechanical–physical–chemical view of nature. This view is very consistent with itself and it becomes more impressive the further we study it. As we climb the mountain of knowledge the landscape broadens. The whole universe appears constructed of electrons, protons and quanta. But allow me to remind you that this is still only a special view. We have throughout excluded from your consideration all questions relating to living things. Thoughts, feelings and decisions, although we have and make them all day long, are not the subject of our studies here. On questions of artistic value, of right and wrong, of politics, of religion, we would remain, as far as we are merely scientists, profoundly ignorant. As

these ideas were not in our premises, so they are not in our conclusions. Because we do not find them in our textbooks some scientists have felt them to be in some way 'unreal'. But that is manifestly an error. The Arts men know better. The state of a man's mind is as much a fact as the state of his books. Admitting then that a physicist is a specialist who, like specialists in arts, has a lop-sided view of life, we must remember that the community needs men who knowing thoroughly the intricacies of their special work, yet are modest and willing to learn about things in general.

Normally Richardson was extremely patient, surely one of the most important qualities for a teacher. Now and then, however, he would lose his temper, a weakness that had affected his whole teaching career. After one such outburst he was so overcome with remorse that he wrote a letter of apology to the student who had annoyed him. At home, too, he could be very angry at times, and would shout at the offending member of the family. Well aware of his weakness, he once took the opportunity of the family prayers at a meal to pray 'Dear God, please help me never to lose my temper again'.

Figure 7.2 The Richardson children, Christmas 1931. *Left to right*: Stephen, Elaine and Olaf.

The house at Castlehead became accustomed to the sound of young people, especially during the school holidays. Olaf, and then Stephen, followed the family tradition by going to Bootham School—Elaine was still too young for boarding school. The boys were keen cyclists and used the gravel path round the house as a dirt track. Although he disliked noise, Richardson was remarkably tolerant and even encouraged such activities. In the summer vacations the house would be invaded by teenage guests, including various members of the Garnett family. One of them brought with him an air rifle. Instead of forbidding it, Richardson set up a rifle range in the attic. The youngsters enjoyed outings to the beautiful countryside of the Firth of Clyde. They would take one of the Clyde steamers, then in their heyday, to a suitable resort, have a swim and then walk several miles to another port of call to pick up the return steamer. When the weather was unsuitable for outdoor activities, they all indulged in elaborate charades. Sometimes they spent the whole afternoon writing and rehearsing a play for an evening performance to the adults. Acting was, in fact, a very important part of the Richardson family life; Dorothy in particular loved the stage and delighted in taking the children to the theatre.

Figure 7.3 Olaf Richardson at Loch Goil, 1935.

Dorothy and the three children still spent some of their vacations at the Garnett home at Seaview but Lewis no longer accompanied them. To him, the Isle of Wight brought back too many memories of the time when Dorothy had been so distraught after a miscarriage. He preferred the peace and quiet of the Scottish countryside where he could walk and bicycle with plenty of time just to meditate, far from the administrative worries of the College. The family spent one summer holiday at an isolated cottage on Loch Goil to which the most ready access was by rowing boat across the loch from Carrick Castle. This proved to be so enjoyable that thereafter they rented annually an equally remote cottage at Knap on the west shore of Loch Long, about 3 miles from the nearest road. It could be reached by walking along a sheep track but they often preferred to row there by dinghy. This was the scene of many happy vacations in which they were sometimes joined by a few friends or relatives who were content to live close to nature.

Figure 7.4 Stephen Richardson with Nansen, 1935.

The teenage visitors in Paisley were not limited to the Garnett and Richardson cousins. I remember, for instance, several encounters with a very clever student named Michael Pocock; his subsequent brilliant career, ended abruptly by death at the early age of 59, culminated in his appointment as chairman of the Royal Dutch/Shell group. Then there were the three charming daughters of the Shanks family (the well known manufacturers of bathroom equipment), who lived across the road. Their presence at the Richardson parties was always welcomed, especially by the boys who were beginning to take an interest in the opposite sex.

Richardson himself never really enjoyed such social activities and would seek refuge in his study at the earliest opportunity. Dorothy was left to organise the party games, a task for which she was admirably suited. In the midst of all the gaiety she would never miss an opportunity to call a halt for a moment in order to bring home a more serious message. I remember a very lively party meal when the room was brightly illuminated by red, blue and yellow lamps. Suddenly Dorothy switched off the blue and yellow lights, leaving everything in a red glow. She then repeated this with the other colours. 'Did you notice', she asked, 'how different everything looked when lit by only one colour? For example, the green leaves looked almost black in the red light. This is how the nations of the world tend to look at each other, each using light of only one colour. What we need is to switch on all the colours so that we can see each other in the full truth of white light.'

As I recall the above incident, I conjure up an image of Dorothy, tall, upright, lean, rosy-cheeked, her hair invariably held in place by a velvet ribbon. Initially she had not been very enthusiastic about living in Paisley; for a time she was 'highly critical of everything Scottish'[20]. In a letter to her niece Peggy Garnett (now Peggy Jay), who was living in Paris at the time, she wrote[21]:

> Do they keep their windows open in France? They don't in Scotland, except the 'educated' I was going to say, but the distinction doesn't hold for the Scottish working folk—nearly all go to good schools and classes afterwards. One can't get windows open in the buses or local trains, other passengers object. But on the 'Royal Scot' (the express train to London) for instance one can.

But she was never one to waste time on grumbling and within a few months she became absorbed in a busy round of social and educational activities. As wife of the Principal of the Technical College, she was often invited to prize-giving ceremonies at local schools. She also entertained the wives of the Governors of the College from time to time. On one such occasion when the good ladies were enjoying their afternoon tea they were astonished to see in their midst an apparition rising through a trapdoor. It was Lewis wearing a dustcap—he had been adjusting a seismograph which he had installed in the cellar under the drawing room. He calmly took a cup of tea and without more ado marched out of the room muttering 'Good afternoon, ladies'.

Dorothy's main concern in the early 1930s was to strengthen the local branch of the League of Nations Union. To this end, she undertook a lot of public speaking on their behalf, supported by material provided by her brother Maxwell, who was still secretary of the Union. In another letter to Peggy she mentioned having eight such engagements in a fortnight including one at the Freemasons Lodge 'They will all take off their regalia before I arrive', she wrote, 'but it will be very exciting. I only hope I may be able to do some justice to the Cause and the Occasion'[22].

As in the Golders Green days, Dorothy's activities with youngsters were not limited to the family and the immediate circle of friends and relatives. She organised a youth club with the object of getting the members interested in international work and especially that of the League of Nations. These Nansen Pioneers, as they were called, gave dramatic performances to obtain public support for the League. Dorothy herself became known locally as 'The Peace Woman'. Lewis once addressed a meeting of the Pioneers; one of the members recalls that he had a mixed reception, some of the audience being impressed while the rest misbehaved[23].

One might wonder how Dorothy found time for so much activity outside her home. The answer is that the College provided her with the full-time services of a resident maid and that the faithful Ellen was still there to look after the children. They all loved her dearly and would turn to her first whenever they needed advice or consolation. On the rare occasions when Ellen was absent, Dorothy's life changed dramatically. She mentioned one such instance in a letter to Peggy: 'I had meant to write for your birthday but, as Ellen was away, I was tied all day and had only time for "business" letters; even then the wee girl [Elaine, age 5] was lonely at whiles'[24].

In June 1933 Richardson wrote to Vilhelm Bjerknes to congratulate him on becoming a Foreign Member of the Royal Society. He mentioned that Dorothy was 'very active in persuading people to respect the League of Nations' and continued: 'She is a great admirer of Fridtjof Nansen's work for the League. So am I. One, rather comical, result is that our son has named his new dog "Nansen" so that his name is frequently shouted in the neighbourhood'[25]. He was a black Labrador, beautiful and intelligent. To demonstrate this to visitors, Lewis would look gravely at the dog and ask: 'What would Nansen do for the League of Nations?' Nansen would then drop down and lie stretched out as if dead! When this was performed to another Norwegian meteorologist, Dr C L Godske, also a great admirer of Nansen, he was almost overcome with emotion; later he reflected that Nansen, a friend of children and animals, would have preferred this form of memorial to an inanimate painting or a piece of sculpture[26].

The cirumstances under which Richardson and Godske met are described in a letter dated 29 September 1936 from V Bjerknes to a Swedish physicist C W Oséen, written just after the former's return to Norway from an IUGG meeting in Edinburgh. Bjerknes wrote of Richardson's 'quite phenomenal achievement' in numerical weather prediction and of his 'very rare mathematical talent' and continued[27]:

> But, regretfully, he has a fault: he is a Quaker. When the Meteorological Office came under the Air ministry he left it for pacifist reasons, and since then he has not had a position which gives him leisure for scientific

research. He is now Principal of a Technical College in Scotland and his duties absorb all his time. He came to the Edinburgh meeting for a brief visit and was glad to hear that my colleague Godske was enthusiastic about his work and wanted to carry his method further using modern observations. He immediately offered to initiate Godske in his future plans—which he no longer expects to be able to achieve himself—and Godske straight away paid him a visit which proved to be very interesting. For my part, I have always wished that one of my collaborators would really familiarise himself with Richardson's methods; this wish has now been fulfilled and time will show how we shall benefit.

Before leaving Edinburgh for Paisley, Godske received a letter from Richardson which read[26]:

In preparation for your visit tonight I have put in your bedroom:
  (i) A copy of my book in the form of loose sheets. Please keep this for annotation.
  (ii) A bundle containing 65 computing forms. I shall be pleased to give you as many of these as you like to carry.
  (iii) A red binder containing my annotated copy of 'Weather Prediction by Numerical Process' with a view to a second edition. I am not willing to part with any of the papers in this binder; but you are welcome to copy anything you like out of it....

Looking back on this visit some thirty years later, Godske wrote[26]:

Having been assistant of V. Bjerknes for many years [he had joined the Bergen group in 1929], I have often heard Bjerknes speak of Richardson as a (impractical) man of genius. As far as I remember Bjerknes had a sincere sympathy for Richardson and I had the feeling that Bjerknes—like myself—considered it a great shame that 'the big Empire' had no better use for Richardson than as a lecturer at a school....

I will never forget him, his kind wife—and his black dog. L.F. was just a 'vicar of Wakefield' type, the ideal of a ruddy whitehaired English country clergyman. The discussions were at the same time inspiring and depressing—depressing because I had the feeling that Richardson had lost contact with science, inspiring because of the many ideas, and the man himself.

Richardson must have found Godske equally inspiring. The following day he wrote to the Cambridge University Press[28]:

This question does not arise at present; but I suggest that you should file this letter where it can be found in twenty years time. A young Norwegian mathematician, Dr. C.L. Godske, at present research assistant to Prof. Vilhelm Bjerknes, is taking a keen interest in numerical weather prediction. At the request of Prof. V. Bjerknes I have explained to Dr. Godske my plans for a second edition of my book, and have provided him with some notes on the project. If then there should, in years to come, be a call for a second edition; and if, for one reason or another, I am then unable to undertake the work, I suggest that you should consider to offer it to Dr. Godske.

The thought that there might be an increased interest in the book in twenty years time was remarkably prescient, for it was in the mid-1950s that numerical methods were being introduced into the daily practice of weather forecasting. Although he maintained his interest in dynamic meteorology throughout his life, Godske's principal field of research had by that time switched to the application of statistics in meteorology. So far as I know, he never followed up the suggestion that he might prepare a second edition of Richardson's book. In 1940 Godske became professor of theoretical physics in Bergen, where he died in 1970.

V Bjerknes' 'F.R.S.-ness', as he termed it in a letter to Richardson, was but one of the many honours he received during his long life—he died in 1951 at the age of 89. One honour, however, eluded him, the Nobel Prize, for which he was nominated in 1922. In spite of the strong international support he received from Sir Napier Shaw and others, his candidacy was probably not considered very seriously by the selection committee for the simple reason that the Nobel Prize in physics must in their view be awarded to a physicist in a very narrow sense of the word. To this day, not a single meteorologist—or geophysicist, for that matter—has become a Nobel laureate. Even without this limitation Bjerknes' chances in the early 1920s could not have been very good for he was up against such outstanding physicists as Albert Einstein and Niels Bohr, who were awarded the Prize in 1921 and 1922 respectively. Bjerknes was clearly disappointed. In another letter to Oséen in 1938 he reminisced about his failure, his main point being that the award of the Nobel Prize would have given international recognition to the Bergen School at a time when it was facing great difficulties[29].

In using the word 'fault' in connection with Richardson's being a Quaker, Bjerknes was surely thinking in terms of its adverse impact on his scientific career in meteorology. Fault or not, he and Dorothy played an active role in the Quaker Meeting in Glasgow. From a tantalisingly brief entry in my diary I find, for instance, that on a Sunday in 1935 'Dr. Richardson opened the meeting with an interesting comparison of science and religion'. For him, religion was not something confined to meditation at a meeting for worship on a Sunday morning. Nor was science an intellectual pursuit, far removed from the practical problems of mankind.

The Richardsons often served as hosts to Quakers from other parts of the country who were visiting Glasgow Meeting. One of these was Corder Catchpool, who had spent several years in Berlin serving in effect as a Quaker ambassador. After speaking about his experiences to an evening gathering of Friends, he had a sleepless night (he suffered from insomnia, probably as a result of the years he spent in gaol for refusing military service after leaving the FAU in 1916). The following morning, hearing of this, Richardson remarked quietly: 'The best cure for insomnia is sleep.'[30]

As mentioned earlier, Richardson himself was not always the soundest of sleepers. His interest in psychology, and possibly a tendency to introspection, led him to analyse his dreams to see what guidance they offered him. Another manifestation of this tendency was that at the age of 55 he tested his own intelligence. He recorded that on the Cattell scale he had an IQ of 140 to 150 which was 'high enough for observational science, but not for the higher flights of pure mathematics'[31]. He also analysed his own thought processes and decided that

> many consequences followed from a tendency of my mental machine *almost* to run of its own accord. This makes possible an intentionally guided dreaming. If the machine ran *entirely* of its own accord, one would have no control, and the dream would be out of touch with reality. If the machine had to be pushed all the way, then thinking would be too difficult to maintain for long. It is the 'almost' condition that is advantageous for creative thinking. I take it to be the essence of the creative temperament. In some ways it is a nuisance; for example I am a bad listener because I am distracted by my thought[31].

He attributed to this same trait his being a bad motor driver, although as we have already seen his FAU colleagues did not share this assessment of his driving ability.

There is yet another example in his biographical notes of Richardson's critical self analysis. 'From infancy', he wrote, 'I had a strong curiosity. William McDougall in his 'Social Psychology' took curiosity, along with its accompanying emotion of wonder, to be an instinct. Along with wonder, one should mention also doubt. To me doubt is mostly pleasant'. After referring to the occasion when he had experienced a doubt so intense as to be distressing (see p 106), he continued[32]:

> But for fifty years I have remained comfortably in doubt as to the existence of a future life, or as to which political party is the best. In respect to the persistence of doubt I used to contrast Sir Napier Shaw with Sir Henry Lyons [another famous British meteorologist]. Sir Napier always had in his mind a copious supply of unsolved problems, which he offered to other meteorologists: Sir Henry either solved a problem promptly, or else threw it out of his mind.

Richardson had been taught at Bootham the validity of the saying *Mens sana in corpore sano* and attached importance to keeping physically fit. He invariably went to work on foot or by bicycle. Long before the days when jogging became so popular, he used to take an early morning run for half an hour. He was a keen walker and loved to teach youngsters how to use a compass and map for finding their way across a moor in the mist. In July 1935 I climbed Ben Lomond with the Richardson family. Lewis was a little short of breath by the time we got to the top, but was clearly in good physical shape. The party included Michael Pocock and a German refugee, Dr Otto Marienfeld, who was interested in my experiences on a six-week cycling tour in Germany earlier that summer. The Richardsons helped him and many other Germans to settle down in this country after escaping from persecution under the Hitler regime. These refugees were extremely grateful for all the assistance they received. Some years later one of them showed her appreciation by presenting to Lewis a German translation of his 90 page monograph on generalised foreign politics. Richardson tried hard to get the translation published, but without success.

Figure 7.5 Resting on Ben Lomond, July 1935. *Left to right*: Otto Marienfeld, Dorothy, Lewis and Stephen Richardson, John Garnett, Michael Pocock. *In front*: Iona Pocock, Elaine Richardson (Traylen).

Like many people of mathematical bent, Richardson was very fond of classical music. He greatly preferred chamber music and took exception to loud symphonies—in fact he disliked all loud noises, probably as a result of his wartime experiences. He felt that in writing works for large orchestras the composer could too easily lose the design and beauty of music in the exercise of power over massed sound. Elaine remembers an occasion when they were listening on the radio to a noisy symphony by a Slav composer. At the end Richardson walked over to the radio, turned if off and said 'Now let's have Peace, by Switchoffsky'! My brother recalls that he once described the slow movement of

Beethoven's seventh symphony (which is one of my favourites) as an 'ode to monotony'. But perhaps the best story of Richardson in relation to music was when he was allegedly learning to play the violin—at an age when few people would take up such a difficult instrument. One day he asked his teacher whether he was making good progress. 'Well', said the teacher, 'you'll never be a Yehudi Menuhin.' 'Will you?' was Richardson's quick rejoinder.

The year 1935 proved to be another turning point in Richardson's life. It will be recalled that in 1926 he had decided to give up meteorology for psychology, but that his subsequent work in psychology did not appear to be specifically directed to problems of peace and war. The reason he gave for this lull in his peace work was that international affairs from 1919 to 1935 'were comparatively tranquil'[33]. Even after Hitler's dramatic rise to power in Germany, many people still hoped that the League of Nations would prove to be an effective mechanism for preventing another major war in Europe. From 1922 onwards, disarmament had been at the forefront of the activities of the League, culminating in 1932 in the opening of the Disarmament Conference in Geneva. When this Conference came to an inglorious end in 1934, disarmament was no longer the catchword; many intellectuals turned their thoughts to rearmament as international tension increased. But not Richardson. To him the failure of the Disarmament Conference meant that it was 'time to reconsider and republish' his ideas on war. From then on research on the causes of war and how to maintain peace was to become his passion.

On 18 May 1935 a letter under his signature appeared in *Nature*. It contained a slight revision of the differential equations first published in his 1919 booklet *Mathematical Psychology of War*. The equation showed that the rate of increase of the 'preparedness-for-war' of a country is proportional to the actual 'preparedness-for-war' of its opponent, after allowances have been made for the effects of 'fatigue and expense' and of dissatisfaction with treaties. The equations implied that mutual disarmament without satisfaction would not be permanent—neither would unilateral disarmament. A race in armaments would occur if the defence terms in the equations predominated. Richardson submitted that 'the equations do describe, at least crudely, the way in which things have been done in the past'.

A few months later *Nature* published a second letter which Richardson suggested was 'relevant to the Naval Conference'[34]. This time he developed from the equations some ideas about the conditions for balance of power and the way in which such a balance might remain stable or become unstable.

As the international situation continued to deteriorate in Europe, so did Richardson redouble his efforts to complete his war studies and to call attention to his results before it was too late. During 1937 and 1938 he wrote a monograph entitled *Generalized foreign politics* (he subsequently regretted that he had not used the more appropriate title *The theory of arms races*)[35]. He submitted the manuscript to three publishers, but the only group to express any interest were the editors of *Psychometrika*, an American journal devoted to the development of psychology as a quantitative rational science—surely an aim which was close to Richardson's heart. They were, however, only prepared to publish the first part of the text, less than a third of the total. Richardson regretfully declined their offer on the grounds that 'the later parts are valuable' and that 'a reader would not follow them unless he had part I at hand'; furthermore, '*Psychometrika* is only to be found in psychological or large libraries'[36]. We might add in parentheses that some years later he was nevertheless only too glad to have some papers published in *Psychometrika*.

Just before he received the reply from *Psychometrika* Richardson made a preliminary enquiry to the Cambridge University Press. A week later he presented an outline of the monograph to a meeting in Cambridge of the Psychological Section of the British Association, after which he received an offer from the British Psychological Society to submit his manuscript to the Cambridge University Press for publication as a monograph supplement to their journal. He informed the Press of this arrangement but they wrote to him a few weeks later to say that they did not feel that his work would be suitable 'for inclusion in their Catalogue'[37]. Richardson responded by return of post to this 'crushing reply'. Had they not forgotten the arrangement with the British Psychological Society? 'I feel, Sir,' he wrote, 'that you owe me some further explanation, for I regard your letter as a very serious rebuff. What is it that irks the Syndics? Do they remember that they made a small loss on my book 'Weather prediction by numerical process' which they published in 1922?'[38].

There had indeed been a lapse of memory and as soon as this had been remedied the arrangements went ahead smoothly and the monograph was published in June 1939. The British Psychological Society contributed £50 to the cost of publication, about a third of the total, and Richardson paid the rest. In the meantime he had been dismayed by what he considered to be the 'many misleading newspaper reports' of his address to the British Association. Of the 45 reports he had seen, only six were good while most of the others 'contained sillinesses of the newspapers own invention.' Some 'contained a plain lie'[39].

The 90 page monograph begins with a non-mathematical section containing the gist of Richardson's ideas. In this he used for the first time the style of a Socratic dialogue which was to become a feature of his subsequent writings. The following extract[40] will serve both to illustrate this style and to let Richardson summarise in his own words what he was trying to achieve.

> CHAIRMAN. 'I now call on Dr Richardson to explain his science of foreign politics.'
> CRITIC. 'Sir! I beg to move the previous question: that we do not waste our time on such an absurdity. How can anyone possibly make scientific statements about foreign politics? These are questions of right, of loyalty, of power, of the dignity of free choice. They touch a little on law, but are far beyond the reach of any science.'
> AUTHOR. 'I admit that the discussion of free choice is better left to the dramatists. But nowadays science does usefully treat many phenomena that are only in part deterministic, witness the many social applications of statistics and the astounding progress of theories as to the probable position of an electron.'
> CRITIC. 'The electron? That surely is irrelevant. On glancing at your summary I see mathematical equations with the symbol t in them. Does t stand for time?'
> AUTHOR. 'You have guessed rightly.'
> CRITIC. 'Can you predict the date at which the next war will break out?'
> AUTHOR. 'No of course not. The equations are merely a description of what people would do if they did not stop to think. Why are so many nations reluctantly but steadily increasing their armaments as if they were mechanically compelled to do so? Because, I say, they follow their traditions which are fixtures and their instincts which are mechanical; and because they have not yet made a sufficiently strenuous intellectual and moral effort to control the situation. The process described by the ensuing equations is not to be thought of as inevitable. It is what *would occur if instinct and tradition were allowed to act uncontrolled*. In this respect the equations have some analogy to a dream. For a dream often warns an individual of the antisocial acts that his instincts would lead him to commit, if he were not wakeful.'
> CHAIRMAN. 'I think we might go on.'

Richardson then explained why with his theoretical approach he could not enter into great detail:

> A rule of the theoretical game is that a nation is to be represented by a single variable, its outward attitude of threatening or co-operation. So the great statesmen, who collect, emphasize and direct the national will, need not be mentioned by name. This is politics without personalities. It must seem, to most newspaper readers, to be a miserably meagre way of treating the subject.

He continued: 'Whether a science should be made in detail or in broad outline is sometimes a question not of truth but of convenience'[41]. A meteorologist, for example, may consider the movement of air by looking at the individual molecules or by taking statistical account of the average effect of eddies. In the same way, a social scientist can choose to enter into detail or deal with generalities[42]:

> Foreign affairs as they appear day by day in the newspaper: the text of the despatch, the facial expression of the ambassador as he comes away from an important interview, the movement of warships, these may be likened to the eddying view of a wind. Whereas the theory here presented may be likened to an account of the general circulation of the earth's atmosphere.

After this entertaining preamble, Richardson elaborated his earlier ideas on how two countries can become engaged in an arms race. This time he pointed out that his equations could equally well describe the opposite of an arms race, simply by reversing the signs of the variables. This would imply a constantly increasing co-operation between the parties, which for two individuals would be equivalent to falling in love. One of Richardson's disciples, Kenneth Boulding (about whom we shall hear more later), suggested that a succinct expression of the mathematics of his arms race model would be the phrase from a popular song 'Lay down your arms and surrender to mine'[43]. For Richardson, one form of co-operation between countries, which has the advantage of being readily expressed in figures, is the amount of international trade. From a test of his equations using data from the 1909–1914 European arms race, he found, as predicted, a straight line fit when the total annual defence budgets of the nations involved are plotted against the annual rate of increase of these budgets. He claimed that the 'mere regularity of these phenomena shows that foreign politics had then a rather machine-like quality, intermediate between the predictability of the moon and the freedom of an unmarried young man'[44]. He extrapolated the straight line and read off the arms expenditure which in 1909 would have corresponded to no annual increase. It amounted to £194 million whereas the actual expenditure had been £199 million—and so the arms race had begun! He then looked for some form of goodwill or co-operation amounting to £194 million and found that the imports and exports between the two alliances were approximately of this magnitude—a very encouraging result.

Richardson remarked that in physics the equations most similar to his arms race model appeared in a paper on the ignition of explosive gases, which he felt was a title 'not inappropriate to the international situation'[45].

In the remainder of the monograph, Richardson considers the defence budgets and international trade of seven great powers from 1922 to 1938 and extends his arms race model progressively to three and then to several nations. These sections were subsequently revised quite substantially in another publication and we will therefore not discuss them further for the time being. The concluding section 'Discussion on applications' is again in the form of a Socratic dialogue of which the tenor can be appreciated from the following samples[46]:

CRITIC. 'So you have got your fine equations finished have you?'

AUTHOR. 'Science is never finished. But I have at least, got an approximation which is mathematically tractable and politically interesting.'

CRITIC. 'And what do you conceive to be the use of it?'

AUTHOR. 'It shows how threats or co-operation interact with grievances and costs.'

CRITIC. 'But surely practical statesmen understand that already; although they express the connexions in plain language, whereas you prefer an obscure symbolism.'

AUTHOR. 'Judging by their public speeches, some understand it and some do not. Many of them in effect assume that what I have called the "defence-coefficients" are negative, whereas the mathematics show that the existing arms race could not have developed unless the defence-coefficients were positive.'

. . . . .

CRITIC. 'I still don't like the fatalistic look of your mathematics. The worst disservice that anybody can do to the world is to spread the notion that the drift towards war is fated and uncontrollable.'

AUTHOR. 'With that I agree entirely. But before a situation can be controlled, it must be understood. If you steer a boat on the theory that it ought to go towards the side to which you move the tiller, the boat will seem uncontrollable. "If we threaten", says the militarist, "they will become docile." Actually they become angry and threaten reprisals. He has put the tiller to the wrong side. Or, to express it mathematically, he has mistaken the sign of the defence-coefficient.'

CRITIC. 'But how can experienced statesmen possibly be mistaken about the sign of an important political effect?'

AUTHOR. 'They are not altogether mistaken. They attend to the immediate effects of fear and they ignore the after-effects, which are of the opposite sign, and in the long run more important.'

CRITIC. 'Fear? The British nation is not influenced by the contemptible emotion of fear.'

. . . . .

CRITIC. 'A serious objection to your description is that it does not mention intelligent aggression planned by a leader as moves in a game of chess.'

AUTHOR. 'That is admitted; for a description of predictable tendencies cannot include anything so unpredictable as moves in a game of chess. Yet the acts of a leader are in part controlled by the great instinctive and traditional tendencies which are formulated in my description. It is somewhat as if the chessmen were connected by horizontal springs to heavy weights beyond the chessboard.'

CRITIC. 'But a silent dictator has a great advantage in playing chess against half a dozen democracies who discuss future moves in loud voices.'

AUTHOR. 'That silence and those discussions are slow in having effect on the instinctive and traditional tendencies...'

. . . . .

CRITIC. 'What answer can there be to force but force? Tell me that!'

AUTHOR. 'If there really were no answer to force but force, then although it is hundreds of years since the time of Robert the Bruce, the relations between Scotland and England might still be describable in partisan language thus: A noble army of Scots fully equipped with the latest means of defence might still be encamped on the north side of the Cheviots facing a dastardly horde of English cruelly armed with their wicked weapons of destruction. But all that hate has evaporated; and not by conquest.'

CRITIC. 'How then do you think that peace ought to be maintained?'

AUTHOR. 'National self-righteousness is not enough. Si vis pacem para justitiam. In the words of R.B. Gregg (1936), "Peace is a by-product of the persistent application of social truth, justice, and strong intelligent love.... The price of peace is the price of justice."'

Richardson's presentation at Cambridge in August 1938 took place in an atmosphere of crisis with fears of an early outbreak of war in Europe over Czechoslovakia. He drew some lessons from the subsequent chain of events and discussed them in an additional section which is included in the published version of his monograph. The fact that the Munich agreement was quickly followed by a rapid increase in expenditure on armaments by all the countries involved upheld, in his view, the main thesis of his work and demonstrated the fallacy of believing that nations can be frightened into peaceable submission by severe threats.

The first review of *Generalized Foreign Politics*, which appeared in *Nature* in October 1939, was by H T H Piaggio, Professor of Mathematics at the University of Nottingham. He wrote of Richardson's daring attempt to deal mathematically with foreign politics, but questioned the validity of his reasoning[47]:

> Some may think that the author has made the first rough approximations in a new branch of science, which can be polished and corrected later. The mathematician, who knows that a small change in a differential equation can sometimes produce a large change in the solution, may be distrustful of the way in which the fundamental equations were obtained. Students of human affairs will assert, perhaps too hastily, that foreign politics cannot be treated mathematically.

Piaggo continued:

> One great defect is that the theory does not include the effect of 'intelligent aggression planned by a leader as moves in a game of chess.' Unfortunately, this is the aspect of the question which is of vital interest to millions to-day.

Writing just after the outbreak of the Second World War, he was doubtless thinking about the recent actions of Hitler.

H G Forder, writing a few months later in the *Mathematical Gazette*, also expressed some doubts about the validity of Richardson's approach but admitted that in his analysis of the 1909–1914 arms race he had found remarkably close agreement between fact and theory. 'If the question were one of physics', he wrote, 'it would be regarded as settled. But in physics we believe that laws hold, in politics we may not unreasonably doubt it, and many more confirmations would be needed to produce conviction'[48].

Richardson took these and other criticisms seriously and did his best to answer them, in his later work.

In August 1939, just a few weeks before the German invasion of Poland, Richardson paid a brief visit to Danzig to see the situation for himself. He described his impressions in articles published in the *Paisley Daily Express* and the *Northern Echo*. In the latter he tried to explain the Danzig crisis in terms of an imaginary situation in the north of England where 'for Danzig we put Tyneside, for Gdynia we put Blyth, for Polish we put Irish and (whew!) for German we put British'. If Tyneside were reabsorbed into England while Ireland retained the thoroughly Irish port of Blyth, would both countries be content? Or, he asked, would the British then want Blyth as well?[49]

He treated the inhabitants of Paisley to a more descriptive account of his personal experiences. In Danzig he had heard German spoken everywhere, even in the Anglo-Polish Trade Bank. But Gdynia, about 15 miles west, was 'thoroughly Polish'. He concluded that[50]

> the Germans believe that the Poles are wickedly claiming large areas that are certainly German; and the Poles believe that the Germans are wickedly claiming the Corridor, which is certainly Polish. Moreover, both sides support their claims by concentrations of troops.

He was clearly not optimistic about the chances of avoiding war.

In the diary which he kept during this two-week trip he recorded several incidents which in his view illustrated the local feelings. In a Danzig hotel, for instance, he asked in German for a room and received an evasive reply. Only when he had explained that he was English rather than German was he told that there were plenty of rooms; 'You see' said the receptionist 'this is a Polish hotel'. He had a similar experience when asking his way in what had formerly been the German-speaking town of Bromberg, which he noted 'the Poles write Bydgoszch, and pronounce 'Bidgosh''[51].

On his way home, Richardson spent a day in Berlin where he met his old friend Corder Catchpool and another well known Quaker, Horace G Alexander, who were trying against all the odds to keep open some unofficial lines of communication between the German and British

Foreign Offices, fully aware that any chance of their being able to influence events was infinitely small. Richardson was amused at the different ways in which the Germans saluted each other. 'The clearest "Heil Hitler" that I heard was shouted by a small boy. In Pomerania grown-ups pronounced it quite quietly, sometimes as "'eil 'itler" sometimes merely as "'ler".... I bade farewell to my landlord with "Heil Hitler and King George" which he answered by "Jawohl"'.

The outbreak of the Second World War within three weeks of his return to Paisley intensified Richardson's passion for peace research and he soon decided that he ought to devote more time to it. On 16 February 1940 he wrote to the Governors of the Technical College[52]:

> I feel that I must make time to prosecute thoroughly researches on the Instability of Peace in continuation of my recent book 'Generalized Foreign Politics'. Accordingly I hereby give notice to leave the service of the Governors three months hence: that is on May 16th. I have chosen this date because then I will have finished the evening classes which I teach personally.

The letter was considered by the Board of Governors on 28 February. Their decision was recorded as follows[53]:

> Having ascertained that his decision is irrevocable, being based on a desire to engage research work of an important character, [the Governors resolve] to accept the resignation, but with profound regret, in view of the outstanding and laborious services rendered by Dr. Richardson during the period of $10\frac{1}{2}$ years, culminating in the Extension Scheme which has now been completed.

In view of the national crisis, the Governors decided only to make a temporary appointment and Mr John Denholm, the head of the physics department and registrar, was accordingly designated as Acting Principal. As he did not wish to occupy the official residence, the Richardsons were allowed to stay on at Castlehead 'during the pleasure of the Governors'.

I have felt it right to give a detailed account of Richardson's resignation because of a rumour which still exists in the College and among some of his friends and relations that the real reason for his leaving was that it had been decided that the College was to be used for war purposes to which he objected on pacifist grounds. In fact it was not until October 1941, more than a year after his resignation, that the Governors approved a scheme for training service personnel in wireless at the College. There is no record of this question having been raised during Richardson's tenure and I have no doubt that the reason for his resignation was as stated in his letter and in the Governors' decision.

Another rumour about Richardson's resignation which persists to this day is that he had just received a legacy which, supplemented by his modest pension, provided sufficient income to enable him to live comfortably in retirement. The facts are recorded in his biographical notes for the Royal Society: 'During the nineteen-twenties I received a legacy of about £7000 from my father. Part of this facilitated research. I have often been anxious lest I were wasting the substance of my wife and children on my personal interests'[54]. Members of his family assure me that there was little left of the legacy in 1940 and they also believe that some of it may have been lost by unsuccessful investments— Richardson was never a good businessman. The income from the remainder could only have added marginally to his pension. One thing of which there can be no doubt is that the Richardsons lived very frugally after his retirement.

Any objection which Richardson may have had to the use of the College for the training of military personnel did not apply to the application of his own scientific knowledge and experience for the training of those engaged in civil defence. In addition to serving as an air raid warden, he gave a series of lectures to Gas Identification Officers, who formed part of the precautions taken by the Government to minimise the effects of any use by the enemy of poison gas against the civilian population. He approved of such precautions, including the issue of gas masks, but was appalled that the training of the GIOs had not included anything about the mechanisms whereby gas is dispersed by the atmosphere.

He began his first lecture by asking the audience to imagine that after an air raid 'there is a dangerous concentration of phosgene in the air. There is no wind, perchance there is fog. The question is: how long is the gas likely to remain? That isn't a chemical question; it is in fact the job of a meteorologist'. He continued with a fable about a doctor of literature who wanted to learn something about chemistry (the GIOs were all chemists). 'In literature', said Richardson[55],

> you can begin almost anywhere, so he took the first two books he chanced to find. One was about the structure of proteins, the other a treatise on electrometric titration. Being disciplined in thoroughness, he read every page, or tried to. He then decided that chemistry was an utterly inscrutable subject, and to the end of his life he never learnt the use of litmus paper. I suspect that some high official in the Ministry of Home Security has approached the subject of turbulence in the same way as our mythical doctor of literature approached chemistry. For how else can I explain official inattention to anything so vital as turbulence?

He then proceeded to highlight the importance of turbulence by showing that without it a cloud containing a lethal concentration of poison gas could take several weeks to become sufficiently dilute to be harmless, whereas with turbulence this time could be reduced to less than a day.

The substance of Richardson's lectures was guidance on how to estimate the rate of dissipation of poison gas and how to make the observations needed for carrying out the calculations. He described various simple devices of his own design, such as a wind vane, a gust meter and a thermometer for measuring the lapse rate (the change of temperature with height). Even in the absence of such instruments he suggested that useful information could be obtained by observing puffs of cigarette smoke or the behaviour of dandelion seeds as they parachute towards the ground—he had not forgotten his own experiments of some twenty years earlier.

Richardson also gave talks to groups of people with less scientific knowledge. Stephen described an occasion when his father had invited all the local residents to his house to discuss air raid precautions. Never had all the neighbours gathered before, and the roar of conversation defeated various attempts by Richardson to start the meeting. Suddenly there was a crash; everybody turned round to find the speaker lying flat on the ground. 'That, ladies and gentlemen, is the correct procedure in the event that you hear a bomb dropping. We will now begin the meeting'[56]. And they did.

Dorothy also served as an air raid warden and, like her husband, she took her duties most seriously. Their son Olaf relates what happened one night when some stray bombs from the Clydebank blitz landed in Paisley[57].

> A bomb exploded in a garden about 250 yards from the house while Dorothy was patrolling. Lewis promptly left the house and raced to the scene but luckily he saw no damage—only a woman standing on the far side of the bomb crater. So he called out 'Go away you stupid woman; a BOMB has exploded!' The reported reply was 'Don't shout at me; go away yourself!' Lewis retorted 'I am a warden so do as I say', to which the woman replied 'I also am a warden—you must be Lewis'. It was of course Dorothy.

After Richardson's death, Dorothy revealed that at about the time of his resignation from Paisley Technical College he had been offered a university chair[58]. Earlier in life he would have jumped at such an offer but by then he had already determined to devote his energies to peace research and could not take on any responsibilities which would divert his attention elsewhere—the offer had arrived too late. Dorothy reported that it was a difficult choice, but I doubt if he went through the same period of intense doubt as in 1920 when deciding whether or not to resign from the Meteorological Office. With great regret—but not too much hesitation—he turned down the offer.

CHAPTER 8

Retirement: Paisley 1940–43

By the summer of 1940 Richardson was at last free from teaching and administration and was thus able to devote most of his time to peace research. His original work in this field had been mainly deductive. He had been led to his arms race model by theoretical considerations and had then deduced from this model the consequences of various lines of action; it was many years later that he had attempted to provide statistical evidence to support his theory. Now he decided to adopt an inductive approach, going from the particular to the general. Believing that what has happened often in the past is likely to happen again, he wanted to ascertain whether a statistical analysis of past wars would indicate any persistent quantitative relationships which might be of practical significance. For each war he needed to know the dates of beginning and end, the causes, the number of casualties and any relevant facts about the belligerents, such as their languages and religions. As he was unable to find all this information in any single existing publication, he realised that he would have to conduct a painstaking search of history books, encyclopedias and other sources and this meant that it was essential for him to have ready access to a good library.

Although Dorothy had long since become reconciled to living in Scotland, she had often discussed with Lewis the possibility of returning to England, preferably to London or Cambridge, when he retired. This would have made it easier for her to see her sister, Hilda, and other members of the Garnett family. Lewis would have been within reach of some of the best libraries in Britain and, in addition, would have enjoyed the mental stimulation of meeting his friends in the Royal Society. On the other hand, Paisley had the advantage of being less affected by the war, and although it seemed to him to be rather far from the main centres of learning it was close to Glasgow with its good libraries, including that at the University. Furthermore, the cost of living would be lower in Paisley, especially as they had the offer of free accommodation from the Technical College, a factor of some importance when living on a greatly reduced income. They decided to remain in Paisley for the time being.

Unbeknown to Richardson, Professor Quincy Wright of the University of Chicago had already embarked on a major objective survey of past wars, the results of which were published in 1942 in his monumental two-volume book *A Study of War*. One cannot but wonder whether it would have made any difference to Richardson's researches if he had been aware of Wright's efforts; would he, for example, have decided to await the publication of Wright's list of wars, thereby saving himself from the endless hours of toil necessitated by the preparation of his own list? It seems more probable that he would not have been satisfied with the criteria adopted by Wright and would still have persisted with his own independent approach. Many social scientists feel that the final outcome of these two individual efforts was more valuable than would have been the result of a single joint study, for in many respects the work of Richardson and Wright turned out to be complementary rather than repetitive.

To illustrate this we need only consider the fundamental question of how to define a war. As Quincy Wright had a legal background, it is not surprising that he attached considerable importance to the legal recognition of warlike action and to the legal and political consequences of each incident. For Richardson this was too narrow an approach. He considered that 'from the psychological point of view, a war, a riot, and a murder, though differing in many important aspects, social, legal and ethical, have at least this in common, that they are all manifestations of the instinct of aggressiveness'[1]. He therefore decided to include in his studies all quarrels involving loss of human life; these he called 'fatal quarrels'.

To avoid the risk of national prejudice, Richardson extended his search to cover the whole world. Although he limited it to the period beginning in 1820, this search would still have been a sizeable operation for a team of research students; for one man working in isolation it was a mammoth undertaking. Among the many difficulties in obtaining estimates of the number of casualties in any conflict was the inconsistency between historical sources. For example, one book states that several hundred Armenians were killed in Turkey in 1895 whereas another mentions a loss of 30 000. Wherever possible, Richardson consulted 'authorities whose bias, if anything, was likely to be opposite'[2] and then took a mean of their statements.

When he was nearing the end of his search, Richardson became aware of the list of wars published by Pitrim A Sorokin in 1937 in his book *Social and Cultural Dynamics*. Sorokin's list, which extends back to 1100 AD, also includes estimates of the number of casualties, but only for the greater nations. Richardson compared the two sets of figures and concluded that there was no reason for him to change his own estimates.

Richardson assembled his information about fatal quarrels in a card catalogue which was a very convenient method for his subsequent statistical work. For publication, however, he had to decide in what order to arrange the cards. He was not in favour of a listing by names in alphabetical order, as some of the smaller incidents had no generally recognised names and in some cases a single incident might have several names, even in one language—for example, the Indian Mutiny, Sepoy Rebellion and Great Revolt. He elected to make a preliminary grouping according to the *importance* of each incident and then to list each group in chronological order. There still remained the question of how to assess the importance. The simplest thing might have been to use the number of fatalities but Richardson felt this might risk giving a ridiculous pretence of unjustifiable accuracy—bearing in mind the uncertainty in the records of casualties. He decided instead to use the logarithm to the base 10 of the number of deaths, which he called the *magnitude* of each quarrel. Thus a war with 10 000 deaths would be of magnitude 4, a riot with 100 deaths magnitude 2, a simple murder magnitude 0, and so on. Richardson likened the use of this scale of ten to the action of a sieve in that it 'retained the reliable part of the data, but let the uncertainties pass through and away'[3]. A good precedent for such a logarithmic scale is provided by astronomers when they talk of the magnitude of a star. Although the stars visible to the naked eye were originally grouped into six magnitudes by the ancient Greeks on the basis of visual estimates of their intensity, the validity of the concept was confirmed in more recent times when it was found that the magnitude of a star depends on the logarithm of its intensity as measured by an accurate instrument.

For magnitudes above 2.5 (i.e. more than 316 deaths) Richardson obtained his facts from works on history and for magnitudes less than 0.5 (fewer than 3 deaths) they were taken from criminal statistics. Between these two limits he found that 'the information is scrappy and unorganised; what there is of it suggests that such small fatal quarrels were too numerous and too insignificant to be systematically recorded as history, and yet too large and too political to be recorded as crime'[4]. These differences naturally affected the rate at which he was able to assemble material for the various types of fatal quarrels. He felt that his collection was reasonably complete for magnitudes 3.5 and above after about a year but for magnitudes in the range 2.5 to 3.5 he was still adding new information many years later. For magnitudes between 0.5 and 2.5 he only succeeded in getting reasonably reliable data on a small number of random incidents while for the category below 0.5 he never attempted to examine individual events.

Towards the end of 1941 Richardson felt that his collection was sufficiently representative to justify drawing some tentative conclusions. His first public announcement was in a brief letter to *Nature* on 15 November calling attention to the simple relationship between the number of fatal quarrels and their magnitude, which he later summarised in the phrase: 'The larger, the fewer'[5]. From his list of fatal quarrels for the period 1820 to 1929 AD he counted the number in each successive interval of magnitude 7.5 to 6.5, 6.5 to 5.5, 5.5. to 4.5 and 4.5 to 3.5. The resulting totals were respectively 1, 3, 16 and 62, in other words very nearly in agreement with the geometrical progression 1, 4, 16 and 64. However, if this progression is continued, it gives 16 384 for the number of murders instead of the observed 10 million. He concluded: 'These remarkable facts call for explanation. A full account of this and cognate matters is nearly ready for publication elsewhere'[6]. But it was not until 1948 that the full account was published in the *Journal of the American Statistical Association*. By that time he had found that the apparent inconsistency in the number of murders could be eliminated by a more refined statistical treatment of the data. He was also able to produce some evidence from studies of banditry and ganging to suggest that his relationship applied also to the magnitudes between 3.5 and 0.5. His conclusion was that deadly quarrels of all magnitudes, from the world wars to the murders, are suitably considered together as forming one wide class, graded as to magnitude and as to frequency of occurrence. He also pointed out that the heavy loss of life occurred at the two ends of the sequence of magnitudes; during the interval 1820 to 1945 AD 36 million had been in the two world wars and 10 million in connection with murders. The total of about 59 million from all fatal quarrels was only about 1.6 per cent of the number of deaths from all causes. 'This is less than one might have guessed from the large amount of attention which quarrels attract. Those who enjoy wars can excuse their taste by saying that wars after all are much less deadly than disease'[7].

Richardson naturally tried to find a theoretical justification for the relationship which he had established between the frequency and the magnitude of fatal quarrels, but never really succeeded. Starting from the thought that 'towns and wars are both manifestations of human aggregation'[8], he discovered, however, an interesting parallelism

Retirement: Paisley 1940–43

between this relationship and the distribution of towns of different sizes. This led to the development of a mathematical equation to describe how a population which was initially distributed uniformly over a country would tend to concentrate in towns under the opposing forces of gregarious attraction and population pressure. He compared this with Jeans' theory of the formation of stars, under the opposing forces of gravity and pressure, from an initial uniform distribution of matter in space. He announced this equation in a letter to *Nature* in December 1941 and added that a fuller account of his theory was ready for publication.

All I have been able to find of this fuller account are two unpublished articles on gregariousness—neither of which seems to be complete—which were amongst the papers left after his death. As mentioned earlier, Richardson shared with his sister Edith a preference for being alone rather than being in a crowd. In a personal biography of his father, Stephen wrote[9]:

> In preparing his entry for *Who's Who*, my father asked the family if he could answer the question about hobbies and sports with the one word 'solitude'. However brief the biographical sketch, his life of 'solitude' should be added. He found social events tiring and confusing and needed a great deal of quiet time by himself.... On one occasion he asked if he might visit his sister for a week end. She replied that she would enjoy his visit but did not wish to talk. They spent a pleasant two days together in silence.

Some of Richardson's thoughts on solitude are reflected in one of his unpublished articles *Gregariousness and its Opposite*. In begins with two anecdotes about real people whom he did not name but whose identities are rather obvious. The first is a conversation between a gregarious woman (clearly his own wife, Dorothy) and her less gregarious husband[10].

SHE   "Oooh! I've been to a *splendid* meeting. The hall was *packed* so that people were standing all round the sides and sitting on the gangways, and the cheering was tremendous!"
HE   "And what did they decide?"
SHE   "I don't know that they exactly *decided* anything. That wasn't what the meeting was *for*. But there were some *thrilling* speeches!"
HE   "About what, for example?"
SHE   "Oh well, its rather hard, you know, to give a matter of fact account of a public speech. But it was a simply *glorious* meeting!"

The second anecdote is about a visit to a non-gregarious woman (obviously his sister Edith), an artist, who

> lives by herself in a studio near a village. Occasionally friends or relatives called to see her and she welcomed them with enthusiasm thus: 'Well, my

dear, I am *delighted* to see you; although, as you know, I hate visitors in general.' She then entertained them thoroughly with sumptious meals and much conversation. This same artist would refer jokingly to her tendency to misinterpret certain well-known phrases. Thus she would say: 'You doubtless know the hymn, and a fine solemn hymn it is too, which goes

> At even when the sun was set
> The sick, O Lord, around thee lay
> Oh with what diverse pains they met
> Oh with what joy they went away.

'But to me', she said, 'the last line has long had another meaning. It is a description of myself saying goodbye to visitors; and the joy is mine at seeing them go away.'

This last remark brings to mind a true story about a little girl aged 3, who shall remain anonymous. Her mother had been entertaining a large number of ladies to afternoon tea and as the ladies were leaving the little girl shook hands with each of them and said politely: 'Thank you for coming. And thank you for going'.

Richardson's article continued:

> W. McDougall included gregariousness in his list of instincts.... It satisfied his definition of an instinct as an inherited or innate psycho-physical disposition... I wish now to suggest that anti-gregariousness has an equal claim to be considered an instinct. For there are people who tend to perceive, and to pay attention to, crowds, even imaginary or hypothetical crowds, and then to experience vague emotions of dissatisfaction, discomfort and aversion together with a tendency to move 'far from the madding crowd'. The outstanding question is whether this disposition is innate.

He felt that some insight into this might be obtained from an analysis of people's dreams. After working for three months in London, the brother of the 'artist' (i.e. Richardson himself)

> had a blissful dream in which he was put ashore, with plenty of biscuits and other stores, on Rockall, an uninhabited island in the Atlantic, to keep watch on the weather. This dream recurred in similar circumstances over several years until the dream-cave in Rockall became quite familiar. As a happy variant of the same theme he was dropped from an aeroplane on the top of Mount Everest, with ample provisions, to keep watch in a snow hut. After three months he was lifted off and was quite pleased to see a London bus again. He interpreted these dreams as caused by *privation of solitude*. In actual life he never experienced any such extreme solitude.

Richardson then referred to the record which the 'solitary artist' had kept of her dreams over a period of eight years. As this included dreams on more than 1000 nights, he had been able to subject it to statistical analysis, especially as regards the way in which the pleasantness of each dream was related to the number of people present in the dream. He had then asked himself what attribute of a crowd she had disliked. He wrote:

> I find no explicit answer; but the following is in conformity with all her dreams and conversation as far as I know them. The objectionable attributes of a crowd were its incoherence, it disorderliness, its too many diverse impacts on her personality. Even a gregarious reader may be able to understand this dislike if he will imagine an orchestra in which each player plays his own tune unrelated to the others.

Was this a premonition of Stockhausen?

In 1941, Richardson sent a copy of the article to his brother Gilbert for comment. Gilbert's reply, using the spelling recommended by the Simplified Spelling Society, began: 'It interests me tu see a haaf-baekt literary produkshun. Dhe mor so as I hav just resently produest wun mieself, and dhen wurkt it up tu a kondishun fit tu publish'. As a 'strieking eksample' of gregariousness he cited 'foot-baul machez wocht bie 10,000, 60,000 or mor' spectators, some of who go 'not ocnly tu se dhe gaem, but olso just to be in dhe kroud and join in dhe uproer'. He contrasted this with the action of St Bruno who in 1084 'establisht himself in dhe wildernes ov dhe Grande Chartreuse'. When Gilbert had visited the monastery there he had been impressed by the inscription on the door of one of the cells 'O BEATA SOLITUDO O SOLA BEATITUDO'[11]. These ideas were incorporated in a later version of Richardson's article but it was still not in a 'kondishun' for publication.

In another incomplete note entitled *An Abstract Formulation of Fashions*, Richardson attempted to develop a mathematical model of the behaviour of people faced with a choice of actions. He began by justifying the use of an idealised model[12]:

> Among natural objects a circle is scarcely to be found. Yet it is the simplest closed curve. Many of its properties were not obvious from the definition, but had to be proved, as in Euclid. The abstract conception of the circle simplifies descriptions of actuality; as for example when it is said that the outline of the sun's disk is more circular than that of the planet Jupiter. Those whose interest in roundness is confined to round fruits, round plant stems, and round pebbles, may feel that theorems on the circle are fussy and irrelevant. Similarly there will doubtless be some practical social psychologists who will take no interest in the following theory. The author proceeds nevertheless.

The theory took account of the common tendency to 'go with the largest crowd' and led to the conclusion that it would be easier to start a new fashion when there is a large number of choices, each with the same number of adherents. As an illustration of his theory he considered the probability of somebody using standard spelling being converted to Simplified Spelling, and vice versa. We shall see that some of the ideas expressed in this note were taken up later in Richardson's work on war moods.

The preceding discussion of Richardson's unpublished notes on gregariousness arose from a reference, in his letter to *Nature* of December 1941, to a *fuller account* being ready for publication. In fact, he did not publish any further results from his peace research until 1944. In the meantime, he continued his statistical analysis of past fatal quarrels and updated his earlier work on arms races, in the hopes of being able to publish all his findings in a book. He was especially keen to have published the information on fatal quarrels which he had so painstakingly assembled in his card catalogue. In the early stages of analysing this information he had thought of indicating which of the belligerent parties was the aggressor but the evidence on this was so conflicting and ambiguous that he soon abandoned the idea. He also rejected a proposal to classify wars uniquely into two groups, civil and international; even if rules could be drawn up to make such a distinction fairly definite, he felt that it would be wrong to single out political loyalty from all the other agencies making for social unity or discord.

In the next stage of his search Richardson made a list of the social relations between the belligerent parties which were mentioned in the history books he consulted. Here he found 'the usual difficulty that brief statements tend to be inaccurate, while accurate statements tend to be complicated and therefore incomprehensible'. His first list turned out to be too simple to be applicable to every war and he therefore revised it to make it 'more appropriate to the data and logically clearer'. 'Having thus muddled through somehow', he finally felt able to recommend his system 'as a good working compromise between incompatible wishes for accuracy and for simplicity.'[13]

The system which Richardson adopted was based on a list of 59 social situations arranged into three groups: (*a*) those expected usually to make for amity; (*b*) those expected to be ambivalent; (*c*) those expected to be usually causes of dislike, suspicion, contempt or arrogance. To each item in group (*a*) he assigned a small Roman letter, in group (*b*) a cursive Greek letter, and in group (*c*) a capital Roman letter. Typical examples are: m: common government; λ: the belligerents had previously fought as allies against a common enemy; R: the belligerents were rivals in trading with third parties. For each fatal quarrel above magnitude 2.5, he noted all the letters corresponding to the social situations which he had found in history books; in some cases there were only one or two letters, while in others there were more than ten.

When Richardson described his list of social situations to a meeting of psychologists in Edinburgh in 1941, several speakers gave examples of wars where a factor which would usually have had a pacifying influence had had the reverse effect. To take account of this valid criticism, Richardson divided the letters assigned to each fatal quarrel into three compartments according to the *actual* influence as distinct from the *expected* influence; the pacifying situations were placed in the first compartment, those with unknown effects in the second, and the annoying situations in the third. It is thus possible to see at a glance whether the expectations were justified in any particular instance.

The above coding scheme was completed by a system of arrows and suffixes, the arrows to indicate the direction of such situations as one belligerent wanting to acquire territory from the other, and the suffixes to indicate the number of years that a situation, such as common government, had lasted prior to the conflict. For multiple wars, the belligerent parties were divided into pairs and separate entries were made for each pair in the form of a matrix. The whole classification system was thus extremely complicated, but much less so than the phenomena that it was designed to describe. It enabled Richardson to compress into less than 80 printed pages the relevant information about more than 300 fatal quarrels above magnitude 2.5 As we shall see, the list was not published in book form until 1960; it was then hailed as 'a remarkably comprehensive collection of historical facts concerning recent wars' and 'a mine of information for social scientists and historians as well as for statistical analysts'[14]; Richardson himself suggested that an appropriate nickname for his summary would be *Potted History Book*[15].

The mere preparation of Richardson's card catalogue and the compilation from it of the classified list of fatal quarrels was no mean achievement; to Richardson it was, of course, not an end in itself but simply a means to an end or, more specifically, a tool to be used in analysing the causes of past wars. We have already mentioned one of the early results of his analysis, namely the relationship between the magnitude and frequency of fatal quarrels. Another result, to which Richardson attached greater importance, relates to the distribution of wars in time. From his data on about 60 wars in the range 3.5 to 4.5 (number of deaths between three and thirty thousand approximately), he counted the number of years in which no war began, in which one war began, in which two wars began, and so on. He found that the resulting distribution (65 years with none, 35 with 1, 6 with 2, 4 with 3, and none with 4 or more)[16] was remarkably close to what is known as the Poisson distribution of improbable events, named after its discoverer, the French peer S D Poisson, who like Helmholtz had studied medicine before switching to mathematical physics. Among the best known examples of a Poisson distribution are deaths by kick from a horse and the emission of alpha-particles from radioactive substances. Richardson gave another example which would surely appeal to a cook.

> Take enough flour to make N buns. Add λN currants, where λ is a small number such as 3. Add also the other usual ingredients, and mix all thoroughly. Divide the mass into N equal portions, and bake them. When each bun is eaten, count carefully and record the number of currants which it contains. When the record is complete, count the number y of buns, each of which contains exactly x currants. Theory would suggest that y will be found to be nearly equal to Y as given by the Poisson formula.

He added, in characteristic humorous vein, 'I do not know whether the experiment has ever been tried'[17].

Seven Weeks' War, 1866

			References
4.6	PRUSSIANS for leadership of German States.	ITALIANS to acquire Venetia.	Dumas. Orsi. Bodart.
AUSTRIANS	— | c i g M_{103} λ_2 | — vi.16–vii.22	— | g I | M_8 L) T) vi.20–viii.12	E 10, 289. E 12, 752.
BAVARIANS	— | c i g | — vii–viii	O	E 12, 806. E 20, 384.
SAXONS	— | c i g M_{103} | — vi.15–vii	O	
HANOVERIANS	— | c i g | — vi.14–vi.29	O	

Results. Prussia acquired dominance. Austria ceded Venetia to Italy.

Figure 8.1 An example of Richardson's matrix from what he called his *Potted History Book*. If written out in full, the information about the Seven Weeks' War (magnitude 4.6) would take at least a printed page.

Richardson later repeated his analysis of the distribution of wars in time using Quincy Wright's list; he obtained the same agreement with the Poisson distribution. Such a result could have been obtained from the hypothesis that there is the same very small probability of an outbreak of war somewhere on the globe on every day. Richardson was careful to point out that this did not however imply any ability to predict the dates of future wars: 'Poisson's law relates only to the total number of years having a given character, and does not tell us the character of any particular year'[18].

The importance of this discovery for Richardson's further peace research was that it provided justification for assuming, as a working hypothesis, that the laws of chance could be applied to historical events. One such application of probability theory was an attempt to determine whether wars have become more frequent. He divided Wright's list into two equal intervals of 215 years from 1500 to 1715 AD and from 1716 to 1931. There were 143 wars during the first interval and 156 during the second; this increase could however be explained away as a chance effect[19]. A similar result was obtained from a statistical analysis of his own list of fatal quarrels above magnitude 2.5, although there was some suggestion that larger wars might be becoming more frequent and smaller incidents less frequent.

A rather lower degree of mathematical sophistication was needed to determine the extent to which different nations had been involved directly in past wars. Richardson described the relevant section of an article about his work as resembling quinine in that 'it has a bitter taste but medicinal virtues'[20]. This description was particularly apt for British readers, for no matter whether he began his count of wars in 1480, 1820 or 1850 AD Great Britain came out as the country which had most frequently been a belligerent, with France and Russia not far behind. After eliminating countries with populations less than a million and those which had not been independent throughout the period, he found only three countries which had not been involved in a fatal quarrel of magnitude 3.5 or greater between 1820 and 1939, namely Persia, Sweden and Switzerland. Of these, Persia had been engaged in a number of wars of lesser magnitude and only Sweden had not taken part in any war at all. Those who believed that war could be prevented simply by suppressing German aggression could not find much supporting evidence from Richardson's table of belligerent countries. He stressed in any case that his statistics were not of aggression but of participation; to attempt to determine which country was the most aggressive would, in his view, have led to an entanglement in 'obscure and controversial distinctions between aggression and defence'[21].

In trying to explain the results of this enquiry, Richardson discovered a marked correlation between the number of external wars in which a country has been involved and the number of its frontiers. In the days of the British Empire, Britain had more frontiers than any other country while France and Russia again followed not far behind. He recognised that it would be better to take into account not only the number of frontiers but also their length and the local density of population; this would, however, involve some laborious calculation and he decided, for the time being, to leave it to others. We shall see how in fact he followed up this idea himself at a later stage of his investigations.

Several references have already been made to Richardson's interest in international languages, especially Ido and Esperanto. The founder of Esperanto, Dr L L Zamenhof, was inspired by the belief that a common language would do much to clear away misunderstanding and hence to avoid war. It is not surprising therefore that Richardson endeavoured to ascertain from his collection of information about past wars whether differences in language have in fact played an important role in international disputes. The circumstantial evidence which he could find could not prove anything. 'What is needed', he wrote 'is a comprehensive survey by counting. For counting is an antiseptic against prejudice. The present author's initial prejudice was that Zamenhof judged wisely'[22].

By neglecting small minority languages, it was easy for him to count the number of pairs of belligerents with the same and with different languages, but this information was not in itself sufficient to answer the question; factors such as the number of people in the world speaking each language and their geographic distribution would also have to be taken into account. What he needed in addition was an estimate of the probability of war between groups speaking particular languages on the assumption that diversity of language had nothing to do with the causes of war—a much more difficult problem.

To overcome the complexities of the real world, Richardson tackled the problem by considering an idealised geography in which the world's population is divided into hexagonal cells, like a honeycomb, with a million inhabitants in each cell all speaking the same language. The cells for any given language were arranged in a form as compact and symmetrical as possible. From this model he calculated the probability that a pair of belligerent cells, selected at random, would have the same mother tongue on two different assumptions; (i) that wars could only occur between adjacent cells, or (ii) that wars could occur between two cells anywhere in the world. These calculations gave two extreme values for the probability that a pair of belligerents would have the same or different languages. The resulting estimates were then compared with the number of wars which had actually occurred between various language groups.

Richardson's calculations indicated that: (i) there were fewer wars in which both sides used *Chineses* (his term for all the languages which, though spoken differently, use the same Chinese ideograms) than would have been expected from the population of Chinese; and (ii) there were more wars in which the opposed belligerents spoke Spanish than would be expected from the number of Spanish speakers in the world. For the other chief languages the statistics neither confirmed nor refuted Zamenhof's belief that a common language would have a pacificatory effect.

Richardson also made an analysis similar to the above to determine the effects of different religions. He began the report of the study with one of his dialogues[23].

FIDOR: So you have written an account of the struggles of nations and have left out Almighty God, who decides their destiny.

AUTHOR: A criticism rather like yours was made, you will remember, by Napoleon of Laplace's *Mécanique Céleste*; and Laplace replied, "Je n'ai pas besoin de cette hypothèse."

. . . . .

SCEPTIC: But what do you mean by the word God?

AUTHOR: If Fidor will permit me to answer that question I wish to suggest a definition that will ease my task of classifying the causes of wars all over the world. At a time when the luminiferous ether was much in discussion Lord Salisbury described it as "the nominative of the verb to undulate." That description, though at first sight comical, had the great merit of directing attention away from puzzling and contradictory theories and on to immediate perceptions, such as the sight of interference fringes. Would it not be wholesome to do likewise for religion by defining God as "the accusative of the verb to worship"? For worship is an immediate experience, a yearning toward something larger and better than ourselves.

SCEPTIC: But with that meaning of the word there are a great many different Gods. Sir James Jeans' "Supreme Mathematician" is quite unlike Kaiser Wilhelm's "Good Old German God."

AUTHOR: From the psychological point of view that is the actual state of affairs. People worship different Gods. And this worship does in part decide their destiny.

FIDOR: I would go so far with you as to agree that there is nothing like a strong Christian faith for giving, in time of battle, that courage which ensures victory. The Mahdi, who fought against Gordon, had a fervent Mohammedan faith; the Taiping rebels, who also fought against Gordon, were upheld by their strange religion; and, strangest of all, the Wangoni persisted in opposing the Germans because they trusted in a medicine that would turn enemy bullets to water.

SCEPTIC (*to himself*): Well, thank God, I'm an atheist.

From his analysis, Richardson concluded that the adherents of the Chinese religions were involved in fewer wars than would have been expected had religion been of no consequence and there were more wars between Christians and Moslems than would be expected from their populations, if religious differences had not tended to instigate quarrels between them. There was also a strong suggestion, not however fully substantiated by the statistics, that Christianity incited wars between its adherents while the Moslem religion prevented wars between its adherents. Richardson commented[24]:

> To anyone brought up, as the author was, in the belief that Christians worship the Prince of Peace, it must be startling to notice the predominant occurrence of wars involving Christians, and especially of those in which Christians fought each other.

In any discussion of the results of Richardson's investigations of the impact of differences of language and religion in relation to war, it must be borne in mind that his statistics refer mainly to the period 1820 to 1939 AD. Regarding the apparent pacifying influence of the Chinese languages and religions, he came to the conclusion that the comparative peaceableness of China prior to 1911 (after which it was less pacific) was the result of instruction, and in particular of Confucian instruction. 'If China could thus be made peaceable,' he asked, 'why not the whole world?'[25]

Richardson used his card catalogue to ascertain the importance of various other factors in causing wars. One of his investigations—which he did not pursue in depth—related to the assertion by Karl Marx that the class-struggle was the essential feature of history. Of the 83 wars on his list above magnitude 3.5 during the interval 1820 to 1939, he could only find 11 for which the history books he had consulted had indicated that there was a marked personal economic inequality between the belligerents. From this he concluded that Karl Marx 'exaggerated the importance of the class-struggle, as far as war is concerned'[26]. There followed a brief dialogue: 'COMMUNIST: That only appears because you have taken your information from bourgeois historians, who have missed the true meaning of history. AUTHOR: I had to use such concise sources as were available. But I should point out that the most reliable observers are usually those who have heard of all the theories, but are not addicted to any of them.'

Richardson went on to check the commonly held view that the fundamental cause of war is economic and found that there were 25 of the above 83 wars for which no economic direct cause had been noted in his list. On this he commented[27]:

> It would no doubt be possible to rearrange the list, or the counting, so as to give more prominence to economics; for the boundaries of the class called 'economic' are not sharply defined. It can be argued that every human affair is economic, because mankind is everywhere dependent on material objects, such as food, land and tools. Similarly it can be argued that all human affairs are sexual, because everyone is born. But the word economic would become useless if its meaning became too wide. Let us avoid such extensions.

By 1942, Richardson had assembled most of the above material—the classified list of wars and the results of various statistical analyses—in the form of a manuscript ready for publication as part of a book. In the rest of the book he planned to include a revised and updated version of his *Generalized Foreign Politics*, beginning with a new chapter entitled 'Armaments and security'. This opens with a quotation from *Punch* of 1842 hailing the discovery of an explosive which was so terrible that it would ensure peace. This was but one illustration of the widely held view that the armaments of a nation promote its security and that their cost should be regarded as an insurance premium. On the basis of this theory, Richardson argued, countries with the largest pre-war defence expenditures should either be able to stay out of the conflict or, if they do become involved, they should have the lowest casualty rates. In this form, the theory could be tested statistically. He accordingly set out to collect the relevant figures for the First World War. This proved to be more difficult than he had expected. 'One might reasonably think', he wrote, 'that the losses of life were some of the main historical facts of World War I. Yet there are several standard histories in which the losses are not stated. Why this omission? The author can only guess that histories, like other books, have to be written to please readers and that statistics of casualties do not please many of them in peacetime'[28]. Using the best available estimates that he could find, he was able to show quite conclusively that there was no relationship between the per capita defence expenditure of a country and the casualty rates in the armed services.

In another new chapter, Richardson makes an interesting comparison between arms races and athletic races, inspired by the thought that 'Whoever coined the phrase "arms race" (was he the editor of the London 'Daily Mail' in 1912?) must surely have been thinking of competition and movement and probably also of athletes running'[29]. Taking as a measure of the speed of an arms race the annual defence expenditure, his earlier work had shown that as this speed increased, the acceleration also increased. In athletic races, on the other hand, after a brief initial acceleration the speed of the runner tends to settle down to a fairly steady pace. By way of psychological explanation, he contrasted the athletes' knowledge of the distance to be run and their previous training with the lack of knowledge of nations regarding the lengths in time and money of an arms race.

In his earlier publications, Richardson had provided evidence to support his theory of arms races from what happened prior to the two world wars. He now addressed the question of whether most wars had been preceded by arms races or whether such races are a recent innovation. He concluded that 'the evidence, as far as it goes, is that only a minority of wars have been preceded by arms races'[30]. He was in fact left with the impression that 'unstable arms races of modern type first appeared between 1877 and 1900'[31]. To him it seemed credible, though hardly proved[31],

> that the modern instability of balances of power is a consequence of the application of science to war material. For, because of the many kinds of complicated material required in peacetime by the fighting services, many people are employed in their manufacture, and their cost is a conspicuous item in the national budget. These activities become known to other nations and are interpreted by them as threats.

The criticisms made by Forder, Piaggio and others on his *Generalized Foreign Politics* led Richardson to re-examine the European arms race of 1908–1914. He warned the reader that[32]

> A thorough re-examination of a thesis that has already been stated in bold outline can hardly be expected to make easy or entertaining reading. Yet the subject is a world event of great importance and so should have close attention. The following treatment is intended to convince the studious critic, whatever be his nationality, rather than to please the passer-by.

He then proceeded to review in much greater detail than previously the defence expenditures of the main European protagonists in the First World War and the extent of their foreign trade. After a more refined statistical treatment than in his earlier work, he concluded that international trade acts as a partial counterbalance to the threat imposed by expenditure on armaments; a good fit between his arms race model and actuality is obtained if the defence expenditures are reduced by one quarter of the trade figures. He showed that power in Europe was so nearly in equilibrium in 1907 and 1908 that[33]

> a small mutual obligement between the opposing groups of nations, such as a reduction of their warlike annual budgets by an amount equal to the cost of two days of the subsequent war, would have prevented the arms race from developing and so, presumably, would have avoided the war altogether.

In extending his investigation on arms races to cover the period leading up to the Second World War, Richardson encountered several factors which greatly complicated his work. For the 1908–1914 race he had been able to consider all the major nations as belonging to one of two groups of countries but at the time when he started to study the arms race of the 1930s the grouping of nations into belligerents was still uncertain and he was obliged to treat the problem as one involving many nations acting independently. When he came to revise his results a few years later he could, with the help of hindsight, have reverted to his earlier two-group treatment. He decided, however, that there were some additional merits in the many-nation equations and that it would therefore be best not to make any changes.

Another complication arose from the greater variability in international exchange rates. Prior to 1914 it had been possible to compare the defence and trade figures of different countries simply by converting the amounts in the national currencies to a common monetary unit. Richardson used the same system for his *Generalized Foreign Politics* even although it resulted in some unreal fluctuations in defence expenditures. He would have liked to have used, instead of gold, a standard based on human effort, but had not yet devised a satisfactory way of doing thus. The problem is well illustrated by the defence budgets of the Soviet Union: when expressed in roubles they show a progressive increase from 1931 to 1938, but when the figures are converted to old gold dollars at the official rates of exchange they reach a peak in 1935 and then fall quite dramatically before starting to rise again. In his later work Richardson overcame this problem by converting the annual defence expenditures into the number of persons who could earn that amount of money in a year if they each received the national average wage. He originally called this the 'warlike worktime' of a country but later changed over to the expression 'war-personnel'. Even this did not satisfy him because of its implication that he was only talking of armed personnel; he therefore coined the word 'warfinpersal' a contraction of 'war-finance per salary'[34], but like some of his other proposed new words this rather clumsy term has never been widely used.

In his 1939 monograph, Richardson had already extended his system of equations to cover an arms race involving three nations and had then considered several special cases. For example, he showed that if two of the nations were pugnacious and one compliant, the aggressive preparations of the pugnacious nations towards the compliant nation would tend to settle down to a constant value, depending on old, standing grievances. The pugnacious nations would continue to interact with each other, drifting in certain circumstances towards war, and in others towards an alliance. The theory was also developed for an indefinite number of nations with applications to particular sets of circumstances. The most unstable configuration would, for example, be one in which all countries were of equal size. Improved communications between nations, such as the development of aviation, would tend to have a destabilising influence. The main changes in the revised version of this section of the monograph relate to the application of this theory to a study of the arms race of the 1930s arising from the use of warfinpersal instead of the actual figures of defence expenditures and trade. According to his results, in any year from 1933 to 1938 pacification 'would have required not only an abolition of so-called "defence" expenditure by all these nations but also some further active obligements between them'[35]. Richardson did not consider that the policy of appeasement pursued by the British government from 1937 to 1939 constituted an obligement; he described appeasement in the following terms 'Open up conversations with likely enemies about matters in dispute, taking care not to give them any valuable satisfaction at the expense of your own country. At the same time, increase your armaments both for defence and attack'[35].

When he introduced foreign trade into his analysis of this arms race, Richardson obtained a more satisfactory fit with his theory. The best agreement was produced by subtracting from each nation's warfinpersal one seventh of the worktime value of its international trade. His final conclusion on this was that, while trade between nations tends to prevent them from quarrelling, it does not have such a big influence as he had earlier believed.

The details of Richardson's theory and analysis are of more interest to economists and statisticians than to the general reader. At first sight, the mathematical sections look rather forbidding but in fact there is little beyond the standard of a university degree. As Richardson himself put it 'Nearly all of the pure mathematics is to be found scattered among the textbooks. I had to collect and arrange it. The application to politics is, I believe original, for it has grown from what I published in 1919'[36]. The gist of what Richardson was trying to say in his arms race theory can in any case be gleaned from the interspersed Socratic dialogues, which also provide some light relief for the serious student. Here is another example[37].

> HUMANIST: Do you seriously believe that you can prove, by mathematics, anything about the behavior of nations?
> AUTHOR: All that can be proved by mathematics is that certain consequences follow from certain abstract hypotheses. But, after such proofs are established, we shall compare the observed behavior of nations with some of the deduced consequences and so form an opinion as to whether the hypotheses are a true description of nations.
> HUMANIST: That seems a fairly modest claim. But the mathematics which I learned at school had an air of dogmatic certainty.
> AUTHOR: The conditional mood becomes tiresome if long continued. So the custom is to follow Euclid by expressing any long theory in the indicative, as if it were true. We must not forget that it all depends on hypotheses.
> HUMANIST: But how do you know what hypotheses to take as your starting point?
> AUTHOR: By intuition, luck, or laborious search. Unless they were promising, they would not appear in print.
> HUMANIST: Many people will distrust the application of mathematics to such a subject as foreign politics because they feel that political affairs are essentially very complicated and interwoven, whereas any mathematical hypothesis is, I suppose, comparatively simple, clear-cut, and definite.
> AUTHOR: I leave out all the details, as in landscape sketching, one might draw a complete wood without showing individual trees or leaves.
> HUMANIST: But, even so, does not the mathematics tend to an oversimplification and rigidity of conception: to drawing the wood as if it were a simple geometrical figure?
> AUTHOR: Yes, it does. But that contrast occurs, in greater or less degree, in all applications of mathematics. It is not peculiar to foreign politics. A circle is a simple definite conception. We see many round forms: the sun, various fruits, some hats, most dinner plates. Very few, if any, of them are accurately circular. Yet the circle is a standard idea.

HUMANIST: But is it worthwhile to be mathematically precise in your hypotheses and deductions when you foresee that there will probably be considerable misfit between your conclusions and the actual state of affairs?

AUTHOR: Yes, it is worthwhile. For when the theories that are obviously wrong have been rejected, the select few that remain begin to be used as standard ideas, by which actuality can be classified and named. If our theory of circles were inaccurate or confused, if, for example, we believed the circumference to be three times the diameter or believed that a plane figure was necessarily a circle because all its diameters were equal, we should become entangled in contradictions when we tried to find out whether a particular wheel was circular.

HUMANIST: But still, you seem to me to be addicted to an extreme degree of precision, far beyond what is necessary.

AUTHOR: Many people prefer to have that somewhat irrelevant precision. A book would mean the same if its lines were set crookedly. A man who buys a house expects his banker to keep a record of the transaction to the precise penny, even if he would have been willing to pay £50 more for the house. We relish our bacon, although the machine that sliced it need not have been so accurate.

Richardson ended his *Generalized Foreign Politics* with another such dialogue, an extract from which was reproduced on pages 165–6. He continued this dialogue in 1941 during the war[38]:

CRITIC: Hullo! Here we are again, both still alive. What have you been doing since we last met?

AUTHOR: Recently I have been studying an explosive.

CRITIC: Really? I am glad to hear that you took so useful a part in our war effort. No pacifist nonsense about that! By now, I believe, we've got something far more powerful than the old-fashioned picric acid or T.N.T. What's it called: damitall? No?

AUTHOR: I have indeed been studying an explosive so powerful that, apart from it, none of the others would do any harm. I refer to the human temper.

CRITIC: You are a naughty joker! But seriously now, I suppose that the war will have considerably changed your views. When we last met, you said that bombing airplanes were a danger to the nation that owned them.

AUTHOR: In a roundabout way in peacetime.

CRITIC: Surely you have now abandoned that outrageous statement. What would have happened to Britain if it had not been for the heroic airmen of her bombers and fighters?

AUTHOR: And which nation had the most bombing airplanes just before the war?

CRITIC: Germany, of course, that was the cause of all the trouble. Fools and pacifists hindered our rearmament. Their criminal folly has been vividly described by Cato in his book called *Guilty Men*.

AUTHOR: But did you expect Germany to be beaten?

CRITIC: Definitely.

AUTHOR: From your statements it follows that the possession of bombing airplanes in time of peace will ultimately prove, in a roundabout way, to have been a danger to the nation that owned most of them. Indeed, without foretelling the future, it may be said that the damage already done to many German cities is an indirect consequence of their prewar Luftwaffe.

CRITIC: But think of the long and terrible struggle that we have been let in for by our initial inferiority in the air.

AUTHOR: What do you suppose would have happened if, by 1935, Britain had developed armaments on the enormous scale of those of 1941?

CRITIC: That is what should have been done. Then we should have been able to hold down Germany and, incidentally, to have got our own way anywhere all over the world.

AUTHOR: So that the whole world would have allied against Britain's overweening superiority! On the contrary, the war of 1939 might never have begun if each side had confined itself to defensive armaments.

CRITIC: Nonsense! Hitler intended war anyhow. If there had been no weapons, we should have fought with fists.

AUTHOR: Then a careful study of grievances, ambitions, and greeds ought to be made. The opinions of historians about such motives have already been summarized, as far as they relate to wars that ended since 1819. Let us next attend to the psychoanalysts.

In March 1946 Richardson completed this conversation with a rather sad admission[39]:

CRITIC: In view of the evidence at the Nürnberg Trial I suppose that you will now admit that Hitler and his associates planned the war deliberately.

AUTHOR: Yes, the evidence of formerly secret documents convinces me of what I was formerly loath to believe.

CRITIC: And with that admission all your fine theory about defense coefficients and what-not blows away in smoke!

AUTHOR: Not a bit of it. For it is firmly founded on numerical fact.

CRITIC: But you cannot say both that the war was caused by mutual interactions between populations and that a small clique planned it deliberately.

AUTHOR: I think that those two statements are compatible, because leaders lead peoples where they are willing to be led.

The foregoing very simplified account of Richardson's peace research brings the story up to the end of 1942 when he started to look in earnest for a publisher. In a few instances, such as the 1946 dialogue, we have deliberately jumped ahead by including some later material where this was felt to be particularly relevant. There may well be other inadvertent use of later results, for Richardson continued to revise his manuscripts for the rest of his life and it is not always clear at what stage the various changes were made.

Many of the revisions, such as those necessitated by the discovery of new information about the wars on his list, involved him in a great deal of additional calculation. In these days of electronic computers it is difficult to appreciate how laborious and time-consuming these computations, made solely with a small mechanical desk calculator, must have been. He did receive some help from Dorothy, who was able to devote more time to him as the years went by. After his resignation from Paisley Technical College, there was no more official entertaining at the Richardson home and, as a result of the war, there were fewer visits from family and friends. Although the faithful and much-loved nanny, Ellen Swan, had died just before the war, her services would in any case not have been so much in demand once the children had started to leave home.

The first to go was Stephen, who joined the British Merchant Service in 1937 immediately after completing his school education at Bootham. He spent most of the next ten years at sea and only saw his parents on the rare occasions when his ship was in a British port long enough for him to pay them a visit. From his earliest days, Olaf had been passionately fond of anything mechanical; on leaving Bootham in 1934 he followed his natural bent and became an automobile engineer. Shortly after the outbreak of war he joined the military and in 1941 found himself with the British Eighth Army in North Africa. He did not see his parents again until after the end of the war in 1945. Elaine was at the Mount School in York (the sister school to Bootham) from 1940 to 1944; she was then at home for a year, working mainly with the Forestry Commission, before going to London to study at the Central School of Speech and Drama.

Figure 8.2 The first family reunion after the war, Kilmun, 1945. *Left to right*: Olaf, Elaine, Lewis, Dorothy and Stephen.

Figure 8.3 Knap Cottage, Loch Long, where the Richardsons enjoyed many summer vacations.

In 1942, the Richardsons decided that the time had come to make a move. They still hankered after settling down permanently in London or Cambridge, but while the war was on they felt it would be wiser to remain in Scotland. As Lewis no longer needed to spend so much time at the libraries in Glasgow, their thoughts turned to the beautiful countryside of the Firth of Clyde. The cottages where they had spent so many happy and peaceful summer holidays were far too isolated for all-the-year-round residence and it was eventually to Kilmun, a delightful village on the shores of the Holy Loch, that they found what they were looking for—a house overlooking the water at a price which they could afford. Although they intended this as only a temporary move, Hillside House in Kilmun turned out to be their last home together.

CHAPTER 9

Retirement: Kilmun 1943–53

As the crow flies—or should I say as the corbie or hoodie flies?—Kilmun is only about 25 miles from Glasgow, but the shortest distance by road is nearly 80 miles, a beautiful drive by winding roads along the shores of several lochs and across the Rest and be Thankful mountain pass. The alternative route now is round the end of the Holy Loch to Dunoon—the nearest shopping centre—and then by ferry across the Firth of Clyde to Gourock or Greenock from where there are good train and bus services. When the Richardsons moved to Kilmun in April 1943, however, the Clyde steamers still called regularly at Blairmore, just two miles from their new home. Even so, it would not have been a very convenient place for Lewis' retirement if he had still wanted to travel frequently to Glasgow. In fact, he and Dorothy made the journey not more than about once a month, usually for a visit to the library or a Quaker business meeting. From now on their social activities were largely confined to the village. In peacetime, Kilmun attracts numerous holidaymakers in the summer months, but for the rest of the year it is a quiet community of only a few hundred, mainly elderly people. During the Second World War, however, many youngsters were evacuated there to be safe from the German bombing of industrial Clydeside. The Richardsons soon organised a youth club, the main activity of which was a weekly gathering at their home. A talk by Lewis on some such topic as the weather or the stars would be followed by party games ('murder' was a great favourite) and then Dorothy would provide a snack of buns and hot cocoa.

Lewis also organised a men's reading group which was attended by such people as the minister, doctor and teacher. They would all read the same book and then meet to discuss and criticise its contents. One of the few local residents with whom he could have serious discussions about his war studies was Mrs Lilian Russell, whom the Richardsons had known before moving to Kilmun—she had, in fact, helped to find a house for them. Among the villagers she was nicknamed the 'monkey woman', because of her somewhat unusual hobby of keeping pet monkeys. In her earlier days she was devoted to the service of unfortunate children and became an inspector of approved schools after the death of her husband during the First World War. Subsequently she ran an orchard in British Columbia and then served for several years in Central Africa with Albert Schweitzer. She became especially interested in the care of those afflicted by leprosy and worked many years caring for them in Nigeria and Thailand. In Strone, the next village to Kilmun, she translated books for Dr Schweitzer and wrote some of her own, including *My Monkey Friends* (1938). By all accounts she was a good conversationalist and quite capable of standing up to Richardson in an argument. She admired greatly his efforts in the cause of peace and there is an unconfirmed rumour that she contributed £500 towards the cost of having his books published.

Figure 9.1 Hillside House, Kilmun, January 1984. The Richardsons' former home now overlooks the American submarine base on the Holy Loch.

Richardson divided his day between research work, odd jobs in the house and garden and some form of exercise, tennis, walking or cycling. He had become an enthusiastic grower of vegetables, initially as part of the war effort but later for the enjoyment of watching things grow and of eating the produce. He would give visitors an opportunity of admiring his tomatoes—or whatever happened to be thriving—before disappearing into his den. A few favoured visitors would be invited into the study to discuss his latest ideas or to be shown some new scientific experiment. There they would see his large collection of books, his filing cabinets and some scientific instruments, all well cared for. Some admired his arrangement for keeping his papers, documents and manuscripts in such good order that he could lay his hands at once on any one he wanted to refer to. Or they might be impressed by the array of spectacles suspended one above the other by the ear loops on a string support in two small cupboards on either side of the ornamental mantelpiece. Richardson was long-sighted and had assembled over the years several pairs of gold-rimmed spectacles, each prescribed for a different working distance, such as 20, 30, 50 or 100 cm. He stuck plaster strips over the bridge and the sides of the frame to relieve the pressure where they contact the skin, and could quickly identify the pair needed for a particular purpose by the length of the strip.

There was a large desk at which Richardson would sit for hours on end, reading, writing or just thinking. When his concentration began to flag he would switch to some manual occupation. There was usually something to be made or repaired for the house but, if not, he always had some new scientific experiment in mind—his inventive streak never left him. During the war, for example, after hearing from Stephen about the harrowing experiences of seamen drifting for days in lifeboats after their ship had been sunk by enemy action, he tried to devise a simple and efficient way of obtaining a supply of drinking water by distillation from seawater. He designed several stills, using either solar energy or the warmth of the human body to evaporate the seawater[1]. He would call on Elaine to help in the testing of his gadgets, but she was a reluctant assistant, not being interested in the physical sciences.

A few years later he constructed a very simple system of rotating pivoted rods to demonstrate the principle of conservation of angular momentum, so well known to our champion skaters. He started to write an account of this for publication but apparently did not get beyond the opening paragraph[2]:

> It is on record that when the setting up of a physical laboratory at Cambridge was under discussion, one of the classical dons objected, saying that if an undergraduate would not believe the word of his tutor—probably a reverend gentleman—but required experiments to convince him, it was a sign that he did not properly appreciate the value of evidence. The attitude of that classical don does indeed now seem archaic. Yet there are still circumstances in which physicists need to estimate the value of hearsay evidence; (i) when, as citizens, they read newspapers or

listen to broadcasts, (ii) when, as students, they are told about observations which they have no opportunity of checking, such as Tycho Brahe's observations leading on to Galileo's reduction and Newton's explanation. The most accurate proof of the conservation of angular momentum is astronomical. As far as I know a neat and accurate laboratory experiment is lacking. The one here described has long been known in the qualitative form; the purpose of the present note is to add the details which make it quantitative.

A good example of the way in which Richardson applied his scientific knowledge to everyday life is provided by the cheap and efficient system of central heating which he installed in the house at Kilmun shortly after their arrival. He set up an anthracite-burning stove in the hall near the foot of the stairs and erected a large-diameter flue-pipe alongside the stairs and through a hole in the ceiling. He once demonstrated this system to me and was especially proud of the fact that the flue was quite cool where it passed through the ceiling; its diameter had been calculated in such a way that all the heat would be dissipated in the lower stretch and none would be lost. To obtain the best results, the doors of rooms in use were kept open and the others had to be kept closed; Dorothy was quick to call my attention to this when I inadvertently left the wrong door open.

Unlike Dorothy, who tended to be rather slapdash about such household chores as washing the dishes, Lewis adopted a scientific approach to all his jobs in the house. When he had washed the dishes they not only looked clean but were clean enough for use in a chemical laboratory. This reflected his phobia for germs, which also led to to decree that nobody suffering from a cold should be allowed to help with the dishes. He kept day-to-day records of his work in the garden, including the weight of any fertiliser used (to the nearest gram). Like many gardeners, he experimented with different kinds of seeds, but in his case the results were recorded with scientific accuracy. One of the most difficult things in growing vegetables is of course to ensure a continuity of crops. In spite of his systematic approach, Lewis was not always successful in this; sometimes there would be a temporary glut. One of Maxwell Garnett's grandsons to this day thinks of Kilmun every time he eats broad beans—he happened to be on a visit to the Richardsons during the broad bean season and was served them in one form or another three times a day!

Dorothy's nephew John Garnett (now Director of the Industrial Society) used to visit the Richardsons regularly at weekends over a period of 4 years while he was working in Glasgow. He recalls their great hospitality and kindness—and also Lewis' methodical way of tackling any problem. On one occasion there were seven adults and a child sitting round the dinner table and Lewis deliberated for several

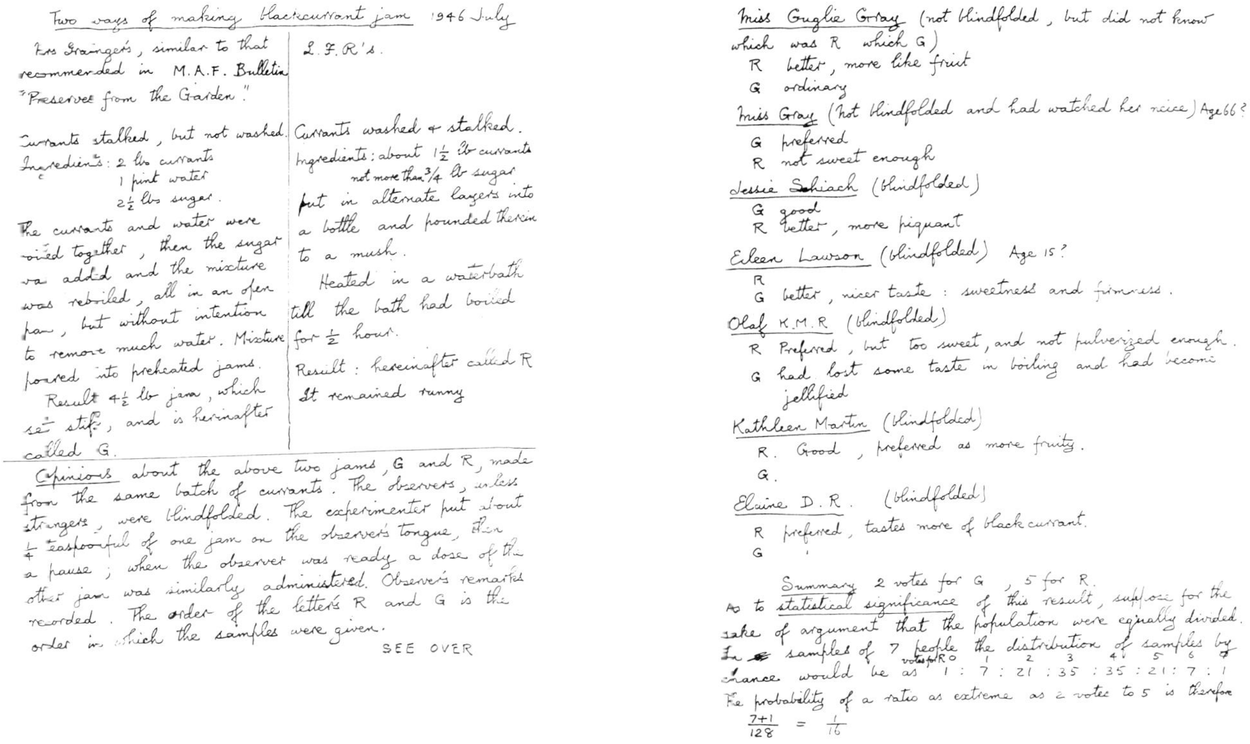

Two ways of making blackcurrant jam 1946 July

Two strangers, similar to that recommended in M.A.F. Bulletin "Preserves from the Garden."

2 L.F.R.'s

Ingredients: 2 lb currants
1 pint water
2½ lb sugar

Currants stalked, but not washed. Currants washed & stalked.

The currants and water were boiled together, then the sugar was added, and the mixture was reboiled, all in an open pan, but without intervention to remove much water. Mixture poured into preheated jam jars, ½ to ¾ jar, which is stiff, and is hereafter called G.

Ingredients: about 1½ the currants not more than ¾ lb sugar put in alternate layers into a bottle and hereafter thrown to a mush. Heated in a waterbath till the bath had boiled for ½ hour.

Result: hereafter called R. It remained runny.

Opinions about the above two jams, G and R, made from the same batch of currants. The observers were blindfolded. The experimenter put about a teaspoonful of one jam on the observer's tongue, then, when the observer was ready, a dose of the other jam was similarly administered. Observers remarked which of the letters R and G is the correct. The order in which the samples were given.

SEE OVER

Miss Guglia Gatay (not blindfolded, but did not know which was R which G)
R bitter, more like the fruit
G ordinary

Miss Grant (not blindfolded and had watched her nieces) Age 16?
G preferred
R not sweet enough

Jessie Schlach (blindfolded)
G good
R bitter, more piquant

Eileen Lawson (blindfolded) Age 15?
R
G bitter, nicer taste: sweetened and firmness.

Olaf K.M.R. (blindfolded)
R Preferred, but too sweet, and not pulverized enough
G had lost some taste in boiling and had become jellified

Kathleen Martin (blindfolded)
R Good
G preferred, as more fruity.

Elaine D.R. (blindfolded)
R Preferred, tastes more of blackcurrant.
G

Summary 2 votes for G, 5 for R.

No statistical significance of this result supposes for the sake of argument that the population were equally divided. In a sample of 7 people the distribution of samples of G would be as 1 : 7 : 21 : 35 : 35 : 21 : 7 : 1
                                                  0   1   2   3   4   5   6 7

The probability of a ratio as extreme as 2 votes to 5 is therefore
$$\frac{7+1}{128} = \frac{8}{16}$$

Figure 9.2 Pages from Richardson's House and Garden Diary, 1946.

minutes on how best to divide the apple pie into $7\frac{1}{2}$ portions. He also remembers Lewis' dissatisfaction with the commercially available jars for bottling fruit; he had found that he could get much better results by grinding the glass lids himself and sealing them with melted cough lozenges. On Sundays they used to go to the local church—the nearest Friends' Meeting House in Glasgow was too inaccessible. For John this was always a nerve-racking experience as Lewis would mutter to himself whenever the minister said something with which he disagreed. One Sunday, Lewis walked out in the middle of the sermon—as he had done many years earlier on the Isle of Wight. It may have been on this or some other occasion that Lewis was particularly upset by the minister's remarks about God being on our side during the war; he commented later: 'That man should never be allowed to preach in a Christian church'. For Dorothy, with her Anglican upbringing, going to church was less of a strain; she once told a friend that in fact she was enjoying the local church because she did not feel the same sense of responsibility as in a Friends Meeting.

Lewis was always polite to the young visitors but, with very few exceptions, he tended to be aloof and never spent much time with them. The exceptions were the youngsters who showed aptitude with their hands or an interest in science. To them he would take delight in teaching the simple elements of woodwork or in talking about his research.

For example, one of the youngsters to visit the Richardson home during the war was David Eversley, who later became Chief Strategic Planner at the Greater London Council. At that time, at the age of 20, he was serving in the army in the Strathclyde area and came to know the Richardsons through Friends at Glasgow Meeting, which he occasionally attended. As a soldier he appreciated at first their hospitality—hot baths, solid meals and good talk. Later, when Richardson had discovered that he had picked up in his army training 'some elementary notions of wiring diagrams, and soldering, and could handle tools under instruction', he was 'admitted to the laboratory, and performed simple tasks whilst Lewis talked'[3]. Although David Eversley could not understand fully what Richardson was doing—apparently conducting some experiments with a cathode-ray tube—he was impressed by his boundless enthusiasm. 'What was left with me', he wrote, 'was not a belief that Lewis' methods were correct or could one day bear fruit, but a bit of his sense of urgency; to pursue research, to accumulate data, to order them to see if there was a pattern'. Eversley's scepticism about Richardson's quantified peace studies became stronger as the years went by—we will refer later to his Runnymede Lecture of 1972 on the more general question of the right use of statistical methods in the social sciences. But his gratitude to Richardson remains:

> What Lewis started in me, and is the cause of great thankfulness, is a curiosity about why complex events occur in the way they do; the search for a better data storage and retrieval system, and above all, the belief that science can be put to the service of mankind in a positive, constructive way, and not only for destruction or for the production of chore-saving gadgets. *Rerum cognoscere causas*, that was the motto of the place where I did my degree. One might have varied it: search for the causes of events, even if you do not ever get the answer.

Another young visitor, who still remembers with gratitude her contacts with Lewis Richardson, was a friend of Elaine, Kathleen Martin (now Kathleen Lewis). She first met the Richardsons in 1940 when her family moved to Castlehead in Paisley. Lewis almost immediately asked her mother where Kathleen would be educated and he subsequently helped to find a place for her at Paisley Grammar School. Whenever Kathleen asked Lewis a question he took great pains to give a full answer. He would often stump off in the middle of a meal to find what he wanted in a reference book. Sometimes he would even set up a scientific experiment for her, to demonstrate a particular point. Kathleen used to help Lewis in the garden at Kilmun and would enjoy chatting with him because he always talked to her like an adult. Sometimes he asked her to help him in the laboratory, especially when a steady hand was needed with an experiment. Later she took a teacher's diploma and then went on to a degree in biology, feeling that she had benefited from Lewis' help in expressing herself clearly. From her I learnt another story about Richardson's days at Paisley Technical College. During a demonstration in the laboratory one day, Lewis knocked over some apparatus, called out 'Botheration!' and then stalked out of the room in annoyance. He returned in ten minutes and apologised to the class for having lost his temper. Some years later he told Kathleen that it was not necessary to swear under such circumstances: it would be equally effective to say 'Potash!' or 'Treacle!'.

Kathleen's mother, Mrs C E Martin, also admired Richardson. Knowing that he had a sweet tooth, she sometimes gave him some chocolates (during the days of sweet rationing); in so doing she took great care not to let Dorothy see, for otherwise the chocolates would have been put away for the next parcel of food for refugees. In a footnote to one of his papers[4], Richardson acknowledges some information he received from Mrs Martin about the frequency of fights involving different numbers of adversaries in a school playground.

One day, Richardson invited Mrs Martin and her daughter into his study and asked them to measure the length of the boundary between two countries, using a pair of dividers. They found that the answer depended on the distance between the points of the dividers; the closer they were together, the longer the measured length of the boundary. Richardson's interest in this question arose from his consideration of the role of contiguity—the extent to which countries have geographical contact with other countries—in providing opportunities for war. We shall see in due course the outcome of his further researches on this subject.

Richardson had always seized any opportunity to report the results of his various researches at meetings of the appropriate scientific body and he continued to do so after the move to Kilmun. He enjoyed both the discussion of his papers during the meetings and the personal contacts with his fellow scientists. At a gathering of the British Psychological Society he met Dr Ranyard West, author of many articles and books on psychiatry and psychology. Prior to the meeting, West had been informed that a paper he was to present on his psychological theory of law and international law would be critically handled by Richardson, of whom he had never heard. 'L.F.R. did speak to it', West told me in a letter[5],

> welcoming the new departure as important. I was 'chuffed' and we were in touch from then on, each trying to clarify the mind of the other and to help him publish work of the validity of which we were ourselves a little suspicious!... I was trying to utilise the Freudian 'discovery' of unconscious motivation and the deep irrationality of man which it revealed (so dangerous to his politics!) to establish on a firmer footing the prerequisites of 'law and order'.

This was of course a very different approach to Richardson's, which was of a more statistical and objective nature. The complementary nature of these two lines of attack led to animated discussions on the rare occasions when Richardson and West could meet and to a further fascinating exchange of views by correspondence. The flavour of their discussions can be sensed from West's recollections[6]:

> We pulled each other's legs a good deal. We each claimed to be terribly 'scientific', he with the weight of his mathematics and I backed by my 'scientific medicine'. Dabbling as I did in unconscious motivation, I was not happy about the unity of some of his categories. He lambasted me as a 'philosopher dealing in broad generalizations'. We both wanted to see a decent set-up for the World Order in 1945 and failed to see it.... He seemed a self-sufficient rather than a lonely man.

Both men experienced difficulty in getting their work published. Richardson gave West the name of a Quaker, Bertram Pickard (whose wife Irene was a cousin of Dorothy), as somebody who might help to find an American publisher for West's *Conscience and Society*, which had already been published in England. Richardson wrote enviously: 'Well you are at least to be congratulated on having your book published in Britain—I wish mine was'[7].

Richardson wrote again in April 1944[8]:

> I was much interested in your discourse in Glasgow on Conscience and Society. Interest in me is apt unfortunately to express itself as criticism. It had often previously occurred to me that we had some Foreign Secretaries who believed in the possibilities of a stable balance of power, and so do not understand foreign politics. Notably Lord Halifax. But then it would occur to me how utterly lost and confounded I might be, if I were miraculously transported into a position of responsibility in foreign affairs.

In 1946, West proposed to Richardson that they should carry out a joint research to establish statistically West's 'Law of Inevitable Prejudice'[9]. Richardson replied that he would be fully occupied for a few months rearranging his list of wars in a better form for a hypothetical printer. He continued[10]:

> After that, I might perhaps be willing to co-operate. The first step would doubtless be to omit the words 'Law' and 'Inevitable' and then to arrange some idea about 'Prejudice' in the form of a question to which abundant answers could be obtained.... At the moment I do not see how to formulate a suitable question, although prejudices are certainly often conspicuous.

There is no further reference to this proposal in the surviving correspondence.

In April 1946, both Richardson and West presented papers at a meeting of the British Psychological Society at Durham, under the presidency of Godfrey Thomson, Professor of Education at Edinburgh University. Looking back on this event many years later, West writes[6]:

> I had brought a statistician along to the meeting and we contributed some joint work—it was on 'selective remembering' in History teaching. I was received with such uproarious ribaldry by a part of the audience that I ended in disgust and withdrew the paper, to the detriment of my collaborator: a bad business!

This occasion also remained fresh in Richardson's memory, for he wrote to West in 1952: 'I am sorry about the fiasco at Durham. I was too excited about my own paper to catch on to yours or to anybody else's'![11]

Richardson's paper contained a summary of the results of his statistical analysis of past wars. It created a favourable impression on at least one listener, Dr C P Blacker, General Secretary of the Eugenics Society, of which Richardson had been a member for nearly 30 years— back to the time when it was called the Eugenics Education Society. Blacker agreed to help Richardson in finding a publisher for his book and Richardson accordingly sent him the following description of its contents for passing on to potential publishers[12]:

> The present title of the book is THE INSTABILITY OF PEACE. It might almost be called WARS, AND HOW TO AVOID THEM. Or perhaps a better title would be HISTORY FOR SCIENTISTS, because I have taken facts from historians, but have digested them after the manner of a scientist. That is to say I have searched for *quantitative objective relations*, and have described many that are not well-known. Although the work is about history and politics, it is written from a world-wide, not from a British point of view. Its style resembles that of a university textbook on physics in so far as unemotional prose is intermingled with numerical tables, with diagrams, and with mathematical argument. But, unlike a physics text, my work contains also some passages of dialogue, which anyone can read. Among historians, journalists, statesmen and literary people, only the small minority who can read mathematics will be able to understand this book as a whole. On the other hand it is likely also to find readers among the large and growing body of engineers, scientists and technicians, who habitually read mathematics and who are concerned about international affairs. I am not in favour of relegating all the hard reading to an appendix; for, if that were done, the appendix would be longer than the rest of the book, and would moreover be disjointed.

Richardson then went on to suggest a way in which the book, which amounted to about 500 printed pages, could be separated into three volumes.

Before approaching any publishers, Blacker sought the opinions of D W Harding, Secretary of the British Psychological Society, and of Godfrey Thomson. Harding responded that although it would be difficult for him to express an official opinion, he personally would be delighted if the work could be published, if only to ensure a more thorough discussion and assessment by competent people. 'It always seems unfortunate', he continued[13],

> that the choice of a statistical medium for one's original work should put up such irrelevant barriers in the way of publication. I have no doubt at all that far inferior work is being published with the utmost ease, simply because it needs less costly production.

Godfrey Thomson replied[14]:

> I met Lewis Fry Richardson for the first time at Durham, but I have known his family and his relations for half a century, especially his brother Lawrence and his brother-in-law, Maxwell Garnett. The Richardsons are a Tyneside Quaker family, well-to-do, indeed wealthy.... I am therefore surprised that Lewis should not be able to publish his book himself, or to get funds from the family to do so, but, on the other hand, it seems clear from his career that he must not have been personally very well off in recent years.... He took a post as lecturer in Westminster Training

College which must have been the only training college in time or space to have an F.R.S. on its staff....

I was very pleased to make his acquaintance at Durham and thought him a charming man, as I knew his brother was, and not at all the crank I had been told he was. He won all our hearts by his own speech, and old Mrs. X who was sitting next to me during his speech kept muttering under her breath 'Isn't he a darling?'—a judgment with which I entirely agreed!

I am, in some small measure perhaps, responsible for part of his failure to find a publisher, for one publisher consulted me on the matter and I had, in all honesty, to say that I thought his book would appeal to very few and would not be a venture which they could undertake without financial loss. That, I think, is true, but it is an extraordinarily interesting book and I hope it can be published. I am sure that it is mathematically sound, although the uses to which mathematics is put in it are most unusual—it is this which has led to his being called a crank.

Thinking that Richardson's main concern was to get some financial support—he had a sum of £500 in mind—Blacker then wrote to an acquaintance in the Rockefeller Foundation to seek a grant. While he was awaiting a reply, a further letter came from Richardson: 'The Dutch meteorologists have a proverb "Either the weather is good or it is interesting". Did anything, good or interesting, come of your kind enquiries concerning the publication of my book?'[15] In due course Blacker told Richardson that his initial approaches had been unsuccessful but that he had not given up. In thanking Blacker for having been so persistent in seeking a patron for his book, Richardson wrote[16]:

I am a little apprehensive of any patron, lest he should have a policy, and expect me to conform to it. For I am convinced that my duty, as a research-worker, is to tell the truth as I see it: and that is not easy, when the subject is so riddled with controversy. What I yearn for is a publisher, not a patron. At the moment I am rearranging the list in a clearer form giving a little more information; and that has to be found in numerous books on history.

The final letter in this exchange of correspondence was in October 1946 after Richardson had been told that all Blacker's efforts had been in vain. He thanked Blacker again and added[17]:

The obstacles to the publication of quantitative research about keeping the peace may I think be classified as follows: (a) Most people feel that the essential thing to do about wars is to win them. There is always lots of money for research with that object.... (b) Most people think that they already know what threatens peace, so that to them research about it seems trivial. They fall into two classes:— (i) Reformers who have a rigid policy, and are interested in research if, and only if, it supports their propaganda. (ii) Feelers who say that the threat to peace 'is all the fault of those cursed ...' The name to be filled into the blank alters from decade to decade. (c) Most people cannot understand quantitative research of any kind. They say 'Why cannot it all be expressed in plain English?' This group includes most of the editors, literary people, and historians who otherwise are knowledgeable, interested and wise about peace and war. (d) In the present 'century of the common man', when everyone can read, there is a tendency for publishers to be so occupied in producing popular books for wide circulation, that they cannot bother with research of any kind....

In the meantime I am busy with small but troublesome amendments to my summaries of about ninety wars. When that is all done—and it will take months—I think of trying one of the University Presses.

In one of his letters to Blacker, Richardson mentioned that he had taken the advice of Dr William Stephenson of the Oxford Psychological Laboratory to offer bits of his results to learned societies. The first outcome of this decision had in fact already occurred when the Royal Statistical Society published in 1945 his paper on the distribution of wars in time. Encouraged by this success, he continued to submit chapters of his book to appropriate societies and journals with the result that a substantial part of his work was published in the course of the next eight years. Some of the material, and especially his annotated list of wars, was not suitable for this form of publication; while pursuing his efforts to have the whole book published commerically, he finally decided to follow a suggestion by the anthropologist, Gregory Bateson, that it should be reproduced on microfilm. For this purpose, he divided it into two volumes, the first of which was issued in 1947 under the title *Arms and Insecurity, alias the fickleness of fear*. A revised version was published on microfilm in 1949 and the second volume *Statistics of Deadly Quarrels* appeared in 1950.

Richardson advertised his first microfilm in *Nature* in 1947 but the response was evidently very poor, for in April 1948 he was only able to report to Ranyard West that copies had been bought by the Library of Harvard University and the Douglas Aircraft Co. West himself had a reluctance to view microfilms and suggested to Richardson that he should approach a well known Scottish publisher. This he did, and reported the results to West. 'I had the honour of an interview (by the managing director). He said they expected books to sell out in the first two years. A slow seller, such as mine would probably be, they could only take if it were subsidized.' The letter continued: 'Even if the subsidy (of about £1000) were forthcoming he did not wish to undertake my book now, as he had 1,500 books in the office waiting to be printed or reprinted. We parted on good terms, but I do not anticipate any agreement'[18]. In this he was not mistaken.

On hearing that West was planning a visit to Dunoon, Richardson invited him to come to Kilmun: 'If it would suit you to stay here ... my wife and I would be very pleased, provided your dates fall before March 28, when spring-cleaning is due to begin'[19]. A few months later, in April 1947 (by when the spring-cleaning had presumably been finished.) Richardson wrote to West again to thank him for a suggestion that public opinion might be stimulated in favour of having his book published by getting the committee of a learned society to adopt an appropriate resolution. Richardson felt that it might be better to persuade a number of individuals to send him a letter of support which he could then pass on to a publisher and he enclosed a draft letter for West's consideration. He reported that[20]

> The present position is that since last June I have remade the summaries of all the wars that have occurred since 1820. That took me till February. The Second World War is now included. A consequence is that I must now revise the statistical analysis of all the wars. I am doing it daily, but it may take another $\frac{1}{2}$ year.

West had apparently asked Richardson if there was any need to have his mathematics checked before the book was published. Richardson replied that he was loath to bother any leading mathematician in view of the care which he had taken to check everything himself. He had in any case 'deliberately subdued the mathematics, kept it humdrum, avoiding the elaborations customary in papers of mathematical interest, some of which may be likened to the cadenza of a musical concerto'. Finally, the one bit which he thought was original had already been seen by a professor of mathematics. 'All this considered', he concluded, 'I take responsibility cheerfully for the correctness of such mathematics as I offer for publication. (Heaps more goes into the waste paper basket!)'[20]

The professor who had so obligingly looked at some of Richardson's mathematics was A C Aitken of Edinburgh University. He was well known as the author of several textbooks, a leading authority on matrix algebra and an incredible prodigy in mental arithmetic. This is well illustrated by the following story[21].

> His children asked him to multiply 987 654 321 by 123 456 789. 'I saw in a flash that 987 654 321 by 81 equals 80 000 000 001, and so I multiplied 123 456 789 by this, a simple matter, and divided the answer by 81. Answer 121 932 631 112 635 269. The whole thing can hardly have taken more than about half a minute'.

In 1952 Richardson wrote to Aitken about a matrix problem in connection with the rejection of one of his papers on the grounds that the mathematics were not sound. Aitken decided that Richardson was completely right and admitted that he had once acted as a referee for Richardson and had been able to ensure that his paper was published as a whole and not in truncated form[22].

Some of Richardson's most interesting correspondence from Kilmun was with Quincy Wright. It started in December 1943 shortly after Richardson had found out about Wright's researches on the causes of war. As usual, Richardson got straight to the point; his first letter began: 'Dear Sir, I have gladly paid the high price for your important book 'A Study of War'. It arrived yesterday and its perusal is to be my chief occupation in the coming weeks'. After a brief reference to Richardson's current work, the letter continued[23]:

> As a preliminary to a systematic reading of your book I have begun in a very self-centred, but perhaps excusable, manner, by looking up all the references to my 'Generalized Foreign Politics', to see whether I agree with them. Yes in the main I do. I think them clear exposition and fair comment; but with three exceptions. Will you bear with me while I explain these to you?

He then expanded on the points in question and concluded:

> However these remarks seem an ungenerous way to welcome your great book. Now that I have got them off my mind I can settle down to appreciate its vast range, wealth of documentation and experienced judgment.

One recurring theme in the subsequent letters was whether it is better to deal only with wars, based on a more or less legal definition, or to cover all kinds of conflict, using the number of fatalities as a means of classification. Richardson explained that he ignored the legal aspects 'lest legality should prove to be controversial'; he had 'sought to form an objective list, to which readers of all nations might be willing to agree'[24]. Wright had used legal criteria partly because they were generally followed by historians and consequently were applicable for studies based on historical writings. He did not see how it would be possible to use a quantitative method for a study going back over any considerable period of history; he thought it probable that the average population of belligerent groups, including the feudal principalities of Europe and Asia and the tribes of America and Africa, must have increased between 100 and 1000 times in the past 500 years. 'There has', he argued, 'been a similar increase in the average losses in wars. Thus an absolute figure of losses to constitute a "war" or a "participant" would not be very meaningful'[25]. Wright nevertheless felt that there was value in Richardson's quantitative analysis and encouraged him to continue. Richardson took up the question again some time later[26]:

> As between our professional habits, yours legal, mine physical, we might perhaps agree on the psychological point of view as a middle. A

psychologist can usually recognize a quarrel, and the parties to it. However, I don't feel that complete agreement between us, as to what definitions should be used, is necessary or even advantageous. Diversity keeps ideas on the move. It is remarkable that our diverse definitions have led to closely similar results as to the Poisson distribution of years....

To overcome his difficulty about the increasing size of potential belligerents, Wright suggested that it might be better to express casualty figures as a percentage of the population; an episode could, for example, be called a war if the total losses were more than 0.1 per cent of the total population and a country would not be regarded as a participant unless its losses exceeded 0.01 per cent of its population[27]. Richardson replied: 'What you say in favour of taking the casualties as a percentage of the population seems very reasonable, and I hope you will have it tried out'[28]. He was less enthusiastic about another of Wright's proposals, namely that the importance of a war might be quantified by measuring the amount of space devoted to it in the leading newspaper of each country of the world. 'Personally', he wrote, 'I have no intention of trying it. A conclusive reason is my lack of power; for it would be a task within the scope of a wealthy organization, with stacks of newspapers, and linguists to read them'[29].

The discussion inevitably got round to the difficulties in getting Richardson's book published. Wright was keen on this if for no other reason than that he wanted to see Richardson's list of wars, but he was not optimistic about the possibilities of having it published in the United States in view of the shortage of paper during the war. Later on, when he had seen some of Richardson's unpublished work, he was to play a leading role in making arrangements for its publication. In the meantime, Richardson continued to send parts of his book to societies and journals, although he was never very happy at this procedure. He expressed his feelings to Wright[30]:

> This scattering of bits of an unpublished book into various journals in different countries would be quite all right if the author were a young man who could look foward to reuniting the scattered portions in 10 or 20 years time. At my age, 67, it seems an unsatisfactory prospect for future readers.

The next of his peace research papers appeared in 1947 under the title *The number of nations on each side of a war*. With his scrupulous attention to the acknowledgment of any suggestions or assistance from others, he stated that I had been responsible for starting the enquiry which led him into this subject. I still have a letter from him telling how this came about[31].

> When you visited Hillside [Kilmun] you raised the question whether the frequency of major wars decreases as powers get bigger and whether the Big Five are particularly likely, judged by their past history, to keep the

peace? We searched in my card index and found preliminary answers 'No' to both your questions. However this discussion started a train of ideas in my mind which has kept me busy ever since.

As the letter was written seven months after my visit, my immediate reaction was to feel embarrassed that a simple question of mine could have led to so much work. The letter continued:

What is the relative frequency of wars of r versus s and how is it to be explained? Here r is the number of nations, or other belligerent groups, on one side and s the number on the other. By now I have finished the empirical discussion of the observations and have made, and refuted, about ten theories. Theory No. XI will I expect survive. The reasons why the earlier theories fail are themselves instructive. Early in this enquiry I felt the need for some formal exercise on a subject where the answers to problems are well known; and so I worked out the probability of encounters between molecules in a gas 2 or 3 or 4 or 5 at a time. For a week I was in despair, because I could not get my answers to agree with standard textbooks. However eventually I acquired confidence, decided that the textbooks had missed something, and finally persuaded a Royal Society referee to the same effect; so that what was begun as an exercise is now being printed for the Proceedings of the Royal Society. So a lot of things came out of your visit here, but by processes so roundabout that I can hardly thank you publicly.

Richardson called attention to the similarities between the behaviour of nations and of gas molecules in a letter to *Nature* (1946) and the complete results of his research were published in due course by the Royal Society (1946) and the Royal Statistical Society (1947). On further reflection I felt somewhat less embarrassed about having asked him an innocent question.

Richardson's main finding in his study of the number of nations on each side of a war was that, of the 91 wars of magnitude greater than 3.5 between 1820 and 1939, nearly half were of the simplest type with only one nation on each side. The number of wars decreased rapidly as the number of belligerents increased and there were only 6 which involved 7 or more nations. He obtained a similar distribution for the 200 wars not marked civil in Wright's list. The next steps were to work out an empirical formula to fit these facts and then to find a theoretical explanation of the formula. To Richardson[32],

The contrast between an empirical formula and a theory resembles the contrast in a factory between an old reliable foreman and an imaginative graduate fresh from the university. The graduate might at first make frightful blunders, but in the course of a decade he is likely to improve the technique more that the foreman ever could.

After testing 13 theories of various degrees of complexity, he came to the conclusion that the best explanation was 'chaos restricted by geography and modified by infectiousness'[33]. By 'geography' he meant the opportunity of war for each country, depending on whether it was a worldwide sea power, a coastal state or a landlocked state. By 'infectiousness' he meant the tendency to join the winning side which he illustrated by the story about the Irishman who asked 'Is this a private fight, or may anybody join?'[34]

In acknowledging the receipt of a copy of this paper, Quincy Wright suggested an additional theory, namely that 'the most plausible hypothesis to permit of the kind of analysis you have undertaken is that the desire of political groups to continue to exist is universal'. He also requested an additional copy of the monograph on *Generalized Foreign Politics*, as his original had been loaned to a Polish Ambassador who 'was so interested in it... that I have not been able to recover it'[35]. Richardson was pleased by this last remark; he replied 'I tried in July 1939 to interest the diplomats by sending a complimentary copy of that monograph to the ministry for external affairs in every State on the globe. Your colleague is the first diplomat of whom I have heard that he has studied my monograph'. Regarding Wright's proposed additional theory he wrote[36]:

> I see no obvious objection to your hypothesis. But I feel that the mental effort leading to either proof, or disproof, of it is likely to be arduous. The first step would be to express 'the desire to continue to exist' in a form so definite that quantitative consequences can be deduced from it. I do not, today, see how even to take this first step. But maybe you, or your mathematical assessors, see how to proceed. I hope that they do....

In this same letter Richardson mentioned that another part of his book, dealing with what he called 'war-moods', might soon be published. It was, he said, a third approach to the study of war, different in its concepts from his arms race model and from his statistical analysis of past wars. The paper was in fact published in two parts in 1948 in *Psychometrika*, the journal which some ten years earlier had offered to publish part of *Generalized Foreign Politics*. Richardson describes the paper as 'a study of the moods of populations before, during and after a war' which 'unites an abstract of historical fact with the psycho-analytic doctrine that the unconscious is important'. He continues[37]:

> Only broad general features of a war are considered, namely the numbers of people concerned, the emotional drives of the majority, and those of a few important minorities. Almost all of the well-known dramatic incidents are ignored. Instead the war is described as a continuous conative process: a smoothed war, like a gale without the gusts.

The first part of the study is based on the attitudes of people in Britain and Germany in relation to the First World War. In the absence of any statistical surveys of public opinion in those days, Richardson had to piece together information from newspapers and other published accounts, and from his own personal impressions. From all this, he produced a graph showing very roughly how the fraction of the populations of the two countries openly in favour of the war varied between 1914 and 1919; a rapid increase in this fraction at the outbreak of war was followed by a slow decline and then an almost precipitous fall at about the time of the armistice. The object was to explain the shape of this graph by a theory of dual moods, the overt and the subconscious, and especially of how people's moods can change. Such a change was likened to 'a change of government in the nation, insofar as a different party to the individual's internal controversy takes over control of his voluntary muscles'[38]. Richardson dealt specifically with three moods: friendly, hostile and war-weary. An individual who was outwardly friendly but inwardly hostile was indicated thus:

$$\begin{pmatrix} \text{friendly} \\ \text{hostile} \end{pmatrix}$$

and so on. Using this notation, he summarised the whole course from a friendly peace through war to a resentful peace thus:

$$\overset{\text{Arms race}}{\begin{pmatrix} \text{friendly} \\ \text{friendly} \end{pmatrix}} - \overset{\text{Outbreak}}{\begin{pmatrix} \text{friendly} \\ \text{hostile} \end{pmatrix}} - \begin{pmatrix} \text{hostile} \\ \text{friendly} \end{pmatrix} - \overset{\text{Attrition}}{\begin{pmatrix} \text{hostile} \\ \text{war-weary} \end{pmatrix}} - \overset{\text{Armistice}}{\begin{pmatrix} \text{war-weary} \\ \text{hostile} \end{pmatrix}}$$

He realised that such a dual mood description could not represent the complicated possibilities of human nature but felt that it 'permits a description less ambiguous than that given by the standard phrase in the British King's speech at the opening of Parliament in peacetime: "My relations with foreign powers continue to be friendly "'[39].

Richardson next assigned a mathematical symbol to the fraction of the population in each dual mood and developed a set of equations to describe the rates of change from one such mood to another. He had wondered whether a well known mathematical theory of the rate of spreading of an epidemic, such as bubonic plague, would be applicable to war fever, but concluded that this could not account for the long persistent phase of a war, followed by its comparatively sudden end. He estimated approximate values of the various constants in the equations from the evidence he had collected on the First World War and then made a variety of deductions which he found to be 'in interesting agreement with the observations'[40]. He stressed that in view of the inadequacies in the observations, only a moderate degree of determinism should be expected: 'the theory can only express what is habitual, traditional, or instinctive; whereas we know, in everyday life, the operation of free choice'.

Another section[41] of this paper is devoted to the effect of gregariousness on war fever; it incorporates many of Richardson's earlier unpublished ideas on gregariousness and fashion. His equations show that the effect cannot be a prime cause of change but only a modifier of changes caused by other motives. In a brief final section he refers to an analogy, pointed out by his friend, Ernest B Ludlam, between the role of an opinionated leader in society and that of an activated molecule in a chemical reaction. Although he felt that this suggestion might prove to be valuable, he himself did not feel inclined to develop it further. He explained his reasons by referring to the interval of 24 years between the statement by Arrhenius of the kinetics of bimolecular reactions and the introduction by Bodenstein of the concept of chains of reactions. 'So a psychologist may excuse himself for taking time for deliberation. The author's theory is not yet at the stage of Arrhenius, for it takes no account of the the different energy of different persons in the same dual mood'[42].

Closely related to the above study of war moods was a brief paper which Richardson read at the Twelfth International Congress of Psychology in 1948. He opened his talk by reciting a number of historical examples of how the object of hatred can change with dramatic quickness and of how a common hatred can have a powerful unifying effect. On the basis of this evidence he asked: 'Is there an irreducible minimum of hatred; and, if so, against what objects had it better be directed?'[43] He then discussed the views of various psychologists and concluded with a reference to Dr C G Jung, who 'recommends, in effect, that we should hate the evil in ourselves'[44]. Each of the speakers at the Congress was expected to announce his affiliation—Harvard, Cambridge etc. Richardson simply indicated that he was from Kilmun. An American who wished to visit him afterwards was overheard asking a colleague the best way to get to the University of Kilmun by public transport[45].

I have already mentioned the unexpected outcome of one of my own visits to Kilmun. I had occasion to go there again in June 1948 and this time we chatted about some experimental work which I was about to do on board an ocean weather ship—these ships are based at Greenock, only a few miles across the water from Kilmun. Richardson suggested that I should repeat an experiment which he had made a few months earlier with Henry Stommel, an oceanographer from Richardson's ideal resort for scientists, Woods Hole. Stommel recalls that in reply to his letter asking Richardson whether he could pay him a visit, Richardson had replied 'Come, but bring some golfballs'. Before he had succeeded in finding any (such luxuries were still in short supply after the war), he received a telegram saying 'Forget the golfballs, they all sink'. When Stommel arrived in Kilmun, Richardson explained that he wanted to take the opportunity of his visit to carry out an experiment on the rate of separation of pairs of objects floating on the sea—of importance in relation to the theory of eddy diffusion at sea. In view of the unsuitability of golfballs for this purpose, he had tried various other materials and had found that pieces of parsnip were perfectly satisfactory—they are easily visible and, as they are almost completely immersed, their movement is little affected by the wind. Richardson had made a very simple measuring rule with which to estimate the distance between the pieces of parsnip—Stommel remembers him walking along to Blairmore pier with this rule tied to his body with string, quite oblivious to the stares of curiosity from passers-by. The outcome was that in only a few days they completed a most successful experiment—a classic of its kind—the results of which were published later in 1948 as a joint paper in the *Journal of Meteorology*. The report began laconically: 'We have observed the relative motion of two floating pieces of parsnip...'. The main conclusion was that the rate of separation of the parsnip floats, as determined from measurements of 45 pairs, was consistent with the $\frac{4}{3}$ law which Richardson had discovered some 20 years earlier for atmospheric diffusion. In those days, I was more interested in developing meteorological instruments than in measuring diffusion and I did not follow up Richardson's suggestion.

Figure 9.3 The parsnip experiment (sketch by Henry Stommel).

The parsnip experiment has given rise to several erroneous accounts. In a recent book, for example, we read that Richardson's solution to the problem of finding suitable floats was 'to buy a large sack of parsnips, which were thrown from one bridge on the Cape Cod Canal while he made his observations from another bridge downstream'[46]. In fact, the parsnips were dug from the garden in Kilmun and were cut into pieces before being dropped into the sea nearby! According to Olaf, his father grew the parsnips in such a way as to ensure that they would have just the right specific gravity for the experiment, but of course he chose parsnips because they naturally floated at about the right depth—even so, he had to adjust the depth slightly by pushing small nails into the parsnips.

Figure 9.4 Blairmore pier on Loch Long, where Richardson and Stommel carried out their parsnip experiment.

On his return to Woods Hole, Stommel carried out some further experiments to check whether the $\frac{4}{3}$ law was applicable to diffusion on the ocean surface for larger scales of motion—the maximum initial separation of the parsnip floats had only been about three metres. His measurements from aerial photographs of pieces of paper floating on the sea near Cape Cod, and later near Bermuda, provided justification for the supposition that the $\frac{4}{3}$ law is satisfied for distances up to at least 100 metres[47]. Stommel's many brilliant contributions to oceanography were recognised in 1983 when he was the joint recipient of the Crafoord Prize. This award was instigated in 1982 by a Swedish industrialist to promote research in a number of fields, including geosciences. Meteorologists and oceanographers are at last eligible for an international award of comparable status to the Nobel Prize. With his love of parody, Richardson might well have said: 'Bjerknes! Thou shouldst be living at this hour...'.

In the course of the next year or two, the theoretical derivation of the $\frac{4}{3}$ law became better known in Britain, thanks to some outstanding work by one of Taylor's younger collaborators, G K Batchelor[48]. The combination of this and Stommel's experiments proved to be too great a temptation for Richardson. He described what happened in a letter to me: 'In spite of my persistent resolution to ignore meteorology in order to study conflict, I have been led into writing a paper on 'Transforms for the eddy-diffusion of clusters' which is now with the printers for the Proceedings of the Royal Society'[49] (it was published in 1952). In this final contribution to the literature of his beloved meteorology—apart from a brief note in the *Quarterly Journal of the Royal Meteorological Society* (1952) replying to some criticisms of his 1920 paper on atmospheric turbulence—Richardson called attention to some limitations in our understanding of diffusion processes and suggested a design for a further experiment which might throw additional light on the subject.

Shortly after my return home from the voyage on the ocean weather ship, I wrote to Richardson to ask if he would care to write a biographical note for the journal *Weather*, of which I was one of the editors. He was not enthusiastic about this but offered instead to prepare an annotated list of his meteorological publications; this might help to remedy the way in which some of his work was being neglected (a paper announcing the theoretical derivation of the $\frac{4}{3}$ law had failed to mention his earlier empirical discovery). He added: 'Of course it is said that each generation must discover the world afresh. How would the hunting instinct be satisfied if all the wild animals were already stuffed and put in glass cases?' He was willing to send a photograph although it seemed irrelevant: 'the inside of my face is more interesting than the outside'![50] In the end I wrote a biographical note to accompany his annotated list. He found it 'embarrassingly personal and flattering'[51] but had showed it to Dorothy, who unlike himself had never been shy, and she had liked it. Both notes were published in *Weather* in January 1949—together with a photograph.

Another example of each generation discovering the world afresh is provided in a letter from Richardson to Sutcliffe regarding a theory which Sutcliffe had just published on the resistance of the Earth's surface to the movement of air. 'First an oddity of publication' he wrote. 'I put the theory, derived from Ekman, in the manuscript of my 1920 Phil. Trans. paper, but the referee (G.I.T.) said it was superfluous; and so ultimately you have had to publish it, after working it out in your own way'[52]. Richardson wrote to F Pasquill in 1949 on a related question, the measurement of the force exerted by the wind on vegetation, on which Pasquill had just published a paper. He sent Pasquill a sketch of an apparatus he had designed for this purpose (see figure 9.5) and said that he was welcome to make whatever use he liked of the idea. By way of explanation he added[53]:

> Nowadays I try to work at the causes of wars; and meteorology comes in as a rather troublesome distraction. Sometimes a meteorological idea, like that of this stress-meter, haunts me. One way to get rid of it is to tell someone else, as I am now telling you. The design is marked 'for criticism'; not that I want to be bothered with criticism, but as a warning to the recipient that several features need further study....

Pasquill told me that in fact he had already in hand a simple development of Sheppard's floating stress-meter (which had set Richardson's thoughts in motion) and that he published an account of this shortly after receiving Richardson's letter[54].

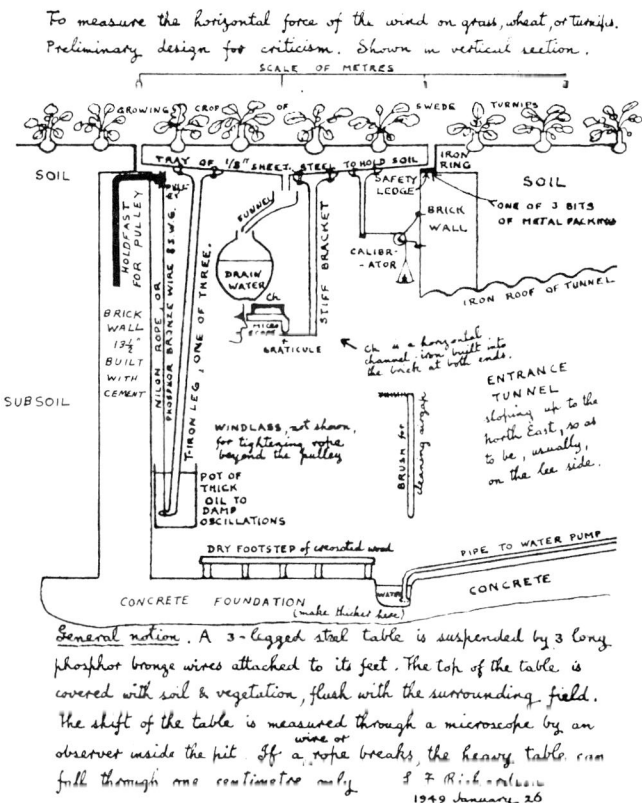

Figure 9.5 Richardson's design of apparatus for measuring wind stress (sent to F Pasquill in 1949).

My next contact with Richardson was in December 1949 when I went to hear him lecturing at a meeting of the Eugenics Society. The initiative for this lecture had been taken by the General Secretary, C P Blacker—perhaps as a gesture of support after having failed to obtain a grant towards the cost of publishing Richardson's book. He wrote to Richardson in June 1949 to suggest that if his work was related to the aims of the Society he might care to present some of his findings at one of its meetings[55]. Richardson was delighted; he was always ready to take any opportunity of talking to different groups of people—in the hope that some of his seeds would fall on fertile ground—and of course he had long been in sympathy with the goals of the Eugenics Society. The contents of his paper were much broader than its title *War and eugenics* might suggest, and the published version (1950) provides a very readable summary of the main results of his peace research. During the actual lecture, he naturally concentrated on a topic of particular interest to the audience, namely the evidence he had found to support the view that intermarriage has a pacifying influence. Amongst the objections he had run into was the response of a bachelor: 'Is it worth while to abolish wars at the expense of ruining all marriages?'[56] A more serious objection was that international marriage on any large and rapid scale was impracticable and could not therefore satisfy those peacemakers who wanted quick results. He nevertheless concluded that it should be encouraged and to this end he suggested the slogan: 'Managers of all countries, unite—by matrimony, in order to hold the world together!'[57] His reasons for choosing managers were that they travel more than most people and represent the more enterprising part of the population; their intermarriage might therefore be expected to minimise any undesirable genetic effects.

After the lecture, Blacker sent the usual letter of appreciation to Richardson in which he indicated his support for the views expressed by the chairman in his vote of thanks. Richardson was somewhat upset and wrote back[58]:

> When I was preparing the lecture I said to myself, 'Now, Lewis, this is a very serious and important subject. You must not descend to levity or jokes.' So when the Chairman thanked me for being informative, I said inwardly 'Thank you, Chairman, I am glad that you think so.' But when he went on to say that I had been amusing, I said to myself 'Alas, Lewis, you have done it again, in spite of good intentions.' And when the Chairman went on to sum up the whole matter in the phrase 'The answer's a lemon', I was deeply discouraged and scandalized. A Chairman of course, like the King, can do no wrong. But when he said 'lemon' I ceased to be his loyal subject.

Blacker replied immediately to say that Richardson had been mistaken in forming the impression that the chairman was trying to express anything but praise and appreciation of his lecture[59].

> It might perhaps have been better if he had said that your discussion of the statistical aspects of war was original rather than amusing; but original ideas are often also amusing, at any rate in the literal sense of that word, that they set one musing or thinking.

Figure 9.6 Richardson addressing a British Association meeting in 1949.

This was certainly not the only occasion on which Richardson felt that he had been unintentionally amusing—and although he complained of having a poor memory, this did not apply to such moments of embarrassment. Six years after his lecture to the British Psychological Society in Durham, for instance, he mentioned in a letter to Ranyard West that 'the audience was difficult, expressing its scepticism by laughing at the wrong time'[11]. Most frequently, however, I believe that his humorous manner of speaking—not to mention his jokes—was fully intentional. He is still remembered by some of our leading psychologists as 'something of a wit' with 'a subtle sense of humour' who often spoke 'with his tongue in his cheek'. At one meeting of psychologists it is reported that he wore his name tag on his back rather than on his lapel and recommended that others should do likewise. When asked why, he replied: 'So that I do not have to embarrass ladies by peering at their bosoms when trying to ascertain their names!'[60]

One person who did not always appreciate Richardson's sense of humour was his brother-in-law, Maxwell Garnett. Their conversations were always rather serious but Richardson sometimes could not resist the temptation to pull Garnett's leg. In a letter[61] to his son Stephen in November 1948, Richardson mentioned a recent visit when he had been talking with 'Uncle Maxwell' about the dangerous situation in Germany, where the Western allies were endeavouring to keep the inhabitants of West Berlin supplied by air in the face of the Russian blockade. From a study of the history of other cities with divided regimes, Richardson had concluded that such a system tended to be unstable and that the divided government of Berlin was an unwise experiment. But, in any case, he felt that the quarrel over Berlin was not about good government but about prestige. Garnett's instant reply was 'not about good government but about Good Faith'. After recounting this conversation, Richardson's letter continued:

> ...he and I had another discussion in which he made a memorable remark. I was comparing communism to the other religions, and pointing out the analogy between the Anglican bishops' condemnation of Darwinism, about 1870–1880, because it was contrary to the Christian faith, and the recent Russian condemnation of the biological non-inheritance of acquired characters because it is contrary to the Communist faith. I ended by saying, in order to provoke him, 'Very dangerous thing: faith.' Maxwell: 'That depends on what you mean by faith. For me, faith is action on an hypothesis with a view to its verification.' L.F.R. 'By faith, I meant believing in disregard of the evidence. Your definition is certainly encouraging. I have heard you mention it before. Did you invent the phrase?' Maxwell: 'I believe so. I don't mind how you define the word, as long as we understand one another.'

Richardson concluded the letter with the comment that faith in the expression 'good faith' has yet another meaning.

Richardson's own faith in God was by no means unshakable. As we have seen earlier, he frequently experienced an intense feeling of doubt and, at least in later life, this doubt extended to the very existence of God—the 'accusative of the verb to worship'. In 1950 he wrote to a friend, Gordon Rogers[62]:

> I join in your prayer (for peace), as far as my godliness permits. There ought to be a religion. But that by law established is to me incredible. Buddhism or Confucianism seem more sensible. I would like to translate customary religious phrases into something that makes sense. For example 'Lift up your hearts' might go as 'Reorganize your hypothalami'.

In his correspondence with Richardson, Rogers had pointed out several mistakes in the microfilms and had offered some constructive criticisms—being a physicist, he had no difficulty in understanding

Richardson's theories. Richardson was most appreciative. He wrote[62]:

> I have arrived at an age when youth and strength appear to have an aura of joy. For example, I am particularly pleased when anybody much younger than myself shows an interest in my work with the hope of improving on it.

In another letter to Rogers, Richardson referred to his arguments with Maxwell Garnett about foreign affairs: '... he is, or was, inclined to regard my theories as 'amusing', but recently has shown more interest in them'[62]. If only this interest had been aroused while Garnett was more active, he might well have attempted to communicate Richardson's theories to a wider public. For this he would have been better qualified than any of Richardson's other relatives or close friends, thanks to his mathematical background, his interest in psychology and international relations, and his considerable experience and skill as a writer—his books on world order are a model of clarity. I am not sure, however, that Richardson would have encouraged such an undertaking, for he firmly believed that scientific discoveries should not be popularised prematurely. 'Any research worker' he wrote 'usually feels that it would be futile and almost immoral to attempt to popularize his discovery until it has been submitted to the few who are capable of judging it critically.'[63] In similar vein he told Ranyard West: 'I am a research-worker, not a popular preacher'.

The problem for Richardson, as for anybody pioneering a completely new field, was to find his peers. He complained bitterly that the subject of quantitative history[64]

> is not at present cultivated under the auspices of any one learned society; indeed it falls between two stools, because the historians, and other students of human nature, are in general strongly averse from mathematics, whereas the mathematicians are apt to regard politics and history as incalculable.

The only learned groups accustomed to consider both mathematics and human nature were, in his experience, the psychologists and the statisticians. He continued: 'Accordingly the best that the author has been able to do in the way of obtaining the attention, criticism, or approval of specialists has been to offer key portions to diverse learned bodies' and of course to 'the comprehensive scientific periodical 'Nature''.

In a letter to Nature in 1947 Richardson called attention to what he considered to be an awkward gap between the Royal Society and the British Academy. The publications of the former are full of quantitative analysis but scarcely any direct mention of human social relations whereas those of the Academy, though 'full of literary grace and social interest', scarcely contain any quantitative analysis. To which of these august institutions should research workers in the social sciences turn? He recognised that the gap was bridged by a number of specialised societies but felt that they lacked 'the breadth and the general authority' of the senior bodies.

One of the learned bodies which Richardson approached was the Royal Historical Society. In 1945 he submitted a paper to them entitled *An Impartial Selection of Fatal Quarrels*[65]. It outlined the object of his war studies and contained his list of fatal quarrels, with an appeal for critical comments from historians. It was within their province, he submitted, to consider such questions as how wars should be defined, what wars have in fact occurred and when each war began and ended. He explained his reasons for using the wider concept of fatal quarrels and his method of classifying them by magnitude. The paper was not accepted for publication.

He expressed his concern about the need for historians to make use of statistics in an article in the 1952 issue of the *British Journal of Sociology* entitled *Is it possible to prove any general statements about historical fact*? The gist of his argument was that historians, such as Toynbee, frequently make generalisations which they illustrate by describing individual historical events. They admit that it will usually be possible to find exceptions but do not attempt to answer the question whether these are sufficiently numerous and important to invalidate the general statement. Richardson thought that to do this it would first be necessary to classify historical events so that they can suitably be counted. To illustrate this point, he told the story of the army contractor who offered to supply pies composed of equal proportions of rabbit and horse. On examination they were admitted to be 'fifty-fifty; one horse, one rabbit!' After describing how he and others had attempted to specify wars in a suitable way for statistical analysis, he showed how the resulting lists could be used for answering questions such as whether alliance in war is persistent. The answer was that the general tendency for alliance and enmity to persist over the period 1917 to 1942 cannot be explained away by chance. While he had greatly enjoyed reading Toynbee's book *A Study of History*, he had not found in it any attempt at providing such statistical proofs of general statements. For example, Toynbee asserts that civilisations have arisen in hard rather than in easy environments; could this statement not be checked statistically? The article concluded with some suggestions for elementary statistical methods which might be included in university courses for history students.

In spite of his aversity to the premature popularisation of scientific discoveries, Richardson accepted a few speaking engagements which gave him an opportunity of expounding his ideas to a lay audience. For example, he addressed meetings of business clubs in Dunoon and

Glasgow and spoke more than once at Quaker conferences, but few of the listeners understood what he was trying to say. He was somewhat more successful in responding to an invitation to write two chapters for a book entitled *Psychological Factors of Peace and War*, which was published in 1950 on behalf of the United Nations Association (successor to the League of Nations Union in which Maxwell Garnett had played such a prominent role). The editor was T H Pear, Professor of Psychology at the University of Manchester, who had shared the platform with Richardson at the Durham meeting of the British Psychological Society. In his introductory remarks, Pear wrote: 'Dr. Richardson reminds us that general statements can be examined, verified or refuted "by collecting the facts from the whole world over a century or more" '[66].

In the first chapter, *Threats and security*, Richardson summarises his work on arms races. He begins by pointing out that threats can have a variety of effects, ranging from submissiveness at one extreme, negotiation or avoidance in the middle, to retaliation at the other extreme. After giving an extensive list of historical examples of these different kinds of response, he postulates that an arms race is an example of a *self-aggravating system*, in the language of operational research, or of *symmetrical schismogenesis*, in the language used by Gregory Bateson to describe the custom of the Iatmul tribe in New Guinea, whereby two men boast alternately, each provoking the other to make bolder claims, until they reach extravagant extremes. This leads naturally to an elementary treatment of the latest version of his arms race model, followed by a brief account of his studies of war moods.

The second chapter, *Statistics of deadly quarrels*, opens with a justification for objective studies of the causes of war in which national prejudices are, at least partly, eliminated. After discussing the different criteria which he and Quincy Wright used in drawing up their respective lists of wars, Richardson gives a semi-popular account of his findings on the distribution of wars in time and on the relation between the frequency of deadly quarrels and their magnitude. He concludes with his answer to the question 'Which nations were most involved?'

Although Pear's book was not exactly a best seller, it certainly brought Richardson's war studies to the attention of a wider audience than his papers in scientific journals. One result of this was that the two chapters were reproduced almost intact a few years later in the well known four-volume book *The World of Mathematics*. The editor, James R Newman, referred to Richardson as 'one of no more than a handful of serious scholars who have in recent years attempted a quantitative treatment of the causes of war, the mechanics of foreign politics, the effects of armament races and kindred matters'. For him, 'Richardson wrote with humour and becoming diffidence; he advanced no sweeping claims for his theories but neither did he undervalue them in false modesty'[67].

At the time of writing the above two chapters, Richardson was still trying to improve his arms race model and was developing his ideas on contiguity. In view of its central place in his peace studies, let us briefly recapitulate how his arms race model had changed over the years. The earliest version, as presented in 1919 in his *Mathematical Psychology of War*, consisted of two equations, which in their simplest form may be expressed in plain language as follows: the rate of increase of warlike activity of a nation is proportional to the actual warlike activity of its opponent, less an allowance for fatigue and expense. In his letter to *Nature* in 1935, he revised the equations slightly by adding a term to account for dissatisfaction with treaties. Then in 1939 in *Generalized Foreign Politics* he introduced a further complication by considering what he called 'submissiveness'. It was this last factor which was to occupy his thoughts almost up to the time of his death.

He gave the following explanation for what to him represented 'a modest increase of detail'[68].

> It seems particularly desirable to include some mention of submissiveness. Nations would not keep up armaments unless they believed them to be effective. The prevailing notion seems to be that armaments are for keeping other people in order. Each nation seems to think, 'If we threaten our neighbours, they will not make trouble.' There must be a considerable body of experience behind this widespread belief. The policy of threats, however unkind it may have been, must have been effective. In the experience of many people it worked in the nursery and in the school playground. It has worked in military experience, in various parts of the world, all through history. The strange fact is that it does not work now [he was writing in 1938] in Europe.... Why is there this contradiction between the theory, by which the maintenance of armaments is defended, and the actual result?

Richardson felt intuitively that when one nation is enormously more populous than the other the minority would tend to be submissive but that this factor would be unimportant in the case of two more or less equal nations. He therefore assumed that submissiveness would be proportional to the difference in the warlike preparations of the two nations. With the addition of a term expressing submissiveness on these lines, his equations showed how a threat from a much larger nation could lead to the submission of a smaller nation.

When he returned to this subject in 1951, Richardson noticed that the revised equations led to some interesting results when applied to the case of two equal groups, namely that an arms race might end without any fighting. The reason was that, as the total expenditure of the two opponents increases, the difference between the individual expenditures increases after a certain time, thus making the role of submissiveness more important. In his letter to *Nature* announcing this finding (1951), Richardson emphasised that although it was a theoretical deduction from his equations he was not aware of any occasion in history when such an event had occurred. Let us hope that before long history will show that an arms race can indeed end without fighting.

Richardson continued to examine the practical consequences of his equations and soon found another result that was completely unrealistic. He therefore tried another hypothesis, namely that submissiveness is proportional to the difference between the potential warlike expenditures of the two nations—the amount they would attain after complete mobilisation—rather than the actual expenditures. Here again the evidence was conflicting, so he turned to a third hypothesis of a purely mathematical nature, much too complicated to be expressed intelligibly in words. This gave a reasonable mathematical description of the changes in the warlike expenditures of the pro- and anti-German groups in Europe over the period 1907 to 1932, but it is not clear whether it had any predictive value.

The results of all these studies were written up in *The submissiveness of nations*, which was accepted for publication in *The British Journal of Statistical Psychology* in the summer of 1953; it appeared in November, two months after the author's death. The following sentences are quoted from the concluding remarks[69].

> The Quakers, Gandhi, and others have tried to persuade everybody to be as non-violent as the exceptional Christian saints. Most people feel the suggestion to be too utterly contrary to their defensive impulses. Could not a middle way be found, with the motto 'defence without provocation'. This motto might well be applied to words as well as to deeds. The existence of the various arms of nation A presumably have different psychological effects on nation B; for example, one may guess that B is much more provoked to retaliate by the existence of A's long-range bombers than by A's land-based fighters. Jonathan Griffin discussed 'defence without menace' in 1936, and now again the problem of the classification of armaments into provocative and non-provocative requires much more attention than it receives.
>
> Submission, except in the sense of defeat, is not a spectacular subject for newspaper headlines; it is not like fighting for freedom. Yet we need to remember that in any organization, whether it be a social club, a scientific society, a business, a national parliament, or the United Nations, a suitable amount and distribution of submissiveness is essential for smooth working. If each member struggled all the time for his own advantage there would be chaos.
>
> The present study has perhaps over-emphasized the bitter submission of the defeated, threatened, suppressed, or snubbed, and has neglected the joyous devotion to an admired leader. To correct the balance let me

instance a classical story illustrating the latter tendency, that of Naomi and Ruth. It is described as an affair of tribal loyalties.

It should be noted that nowhere does Richardson mention 'deterrence', the factor which is often used in defending nuclear weapons. Shortly after the dropping of the first atomic bomb on Hiroshima I wrote to him to suggest the addition of a new term to his equations to take account of the deterrent effect of these new weapons of mass destruction. He did not agree; the reaction of other nations would simply be 'we must have atomic bombs, because they have'[31], a situation adequately catered for by his equations.

So much then for Richardson's final views on submissiveness. Let us now turn to the other subject which continued to occupy his thoughts almost until his death—contiguity. A simple way of bringing out the importance of this topic would be to try to answer the question of whether two countries with a common frontier are more likely to fight each other than two countries with no common frontier. This question of contiguity had been a recurring problem in Richardson's war investigations. For example, in trying to explain why some countries had been involved in more wars than others, he discovered the role played by the number of frontiers of each country. Then, in his studies of the effects of different religions and languages, he had devised a way of dividing the world's population into honeycomb cells each with a million inhabitants, in order to assess the opportunity of war of the different religious and linguistic groups. Again, in investigating the number of nations on each side of a war, he found that much depended on the geographical nature of the nations. Finally, in his paper on *War and eugenics* he reported the preliminary results of a study which suggested that the number of small wars was less than would have been expected on the grounds of contiguity; this could be explained by some kind of pacifying influence, possibly common government.

Richardson tried to draw together all these findings in a paper published in 1952 *Contiguity and deadly quarrels: the local pacifying influence*. He started by grouping the deadly quarrels in his list into three groups; foreign contiguous, foreign non-contiguous, and civil. He then counted the number in each group for each range of magnitude and found that far more of the foreign wars had been contiguous than non-contiguous and that the ratio of the number of civil wars to the total number of wars increased progressively as the magnitude decreased. His next step was to examine the relation between the magnitude of a war and the size of the population of the lesser of the two belligerents; this showed that the suffering, i.e. the percentage of casualties, had been much greater in large wars than in small wars, which again suggested that some kind of pacifying influence was at work for the latter.

The above findings were derived directly from the historical facts in his list of fatal quarrels; no theoretical ideas had yet been introduced. To provide a more convincing answer to the question as to whether common government has a pacifying role, Richardson wanted to determine the theoretical opportunities for war, both civil and foreign, depending on the distribution of population. For this, he needed to divide the population of each country into cells with equal numbers of inhabitants and then to count the numbers of cell edges which were (a) wholly within a nation, (b) shared with another nation, and (c) adjacent to an uninhabited area, such as sea, polar ice or desert. Furthermore, he had to repeat this exercise for cells of different sizes. The first step was to prepare a catalogue of territories arranged in order of population; although he limited this, for practical reasons, to territories with at least 10 000 inhabitants, it proved to be a time-consuming task. The next step was to divide up each country with a sufficiently large population into cells of a million persons, making each cell as compact as possible and with not more than six sides. He then counted for each of these countries the numbers of cell edges of categories (a), (b) and (c). On the basis of a theory which he had developed specifically for the purpose, Richardson next estimated the numbers of each category of cell edge that would be found with smaller cells, each with 100 000 persons, and again with 10 000 persons. He also estimated the number of cell edges for cells of ten million persons. For the smaller cells he included the additional territories from his list with a sufficient number of inhabitants. He was now ready to compare the geographical opportunities for quarrels of different magnitude with the historical facts. The result confirmed his earlier impression that common government, or something closely connected with it, has a pacifying influence for wars of magnitude less than 4.5 (number of killed between 3000 and 30 000 approximately). Although this study was limited to 'what has actually happened' rather than 'what ought to be done', Richardson hinted elsewhere that his findings might be relevant to discussions about whether world government would lead to a reduction in the number of wars[70]. He took up the subject again in a Socratic dialogue between a critic and an advocator of federal government[71].

> FEDERATOR: Clarence Strait seized on the most essential fact when he stated that the League of Nations failed because it was an organization of governments to maintain their national sovereignty and independence, the very claims that needed to be restricted. When we shall have a Federal Union, that is to say a world government elected by the people, affairs will go much better.
> CRITIC: We shall merely have most of the old troubles under new names. There will be wars, but they will be called rebellions or civil wars.
> FEDERATOR: But history shows that peoples whom habit has joined seldom split asunder. Civil wars have been rarer than international wars.

CRITIC: That assertion would be important if it were known to be true. But it needs to be tested by statistics.

In developing his theory on the way in which the numbers of cell edges of different categories depend on the size of the cell, Richardson became involved in some original mathematical studies of a topological nature. One of the most interesting aspects related to measurements of the lengths of frontiers or coastlines, on which it will be recalled he had experimented with some of his visitors. With a unit length (the distance between the pointers of the dividers in his tests with the Martins) equivalent to 200 km, the length of the west coast of Britain comes out at 1180 km, but with a unit of 10 km the result is 2931 km, and so on. The length of the coastline continues to increase, apparently without limit, as the unit length of measurement decreases. In the absence of an international agreement on the scale to be used when measuring a frontier, it is not surprising that Richardson found large discrepancies in the published statements on the lengths of frontiers; for example, the length of the common frontier of Spain and Portugal is given as 987 km by Spanish authorities but as 1214 km by Portuguese. The most remarkable result was that for any natural frontier or coastline there is a linear relation between the logarithms of its measured length and of the unit of measurement. The slope of the straight line gives an indication of the ruggedness of the coastline. Richardson offered no explanation for this relation, simply noting it as a curiosity.

The detailed account which Richardson wrote of the above mathematical work was found among his papers after his death. It was published in 1961 in the *General Systems Yearbook* under the title *The problem of contiguity*. We shall see that it was to make a considerable impact on the mathematical community.

Another paper which Richardson had completed in a form ready for publication but was never published was the third, and most complete, account of his proposals for basing the voting in international organisations on something more logical and realistic than what he called the 'old diplomatic pretence that all sovereign states are equal'[72]. He was disappointed that the United Nations had adopted the same voting system as its predecessor, the League of Nations, namely that each member state had one vote. But, he pointed out, the United Nations 'may in its turn destroy itself, and forthwith arise like the mythical phoenix from its ashes. Proposals for reform, such as the present paper, should be published in advance of any such crisis'. While recognising that the 'fictitious equality' is moderated in practice by rules of procedure and by commonsense, he nevertheless advocated a system whereby the voting strength of each country would be a measure of its international importance, calculated on the basis of several readily ascertainable ingredients. Earlier he had thought that the armed strength of a nation might serve as one of the ingredients, but he now ruled this out on the grounds that this would only 'foster the very nuisance, namely a threat of violence, which a reasonable system of voting strength should be designed to avoid'.

He finally recommended a combination of the following ingredients: money contributed to the international organisation; number of geographical contacts; number of inhabitants; international trade. His choice of geographical contacts or frontiers was based on his finding from his peace research that there is a strong correlation 'between a state's number of frontiers and its number of external wars'. For each of the 60 member states of the United Nations at the end of 1951, he collected the necessary statistics and expressed the numerical value of each ingredient as a percentage of the total of that ingredient for all the states. He then calculated for each country a voting index which was simply the mean of its percentages for the several ingredients. The values of the index varied from 0.23 for Iceland to 17.46 for the USA. He suggested that his index was 'a systemization of commonsense about voting strength'.

Richardson submitted the paper to Unesco and was only too willing to accept their conditions for having it published. Shortly after his death, however, Dorothy received a letter from Unesco (who by this time had had the article translated into French) stating that some unexpected difficulties had arisen and that they could no longer publish the article. Several further attempts were made to find a publisher, but without success.

Perhaps the most important of Richardson's unpublished papers was an article *Hints concerning International Organizations*, the story of which was related to me by Carl C Lienau. Looking back on the 79 years of his very varied life, Lienau described himself to me as being—like Richardson—a maverick, in the non-pejorative sense. On reading about Richardson's peace work in 1939 he felt that he was a kindred spirit and from then on he read all of Richardson's articles that he could lay hands on; being a statistician, he was able to understand the mathematics. In 1943 he wrote that Richardson 'is typical of a new tribe of scientific nomads. Such people are very inconvenient for the preservation of certain fictional boundaries between the sciences.... Quite a meddler with high human organization, this Richardson'[73].

In 1950, Lienau sent Richardson a discourse, based mainly on the latter's *Generalized Foreign Politics*, which he had presented at a seminar in Columbia University. He enquired if Richardson was willing to 'consolidate' his work in the form of a chapter for a book on *Measures of Organization* which he was planning to edit. Richardson replied[74]:

I am much honoured by your invitation... but I am not yet at all clear as to whether what I might be able to write would fit into your scheme. The only bit that I ever published about organization was a one-page joke about a professor of *dis*-organization. If you wish some comic relief, it might do. Or if I am to attempt a serious chapter, it might have to begin somewhat thus: '*Disorganization* by L.F. Richardson. Just as our appreciation of goodness is heightened when we notice the existence of badness, so our understanding of organization may become keener when we notice how it may go wrong....'

Lienau was not discouraged, and my March 1951 Richardson had begun to write a serious chapter. In informing Lienau of this, he wrote[75]:

The JOKE, if it ever reaches you, will be under my pseudonym of *Dafry*. Please keep the writings of that foolish person Dafry quite distinct from those of the serious-minded L.F.R. Today however I feel like the Scotsman who 'aye jokit wi' diffeeculty'.... I have never felt it to be my métier to write about organization. In the phraseology of Thomas A. Edison, 95% of perspiration will be no good if the 5% of inspiration is lacking; and I am not sure that I have the latter.

The chapter was despatched in April; writing it had to Richardson been 'like packing a small bag for a long journey... an exercise in leaving out'. He asked Lienau to 'resist the temptation to print editorial alterations, as if they were my work, without my consent' and continued: 'Some editors do that, and it always infuriates me. On the contrary I regard editorial objections as perfectly fair game if the reader can see that they come from the editor, not from the author'[76].

Lienau told me that the publication of his book had been 'aborted for failure of funding'[77]. If Richardson had still been alive at the time, he might have endeavoured to have his chapter published elsewhere, for it contains several ideas not to be found in his other publications. Let us make amends by quoting a few paragraphs[78].

In the nineteenth century the world could not be organized as a whole, because of the slowness of messages and of travel. Modern radio and aircraft have greatly speeded interaction, yet the world is still not organized as a whole. The chief remaining obstacles appear to be in human nature.

Every organizer uses information. If his stock of information is copious, accurate and well-arranged, he may claim that his process of organization is scientific. Every organizer must also make decisions about value, distinguishing better from worse, good from evil. The process of organization is an interaction between knowledge and judgments of value, and therefore it cannot possibly be reduced to social science alone. This puts me in a difficulty. The editor has asked me to write this chapter because I have published various researches on the causation of wars. In those articles I have tried to be objective and impersonal. But now, when writing about organization, I must inevitably take sides, by saying what I personally think better, what worse. One cannot discuss world-organization without indicating what sort of world one would like. Yet in order to avoid excessive personalism, I will allude to several ideals, as a reminder of how diverse they can be.

Ideal A A world in which there is plenty of variety, excitement, and disorder.

Ideal B A world where love, joy, peace, long-suffering, gentleness, goodness, faith, meekness, and temperance prevail everywhere.

Ideal C A world in which there are plenty of opportunities for the able and energetic to become millionaires.

Ideal D A world in which no one suffers poverty.

Ideal E A world designed to make the next generation better than this one, by selective breeding or otherwise.

Ideal F A world where nearly everyone loyally follows the leader, while dissident minorities are suppressed.

Ideal G A world where friendly cooperation abounds, where competition and controversy are permitted, but fighting is forbidden.

Having mentioned these various ideals, I go on to discuss the difficulties of attaining only one of them, namely G, the last. Allow me to take, as a working model or microcosm of the last world-ideal, the British House of Commons. It has been said to be 'the best club in London'. To dine occasionally with political opponents is regarded by some members as a duty. Yet the House is also the scene of almost daily strenuous contention. In foreign policy and in defence, the opposition have often supported the government; but on other problems the opposition generally try to modify, to delay, or to frustrate governmental action....

A never-mentioned, but very influential, feature of the parliamentary situation is the fact that the opposing parties do not keep party armies. This appears the more remarkable when it is compared with the declaration by Mr. Anthony Eden that, in foreign affairs, diplomacy needs to be supported by armed force. On the rare occasions when a member of Parliament has descended to fisticuffs he has been ordered by the Speaker to leave, and, if he refuses to go, he has been forcibly removed by the Serjeant-at-Arms. These are officers of the House of Commons as a whole, not of one party. How different is international behaviour!...

I write with diffidence about what ought to be done, for I suspect that the connection between social science and practical politics may be very loose. That is why this chapter is called 'Hints'.

In the following pages, Richardson summarizes his work on arms races, submissiveness, the local pacifying influence, the role of common government and so on, leaving out, however, the distribution of wars in time—perhaps he attached less importance to the Poisson distribution than he had earlier. He ends the chapter with a section entitled *Interim precepts for pacification*:

> Friends ask me for definite practical recommendations. I am reluctant for the following reasons. Social science, however much it might be developed, could never tell us exactly what we ought to do: it could warn us off some actions, and suggest that others are harmless; but a wide range of free choice would always remain open. Similarly dynamics does not specify the machines that ought to be made. Moreover dynamics has persisted as enduring truth while machines have been scrapped and replaced by new designs. Social science should persist, and in order that it may do so, it should be kept distinct from ephemeral practical policy. However, with that caution, here follows an attempt to be practical:
>
> (1) If you have the affection and the courage which go to make saints and heroes, try Gandhi's method.
>
> (2) If you find it impossible to love your enemies, try, as next best, to understand them. One good way to get under their skin is to read the novels and plays which they enjoy.
>
> (3) Don't object if your relatives wish to marry foreigners. Such bonds may help to hold the world together.
>
> (4) Don't ban goods merely because they are foreign. Trade is a mild pacifier.
>
> (5) Develop loyalty towards world government.
>
> (6) Always remember that some of the 'defence' preparations, which your nations might intend to be purely for defence, are certain to appear to some other nation as a dangerous threat against which they must make counter-preparations. Nevertheless, some purely defensive preparations do exist, for example air-raid precautions do not alarm unaggressive foreigners.
>
> (7) Because research on the science of pacification is now in progress, look out for new and better techniques.

In justifying his search for pacifying influences, Richardson mentioned that he and his wife did not always have identical views:

> My wife holds that it is wrong to disturb anyone's fundamental beliefs unless one has something better to offer instead. On the contrary I think that people will not bother to search for, nor even to attend to, new truth until they have become somewhat dissatisfied with traditional statements. This domestic controversy was illustrated one day when my wife missed a train by relying on a time-table which was placarded in a railway station. She complained about it to the station master. He said he was sorry, he knew that the placard was out-of-date, but he had nothing better to put in its place.
>
> A time may be coming when statesmen will resemble that station master in so far as they have personally ceased to believe in the Roman proverb

'si vis pacem para bellum', and yet do not possess any alternative belief which could be offered to their peoples to guide their conduct and to make them feel secure. Let us therefore look around for any pacifying influences.

Richardson told Lienau that he had shown the chapter to a friend who was formerly editor of a newspaper. The friend considered that although the style was unsuitable for the ordinary newspaper reader, it might pass for university graduates and research workers; to him the high spot was the story about the station master.[79]

During his retirement, Richardson rarely allowed himself to be diverted from his main task of studying the causes of war. Some of his publications in his later years might appear at first sight to be digressions, but on closer examination it will be found that with few exceptions they were closely related to his central activity. For example, his 1946 paper on the probability of encounters between gas molecules was, as we have seen, a preliminary 'little mathematical exercise' in preparation for his 1947 paper on the number of nations on each side of a war. Similarly, a mathematical paper in the *Philosophical Transactions* and another in the Statistical Section of the *British Journal of Psychology*, both published in 1950, were simply reporting on techniques which he had had to develop in the course of his war studies. We have already mentioned his brief return to the study of diffusion, largely inspired by the visit to Kilmun of Henry Stommel. On the psychological side, there is only one paper from this period of his life which stands out from the others as having no bearing on peace research, namely a report published in 1952 under the title *Dr S.J.F. Philpott's wave-theory*. The origin of this is of some interest.

For more than 20 years, Philpott had been conducting research on the fluctuations of human output when performing routine tasks, such as making elementary arithmetic calculations. For each individual test he plotted a 'work curve' in which the output is plotted against time. The individual work curves did not show much evidence of periodicity but when several hundred curves were added together (in this case with the output plotted against the logarithm of the time), there was a tendency for the peaks and troughs in the resulting 'grand-total' curves to occur at regular intervals. By the use of a somewhat debatable statistical method he deduced from this that there was evidence for the existence of mental waves with a period of about 4×10^{-23} seconds[80]. He further claimed that his results had been corroborated by subsequent experiments, including some made by his students—Richardson had of course himself studied under Philpott when working for his degree in psychology. To many, including Richardson, it seemed incredible that waves of such a short period could be detected from tests in which the output of the person being tested was only measured at intervals of about 5 seconds. Some psychologists shrugged off Philpott's work as an eccentricity or thought that he had been trapped by the delights of statistics. He was, however, known to be a thoroughly honest man and was highly respected by his peers—at one time he was president of the British Psychological Society. Few of his associates had sufficient statistical expertise to examine his methods critically and, for this and other reasons, nobody had seriously challenged his findings. In the early 1950s the editor of the *British Journal of Psychology*, D W Harding (who had in 1946 expressed a favourable view regarding the publication of Richardson's book), felt that someone should pay Philpott's work the compliment of a thorough assessment and suggested to Richardson that he would be well qualified for the task[81].

Richardson accepted the job, but with some reluctance. Quite apart from his desire to devote his energies to his war studies, he did not wish to do anything which would upset Philpott, with whom he had maintained a cordial relationship for nearly 25 years—in 1944, Philpott had spent a few days at the Richardson home in Kilmun. He began his article with a tribute to the great store of experimental data which Philpott had accumulated but rapidly embarked on a highly critical analysis of his statistical methods. In his view, Philpott's extrapolations were far too risky, the spikiness of his grand-total work-curve might be mostly random, and to search among it for periodicity might be fruitless. He complained that Philpott had not made adequate tests of the significance of his results and showed that other numerical values of the period seemed likely to fit the curves just as well as the value given by Philpott. The proper thing to do with his value would be to 'consign it to oblivion'[82]. Finally, Richardson suggested that more useful conclusions might be derived from Philpott's data by studying the individual work-curves rather than the grand-total curves.

Philpott saw Richardson's paper before its publication and was given an opportunity to reply. He began: 'I was more than delighted when I knew that Dr. Richardson had marshalled some of the objections to my theory of the curve. Controversy is one of the main processes in the advancement of science'[83]. He then made some comments of a general nature before proceeding to answer Richardson's objections point by point. He accused him of contradicting his own statement and, in one instance, of being 'quite off the mark'. As regards the question of extrapolation, he felt that Richardson was 'worse than I am'. Philpott was still clearly convinced that his results were valid.

Richardson's comments on Philpott's reply were also published in the same issue of the *Journal*. While accepting some of the criticisms he did not change his fundamental objections to Philpott's conclusions and remarked that he was 'content for the arguments to "go to the jury", which consists of readers of this journal'[84]. Philpott replied once more and again questioned the validity of some of Richardson's arguments. He concluded by repeating his thanks to Richardson and by paying him the following tribute[85]:

> I have known him for many years. I am never quite sure when he is acting as counsel for the prosecution, trying to get a conviction and an exemplary sentence, and when he is acting as devil's advocate before an approaching canonization.... But whether in the one capacity or the other, one can depend on him to put up a good case.

Harding recalls that when it became clear that it would be necessary to publish this series of replies and rejoinders[86]

> Richardson half jokingly suggested that I should use the terms used in the Scottish courts of law—a series of rejoinders and retorts. I thought that they would be too obscure for an international journal, but that they had an attractive archaic sound. I imagine this was characteristic of Richardson's intelligent playfulness.

Harding also told me that he had set the thing going before Philpott became seriously ill, or at least before he knew of it. In fact, Philpott was very ill before writing his final rejoinder and died shortly after.

This digression from Richardson's peace research—about which he would presumably have had second thoughts had he been aware of the state of Philpott's health—started soon after a visit, in the summer of 1959, to his beloved Cambridge. Although he and Dorothy had become very attached to their home in Kilmun, they were at that time still planning to settle down permanently in England. Prior to their departure for Cambridge they accordingly placed Hillside House in the hands of a local house factor with view to selling it. The occasion of the visit was to attend the wedding of their daughter, Elaine, who by this time had completed her studies and was employed by the Norfolk Education Committee as a peripatetic teacher of speech and drama. This necessitated much travel in the area around Downham Market where she met her future husband, Michael Traylen, who was working at the local meteorological station. The ceremony at the Friends Meeting House was a solemn but happy affair. After it was over, the Richardsons stayed on in Cambridge for a few weeks to give Lewis a chance to consult some books in the University Library. They were shocked to find that the cost of living was so much higher than in rural Scotland and regretfully concluded that they could not possibly afford to live in Cambridge without some additional income. Perhaps they were not too disappointed when they returned home to find that nobody was interested in buying their house—even a year later they had not had a single enquiry.

In July 1953 Richardson saw in *Nature* an announcement that his old College in Cambridge was offering some fellowships for which he might be eligible. After receiving a moderately encouraging response to an initial enquiry, he submitted a formal application in mid-September. 'At present', he wrote[87],

> I am revising 'Statistics of Deadly Quarrels' with a view to a second microfilm edition, and in the hope that it may ultimately appear as a printed book. The larger wars, since A.D.1820, are already listed and analysed statistically; but the list of the smaller pugnacious incidents, in each of which about 1000 people were killed, is far from complete; and the search for these small incidents requires a library larger than those in Glasgow.

He continued:

> Even a term's search in the University Library at Cambridge would probably yield a useful crop. Research, as you know, is like climbing an unknown mountain: at the foot one thinks that one sees the summit; but, on reaching it, one beholds a higher peak beyond. My enquiry into the general causes of wars is likely to continue as long as I have strength and ability. In 1940 I estimated that I might have it finished by 1943; yet it still goes on.

Under the heading 'Qualifications' he wrote:

> Broadly speaking, my qualification is a tendency to find out, to arrange, and to publish; when that is achieved, I soon forget. Reputations are usually out of date: mine is connected with activities that I no longer engage in, and results that I have partly forgotten.

The fellowships would normally have been offered for research extending over three or four years, but Richardson suggested that for a person of his age 'it would be presumptuous to plan ahead for more than one year at a time'. In making this reservation, he no doubt had in mind the serious heart condition which had affected him for more than a year. He had in fact never recovered completely from a bad fall in 1949 when 'a stepladder broke under me, and so I could do no gardening for 9 days'[88]. Although he had been warned that his heart trouble might recur at any time with fatal consequences, he hoped that he would be spared long enough to complete the revised version of his microfilm. His desire to work on right to the end may have influenced his decision not to go into a nursing home where he could have received prompt attention in the event of another attack. Towards the end of September 1953 he wrote to Quincy Wright that he was 'very old and tired'[89]; three days later he died quietly in his sleep of a heart attack. In reporting his death to the Provost of King's, Dorothy wrote[90]:

> He loved his college with so real a love; and often regretted that he had to give up, or refuse, scientific appointments wherein he might have

brought honour to his college, because he feared his researches might be of use in war.... It was such a happiness to Lewis to be in Cambridge August 1950 when our daughter was married there. He was dreaming of King's College the night before he died and dreamed he was welcomed there by old friends and colleagues.

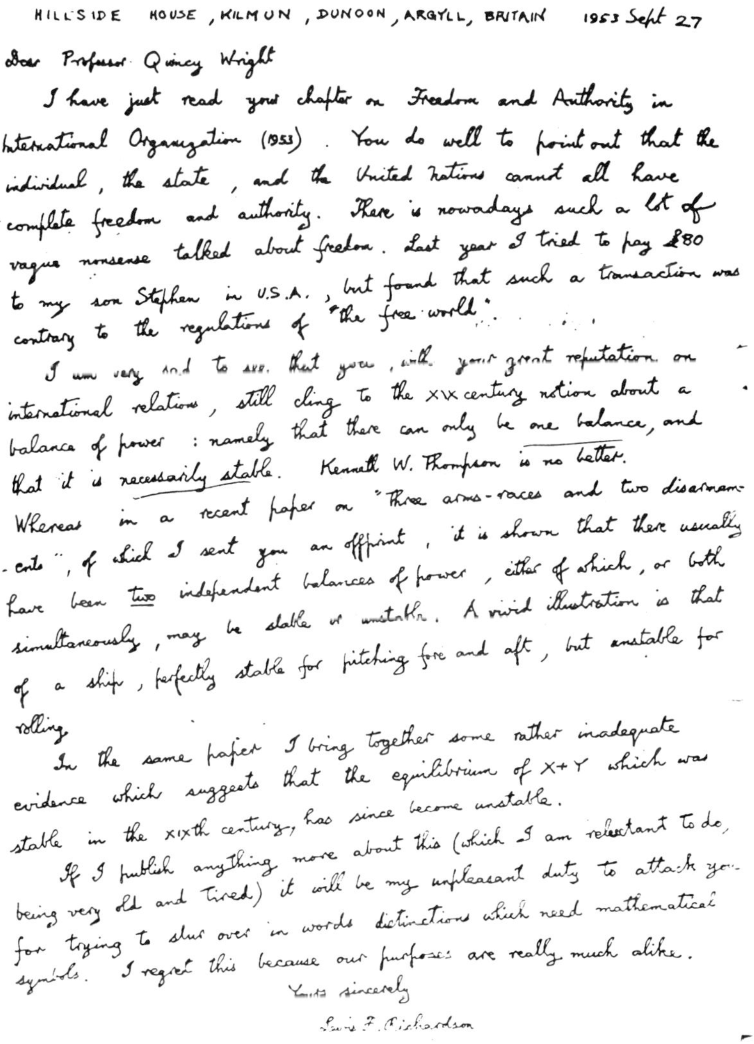

> HILLSIDE HOUSE, KILMUN, DUNOON, ARGYLL, BRITAIN   1953 Sept 27
>
> Dear Professor Quincy Wright
>
> I have just read your chapter on Freedom and Authority in International Organization (1953). You do well to point out that the individual, the state, and the United Nations cannot all have complete freedom and authority. There is nowadays such a lot of vague nonsense talked about freedom. Last year I tried to pay £80 to my son Stephen in U.S.A., but found that such a transaction was contrary to the regulations of "the free world".
>
> I am very sad to see that you, with your great reputation on international relations, still cling to the XIX century notion about a balance of power: namely that there can only be one balance, and that it is necessarily stable. Kenneth W. Thompson is no better. Whereas in a recent paper on "Three arms-races and two disarmaments", of which I sent you an offprint, it is shown that there usually have been two independent balances of power, either of which, or both simultaneously, may be stable or unstable. A vivid illustration is that of a ship, perfectly stable for pitching fore and aft, but unstable for rolling.
>
> In the same paper I bring together some rather inadequate evidence which suggests that the equilibrium of X+Y which was stable in the xixth century, has since become unstable.
>
> If I publish anything more about this (which I am reluctant to do, being very old and tired) it will be my unpleasant duty to attack you for trying to slur over in words distinctions which need mathematical symbols. I regret this because our purposes are really much alike.
>
> Yours sincerely
> Lewis F. Richardson

Figure 9.7 Richardson's last letter to Quincy Wright, written a few days before his death.

Richardson's body was cremated at Maryhill on 5 October and the ashes scattered in the garden there. On the same day Friends attended a very happy memorial service in Glasgow where, according to the official minute[91],

> tributes were paid to the life and character of Lewis Richardson. Mention was made of the integrity of his professional life, of the sacrifices he had often made for the sake of his sense of truth and righteousness. Along with his outstanding intellectual gifts, went courage and gaiety and he was filled with a deep desire that we should live and work so that future generations might enjoy a better world than that in which we find ourselves today.

CHAPTER 10

Epilogue

Richardson's death was duly reported in the press and in the journals of the scientific societies with which he had been most closely associated. In *The Times* the event would have passed off with but the briefest of obituaries[1] had it not been for Ernest Gold who wrote a more substantial supplementary note[2]. There was nothing in the original to suggest that Richardson and his work would not soon be forgotten but to Gold it seemed that thanks to the Richardson number his name would live 'so long as there is a turbulent atmosphere and men and women able and willing to be initiated into its secrets'. Gold had known Richardson personally from the time when they were both working in the Meteorological Office—they had also been living near each other in Hampstead for several years. Being a Fellow of the Royal Society himself, he was an obvious choice for the task of writing the obituary notice for the Society; he proved also be be an excellent choice, for the 20-page notice, published just over a year later, is a very readable and scholarly account of Richardson's life and work[3]. I have already had several occasions to refer to it in earlier chapters.

The local papers gave more glowing accounts of Richardson's achievements than *The Times*. The *Paisley and Renfrewshire Gazette* naturally spoke of his valuable work as Principal of the Paisley Technical College 'where his brilliance was reflected in the high standard of the students passing through his hands'[4]. The *Dunoon Observer and Argyllshire Standard* referred to his home in Kilmun where he and his wife 'kept open house for young people'[5]. The obituary in the King's College annual report of November 1963 related some incidents from Richardson's undergraduate days, largely based on the personal recollections of the author, Sir John Sheppard, a contemporary of Richardson at King's. The obituary also contains a factual account of

Richardson's career, including his work in the Meteorological [sic] Department of the National Physical Laboratory. But there was no suggestion that he had made any outstanding contributions to science for which he would long be remembered[6].

An anonymous notice in *The Friend* concentrated on Richardson's associations with the Society of Friends and remarked that it was while he and Dorothy were living in Paisley that 'many young people ... were enriched by their interest and enthusiasm'[7]. As one of the young people in question, I added a personal tribute[8] and further appreciations were written by Dorothy[9] and by Lewis' former FAU colleague, Tom Ellis[10]. The *Bulletin of the British Psychological Society* paid this tribute to Richardson[11]:

> Although he came late into the field of psychology, his background of mathematics and his personal attitude to the problems of society always made his interventions in discussions either at the meetings of the Scottish branch of the British Psychological Society or at those of the British Association a stimulus to those who were thinking along more conventional lines, however much they may have disagreed with him. In spite of ploughing a lonely furrow, he was always tolerant to those who differed from him and always seemed to preserve a Puckish sense of humour towards their apparent ignorance. He has been described by a fellow undergraduate as a 'rock and flew his colours with superb audacious gallantry.' Those of us who only knew him in later life will agree with his wife's estimate that the 'audaciousness' became an understanding gentleness, but the superb gallantry increased.

In the *Westminster Club Bulletin* Richardson was perceptively described as a 'man of outstanding ability' who 'was absorbed with problems of a scientific nature which lifted him above ordinary mundane affairs' but nonetheless 'always courteous and friendly'[12].

As might be expected, Richardson's death received extensive coverage in the meteorological literature. An anonymous writer in the *Meteorological Magazine* spoke of him as 'possessing one of the most able, original and versatile minds ever devoted to meteorology' and expressed the view that he 'will always be remembered for his pioneer work on the forecasting of the pressure distribution by transforming the equations of motion into finite-difference equations'[13]. The obituary in the *Quarterly Journal of the Royal Meteorological Society* was by David Brunt. His regard for Richardson's work on numerical weather prediction had evidently increased since the days when he had all but failed to mention it in his own textbook on meteorology. He remarked that at the time when Richardson attempted to make such a forecast[14]

> the facilities offered by the modern high-speed computer were not available, while the available observational material, especially in the

Epilogue

upper air, was insufficient to give a complete picture of the state of the atmosphere.

He continued:

> Today's network of surface observations, and our much improved methods of measuring winds in the upper air, offer a much more complete picture of the atmosphere at any given time, while the modern electronic computing machine can carry out in relatively few minutes computations which by ordinary methods would require thousands of computers over many months. Thus, the work of Richardson on the use of numerical methods in predicting tomorrow's weather is now to be regarded as likely to be fruitful in the near future.

To my mind the most imaginative obituary of all was by Brunt's successor as professor of meteorology at Imperial College, Peter A Sheppard, perhaps the most lucid meteorological writer of his day. In an article in *Nature* he referred to Richardson as the author of *Weather Prediction by Numerical Process* and *Generalized Foreign Politics* and continued[15]:

> These mentally adventurous works stamp the man, and of him it may be truly said, as Wordsworth said of Newton: '... a mind for ever/Voyaging through strange seas of Thought, alone.' For it was given to Richardson to be way out ahead of his contemporaries in the effort to mould experience to scientific form. Some have now caught up, or are catching up, with the meteorological research which Richardson did thirty years ago; the fate of his pioneering efforts to form a science of international relations will perhaps not be known for a still longer time.

None of the obituaries mentioned the tremendous support which Richardson received from his wife, Dorothy. This was remedied in a biographical note which Stephen wrote about his father in 1957. He describes how she had complete belief in Lewis' judgment and devoted herself to 'furthering his work, acting as his listener and questioner when he wanted to talk about his ideas, and quietly correcting the spelling and punctuation in his manuscripts'[16]. Elaine remembers her father becoming 'a crusty old man' who used to take out some of his discontent on his wife[17].

> I used to get extremely angry at the way he would speak to her. I was also angry at the way she took it and felt she should stand up for herself. She was a woman of no mean brain but she had dedicated her life—after bringing up her family—to Lewis' work and clearly felt that this was the correct thing to do

Some of our great pioneers and original thinkers may be self-sufficient to the extent of not needing such devotion and dedication by their spouse. Lewis was not one of these. He desperately needed somebody in whom he could confide and who could help him to overcome disappointment and frustration and thereby achieve what he believed to be his mission in life. He was indeed lucky to find what he needed in Dorothy.

After devoting more than forty years of her life to Lewis, Dorothy was naturally shattered by his death. As soon as she had recovered from the initial shock, she decided that the sensible thing would be to sell Hillside House—which was much too large and inconvenient for a single elderly person—and to set up home in England with her sister Hilda, who was suffering from angina pectoris and needed somebody to look after her. Before doing that, however, she had to dispose of Lewis' library and scientific equipment, not to mention the vast pile of correspondence and notebooks he had accumulated over the years. She presented about 50 books to Westminster College and a similar quantity of more technical books was purchased by Paisley Technical College together with most of the scientific apparatus. Other books were distributed to members of the family and close friends—I myself inherited about 20 of his meteorological books. But there remained the most important material, the bound volumes containing his original manuscripts, various notebooks of a more personal nature and all his files of correspondence. In the hope that some day a publisher would be found, she clearly had to keep the typescripts of the two books which had been microfilmed. What to do with the rest?

Dorothy wrote to me in December 1953[18]:

> There is much correspondence between Lewis and his juniors ... giving help and advice over their papers and sometimes gentle criticism. These men are now F.R.S. and or Knights, Directors of Education or Heads of Scientific Societies and Institutions. Other, jolly correspondence with his colleagues of the same age is a joy to read.... There is half a great file drawer full of meteorological correspondence. As far as I can judge most of it is out of date. But before I destroy it, I would be glad if some meteorological man could look at it.

To complicate matters, reprints of scientific papers, sometimes accompanied by letters from people who had not heard of Richardson's death, kept on arriving. Dorothy wrote to me about one from a Japanese meteorologist who had previously been in disagreement with Richardson about an article on eddies. 'The letter tells Lewis that the writer was "entirely wrong in the matter and Dr. Richardson was correct"... I do not know what to do with these papers.' The problem of how best to dispose of all this material was beginning to drag and Dorothy added despairingly: 'I am so tired and rather oppressed in thinking that I cannot do it alone'[19].

Olaf was the only member of the Richardson family who was able to give much help to Dorothy during this difficult period. On being demobilised from the army in 1946, he had gone back into business as a motor mechanic and by the time of his father's death he had a garage of his own near Kilmarnock. He visited Kilmun frequently at the weekend with his wife, Alice, who came from that neighbourhood. He later became interested in caravan sites and finished his career as secretary of a national caravan association; he died in 1983. His brother Stephen was living in the United States of America. After obtaining his master mariner's certificate in 1946, he had switched to academic life with encouragement from his future wife, Peggy, daughter of Newton Henry Black, a professor of mathematics at Harvard. Stephen graduated at Harvard (BS *magna cum laude*) in 1949 and then spent seven years on the staff at Cornell University. He is the author of more than 80 scientific papers, mainly on field methods in the social sciences and on mental retardation in children. The youngest of the three Richardsons, Elaine, was still living in Norfolk in 1953, but the following year she left for a three-year stint in Gibraltar where her husband was engaged in making upper-air observations at the meteorological station—I myself had left Gibraltar just a few months before their arrival to join the staff of the World Meteorological Organization in Geneva.

Dorothy sought advice from some of Lewis' former colleagues, with the result that about 30 thick volumes, mainly of lecture notes in physics and mathematics, were handed over to Paisley Technical College for permanent retention. A small quantity of files and correspondence was retained by members of the Richardson family but the vast majority of the papers were ultimately consigned to a bonfire.

Finding a purchaser for Hillside House continued to be difficult and it was not until early in 1956 that Dorothy succeeded finally in disposing of it. In the meantime, she had visited Stephen in Ithaca and Elaine in Gibraltar, greatly enjoying the company of her grandchildren. Together with Hilda, she set up house at Sonning on Thames, which she described as an old world village in very beautiful country near Reading. As a result of a serious illness at about the time of the move from Kilmun, she could only think slowly and often made mistakes. She was nevertheless able to live a fairly normal life and was very happy to be with her sister; she also appreciated being within easy reach of a Quaker Meeting for a change. Her greatest joy was to entertain her great nieces and nephews when they came down from London for the day. This peaceful episode proved to be tragically short. On 9 November 1956, barely six months after her arrival in Sonning, Dorothy walked up to the main road to meet a ten-year old great niece who was coming to see her by bus. Before the bus arrived, Dorothy's body was being carried away in an ambulance—she had stepped onto the road without looking and had been instantly killed by a passing lorry.

Shortly before this accident, Dorothy told me of her conviction that

Lewis' work would be better appreciated in the future. In the case of his research on numerical weather prediction, which at the time of the publication of his book in 1922 had been assessed as quixotic and unrealistic, she was aware that some encouraging steps had already been taken, before Richardson's death, towards the realisation of his dream that 'someday it will be possible to advance the computations faster than the weather advances'[20]. It will be recalled that the chief obstacles to the successful application of numerical methods in weather forecasting were the lack of upper-air observations and of means for making the calculations with sufficient speed. The first of these obstacles was removed, or at least reduced, by the development of a worldwide network of upper-air stations following the first successful radiosonde ascent in the late 1920s. Initially, the network grew very slowly but the pace accelerated to meet the needs of the military during the Second World War. The earliest systems only gave measurements of pressure, temperature and humidity but techniques were then developed for observing also the upper winds, first by triangulation using radio direction finding, and later by radar, which is more accurate. By 1945, there were several hundred combined radiosonde/radiowind stations in the world, each making observations up to heights of 20 km or so once, twice or even four times a day.

In parallel, great progress was being made in the design of fast computers. Of particular interest in the present context was the Electronic Numerical Integrator and Computer (ENIAC), the first multipurpose electronic digital computer, which came into service in December 1945. Although it was very slow and cumbersome compared with modern computers, it represented a big advance on its predecessors and offered to meteorologists, for the first time, the possiblity of carrying out at least some simple calculations fast enough for operational use in weather forecasting.

In 1946, the great American mathematician John von Neumann submitted a proposal to the US Navy for 'an investigation of the theory of dynamic meteorology in order to make it accessible to high speed, electronic, digital, automatic computing'[21]. The theoretical study was considered necessary in order to simplify the basic equations, as used by Richardson, in such a way as to make them suitable for processing by computer. The project was duly funded and in August 1946 von Neumann convened a conference at the Institute for Advanced Study in Princeton to enlist the interest and support of the meteorological community. The participants included Jule Charney, a young American scientist who was destined to play a leading role in subsequent events. His imagination was fired and he worked on the problem during a visit to Oslo in 1947 and 1948, after which he went to Princeton to become leader of the meteorological group. One of the main simplifications was to treat the whole atmosphere as a single layer and to apply the equations at the level of 500 millibars, which roughly divides the atmosphere into two equal parts by weight. Other modifications were introduced to avoid difficulties which would have been encountered with Richardson's method.

By March 1950, all was ready for the first ENIAC test of the simplified equations using real meteorological observations. In spite of the very substantial simplifications involved, the results were very encouraging. They showed that even with such a crude model it was possible to predict the large-scale weather pattern with an accuracy comparable to that attainable by the traditional empirical methods, and of course greatly superior to what Richardson achieved with his much more complex set of equations. This shows the wisdom of Charney's decision to begin with a simple model and only to extend gradually to more realistic and hence more complicated versions. It should be pointed out that the ENIAC product consisted of a forecast of the flow pattern at the 500 mbar level; it was not a weather forecast in the sense understood by the ordinary citizen and would certainly not tell Mr X whether it would hail on his field! But in the hands of an experienced meteorologist, the 500 mbar chart is an important tool in making a local weather forecast, as the smaller scale events in which Mr X is interested are dependent to a high degree on the large-scale developments.

The 1950 ENIAC experiment was soon followed by further tests, both in the United States and in several other countries, such as Britain, Norway and the USSR, where research was already under way to find the best form of the atmospheric equations for numerical solution. Within ten years, the first operational system for producing weather forecasts by computer in a national meteorological service had been introduced. Nowadays meteorologists throughout the world employ numerical prediction methods involving much more sophisticated models and using the most powerful electronic computers that man has yet devised.

Happily, Richardson was aware of this turn of events before his death. In 1952, he sought information on the subject from the British Meteorological Office, which put him in touch with Charney, who independently had already sent him some reprints of his relevant scientific papers. In his letter of thanks to Charney in June 1952, Richardson wrote[22]:

> Allow me to congratulate you and your collaborators on the remarkable progress which has been made in Princeton; and on the prospects of further improvement which you indicate.
>
> I have today made a tiny psychological experiment on the diagrams in your *Tellus* paper of November 1950 [this was the paper announcing the results of the first ENIAC experiment]. The diagram *c* was hidden by a

card, which also hid the legend at the foot of the diagram. The distinctions between a, b and c were concealed from the observer, who was asked to say which of a (initial map) and d (computed map 24 hours later) more nearly resembled b (observed map 24 hours later). My wife's opinions were ... (that) .. d has it on average, but only slightly. This, although not a great success of a popular sort, is anyway an enormous scientific advance on the single, and quite wrong, result in which Richardson (1922) ended.

So far I have only had time to glance at your five papers. To comment on them now would be rash; but to defer comment would be to risk never making any; for I have other urgent duties.

Richardson sent Charney some offprints including one on which he commented: 'I think it is very like Southwell's 'relaxation method': he maintains that they are quite different. My method was first published in 1910: his 25 years later'. He concluded: 'Again thanking you for your papers, I must now resist a strong temptation to read them all thoroughly'[22].

Charney responded in 1953 by sending Richardson another reprint, but it arrived after his death. On hearing the news from Dorothy, Charney wrote a letter of condolence to Stephen Richardson. He referred to Lewis' 'characteristically generous and kind remarks about our work on numerical forecasting which in a very real way was based on his pioneering efforts in this field'. The letter continued[23]:

> I had not the good fortune to know your father personally, and this I deeply regret, but since my first acquaintance with meteorology and numerical mathematics I have been greatly impressed by the scope and originality of his work and have profited much from it. It is still a pleasure to turn to his great book "Weather Prediction by Numerical Process" to be again charmed by his witty and penetrating remarks. Later when I heard of his qualities as a man as well as a scientist my respect grew even more. As always he went straight to the heart of the matter, and the matter was the prevention of war.

Charney later became a professor of meteorology at the Massachusetts Institute of Technology where he continued his brilliant scientific career until his death in 1981 at the early age of 64. He was recently described as one of the most important leaders in meteorology of this century.

With the growing success of numerical weather prediction came an increasing appreciation of Richardson's contribution to the subject. No longer was he regarded as an eccentric visionary full of original ideas of no practical value. His book had become monumental and epoch-making while he himself had become a genius, 30 years ahead of his time. He was even listed along with Helmholtz, Bjerknes and von Neumann as one of meteorology's 'intellectual giants possessing Da Vinci-like versatility'[24]. During his life, Richardson had received few honours apart from his FRS. None of the scientific societies to which he belonged had bestowed on him any of the awards reserved for their most distinguished members. Now, however, his name received posthumous recognition. In 1960, for example, the Royal Meteorological Society instituted the annual L.F. Richardson Prize, consisting of a book and a modest sum of money, for the encouragement of young meteorologists; it is awarded for meritorious papers contributed to the *Quarterly Journal* by members under the age of 28. It should not be confused with a proposal made in 1957 by an American meteorologist, James E McDonald, that the Society should establish a Richardson Prize in recognition of 'outstanding contributions to the poetic interpretation of meteorology'. 'For who', he asked, 'has condensed into few lines deeper meteorological truth than L.F. Richardson did in the famous vorticity rhyme?'[25]

Following a proposal by a Quaker meteorologist, E Wendell Hewson, *Weather Prediction by Numerical Process* was re-issued in 1965 as a paperback by Dover Publications. Although more than 3000 copies were printed, as compared with the 750 copies of the original, this time the book was out of print in less than ten years. The only difference from the original, apart from the cover, is the addition of an interesting seven page introduction by Sidney Chapman, who it will be recalled had been one of the book's reviewers in 1922. Dorothy often complained that Lewis' pacifist views had disqualified him from obtaining a university post in the aftermath of the First World War. But Chapman, who had also been a conscientious objector, was appointed as a professor at Manchester University in 1919, so this cannot have been the only reason. Chapman reports[26] that he himself had urged the appointment of Richardson as professor of meteorology at Imperial College when the post became vacant in 1934, but as we have seen the successful candidate was David Brunt. By the time of the Second World War, Hitler had changed Chapman's views on pacifism and he served as Deputy Scientific Advisor to the United Kingdom Army Council. Chapman's outstanding scientific career continued until his death in 1970 at the age of 80.

It is interesting to compare the reception of the Dover reprint with that of the original 1922 edition. With the benefit of hindsight, the reviewers were able to refer to Richardson's genius and vision and to the unique nature of his book; some went so far as to claim it as the first textbook on dynamical meteorology. It was recommended to applied mathematicians and meteorologists, both as a classroom text and as a source of inspiration for future avenues of research. The importance of the occasion was underlined by the appearance in the *Bulletin of the American Meteorological Society* of George Platzman's scholarly 37-page article on Richardson—this journal rarely publishes articles even as long as ten pages! It had been triggered by a request to review the reprint, but as the author himself expressed it, '...whereas the conventional brief format was intended, what emerged is manifestly very different!'[27] The article gives an excellent retrospective view of Richardson's work on weather prediction and should be read by every meteorologist interested in the subject. Some of Platzman's comments were based on a file containing the correspondence which Richardson had received from his distinguished contemporaries after the publication of the book, and also his manuscript notes for a possible second edition, presumably an up-dated version of the 'red binder' which he had left in Godske's bedroom in 1936. The file was later handed over to the Royal Meteorological Society for their archives but was tragically lost, along with some other irreplaceable Richardson papers, at the time of the removal of the Society's headquarters from London to Bracknell in 1971—perhaps it too will turn up some day under a 'heap of coal'.

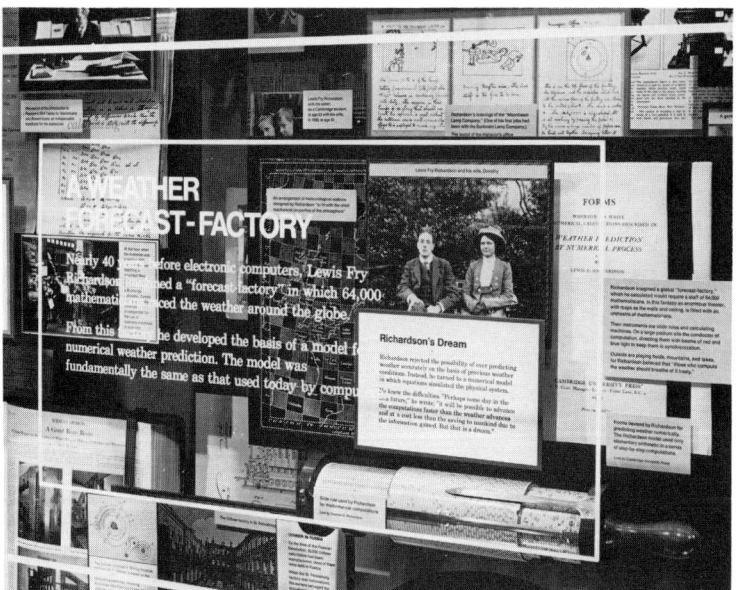

Figure 10.1 The Richardson section at the exhibition *A Computer Perspective*.

At an exhibition 'A Computer Perspective' mounted in the United States in the early 1970s, Richardson was featured as a central figure in the story of the development of computers. The Odhner computer which he used for many of his meteorological calculations can be seen in the photograph of the Richardson exhibit (figure 10.1).

What was perhaps the climax in the posthumous recognition of Richardson's meteorological achievements occurred on 6 October 1972 when a big extension to the headquarters of the British Meteorological

Office was opened by the Prime Minister, Edward Heath. The extension had been built to house the central forecast office, including a powerful new computer for use mainly in weather prediction. In spite of Richardson's somewhat chequered career as a civil servant, the Director-General of the Meteorological Office had decided to name the new building the Richardson Wing. In welcoming the guests, Mason referred to Richardson as 'one of the greatest meteorologists of his and any other generation' and went on to describe how it had become possible to put his numerical forecasting methods into practice[28]:

> Since 1965 all the forecasting for aviation has been done almost entirely by computer with significantly better results than had been achieved by the forecasters when they had to depend upon their subjective judgments and experience.

But he was careful not to claim that all the age-old problems of meteorology had been solved and that the Prime Minister would never be given an incorrect forecast. 'Meteorology' he said,

> like politics, is an exposed profession; several times a day we have to make decisions with inadequate time and data and if we make a mistake 50 million people here and many more millions abroad know about it almost instantly with no possibility of our hiding it. This is a harsh but salutary discipline, and we take considerable comfort from realizing with politicians and economists, that because we shall never be completely infallible, we shall never be entirely dispensable.

Figure 10.2 A C Thomas demonstrating the new computer after the opening of the Richardson Wing. *Left to right*: Dr B J Mason, Edward Heath, N Bradbury.

Figure 10.3 Glass engraving of Richardson in the entrance hall of the Richardson Wing, Meteorological Office Headquarters.

Mason has the well earned reputation of being able to give a speech word perfect without using any notes, and this was no exception. In congratulating him on the excellent way in which he had delivered his speech, Edward Heath added: 'I think I can say that you are the only person besides General de Gaulle whom I have heard deliver a speech absolutely word for word without a piece of paper in front of him. This I envy'. Mason later told me that as a result of this remark he had received several letters addressed to him as 'Mon Général'! In more serious vein, the Prime Minister continued[29]:

> It was a happy inspiration to name it after Lewis Richardson. He was one of those men whose names are not widely known to the public but whose work as you have described it has revolutionised part of their everyday lives. I don't really ... believe that providing an additional 64,000 people would have provided Sir William Armstrong [head of the civil service] the least problem at all, it is only when I tell him to cut the Civil Service down by 64,000 people that the problems really arise.... Perhaps we are

all wondering what Richardson himself would think if he could see today the realization of what even to him, must have been something of a fantasy. I think that Richardson would have been reassured if he had been able to see how the uses of meteorology have been developed for peaceful purposes.

I personally am quite sure that Richardson would have been as pleased as the members of his family who were present to hear the Prime Minister; he would have been even more delighted if there had been some indication of support by the politicians for his ideas on the causes of war.

Let us go back and see what had been happening in the meantime to the results of Richardson's peace research, which at the time of his death could only be found on microfilm or, in part, scattered over a wide variety of journals. Under the circumstances, it is not surprising that his work was relatively unknown; few outside the immediate circle of his friends and correspondents, such as Quincy Wright and Carl Lienau, appreciated what he had been trying to do during the thirteen years since his retirement. Even Nicholas Rashevsky, a professor at the university of Chicago in close touch with Wright, failed to mention Richardson when publishing, in 1947, his own mathematical theory of war, which was, however, on different lines from Richardson's.

A key role in stimulating interest in his father's work was played by Stephen Richardson, who in 1954 was invited to spend a sabbatical year, along with thirty other academics, at the Centre for Advanced Study in the Behavioral Sciences at Stanford University in California. Thanks to the generosity of the Ford Foundation, the participants—sociologists, anthropologists, economists, psychologists etc—were free to talk and write about anything in which they were interested. When Stephen spoke to the group about his father's peace research, several of them were inspired to read the microfilms which he had brought with him. Kenneth Boulding, the distinguished Quaker economist and poet, recalls how this drew him into the peace research movement[30], which was already nascent in the United States. The earliest journal in the field had in fact started in 1952 under the rather formidable title *The Bulletin of the Research Exchange on the Prevention of War*. Further inspiration came in 1955 from the publication of a book *Towards a Science of Peace* by Theodore Lentz, whom Boulding has described as the father of the peace research movement, the grandfathers being Quincy Wright and Richardson. After the year at Stanford, Boulding returned to the University of Michigan where he became involved with other members of the Stanford group in founding the *Journal of Conflict Resolution*. This was quickly followed by the establishment of the Center for Research on Conflict Resolution. Here at last was a group of people capable of judging Richardson's work critically, something that he had felt to be essential before making any attempt to popularise his discoveries. Here, at last, was a journal to which he could have submitted all his peace research papers for publication. Here, at last, was a bridge between the war historian and the scientist interested in the objective study of war.

In recognition of Richardson's pioneering efforts, the third issue of the *Journal of Conflict Resolution* (September 1957) was entirely devoted to his life and work. The main contribution was a 50-page essay by Anatol Rapoport, also of the University of Michigan. Under the title *Lewis F. Richardson's mathematical theory of war* he did for peace research what Platzman was to do later for meteorology in his article in the *Bulletin of the American Meteorological Society*; it should be read by everybody interested in Richardson's war studies.

The essay begins with a fascinating analysis of Richardson's philosophical orientation, psychological and ethical convictions, and scientific strategy. I am not sure that Richardson himself would have realised that his view of large-scale human behaviour derives from Malthus, Darwin, Marx, Toynbee, Pareto, and the mathematical economists; he would have accepted more readily that his psychological–ethical orientation is 'traceable partly to the thinking of Jesus, Tolstoi, and Gandhi and partly to that of the pragmatists'[31]. Rapoport gives a particularly lucid analysis of the deterministic philosophy and its application to large and small scale physical and historical phenomena. Turning more specifically to Richardson, he devotes the next section to the statistics of deadly quarrels, concentrating on the relationship which Richardson discovered between the frequency of quarrels and their magnitude. He then discusses the causes of deadly quarrels and presents a useful consolidation of the various pacifying influences which are analysed in various papers by Richardson; he pays most attention to the effects of common government. This leads to sections on Richardson's arms race model and his work on war moods. Having thus presented, for the first time, a thorough survey of the bulk of Richardson's investigations, Rapoport indulges in a critique of his techniques, and indeed of the whole question of the validity of the application of a mathematical approach to the study of human behaviour. At one extreme he postulates that researches of this kind may prove to have only historical value. More optimistically he argues that 'the findings of game theory and systems of differential equations are the two extremes bracketing the yet unknown theoretical method suitable for the study of human behaviour'[32]. He concludes with the thought that the greatest value of Richardson's work may rest in the way in which it inspires others to follow on where he left off. He suggests that whether his successors are able to construct a good mathematical theory of conflict or not, they[32] will be treating conflict with the objectivity of the mathematician, which is the most complete objectivity achievable by man. And this activity may of itself have a salutary effect on men's minds. After all, the most important achievement of celestial mechanics was not so much in its utility for calendar-making as in its liberating influence on the human mind. Similarly, 'astronomical' methods applied to the investigation of human affairs, if applied at a rate comparable to the growth of mathematical astronomy at the time of its inception, may yet free us from the compulsions which have been driving us towards destruction.

This brief summary of Rapoport's article does not do it justice but will, I hope, encourage more people to read it for themselves. They should also read, in the same issue of the *Journal*, the personal biography of his father by Stephen Richardson; I have used it liberally in writing this book. Finally, the *Journal* contains an annotated bibliography of Richardson's war studies which he himself had prepared.

This growing recognition by American social scientists of the value of Richardson's microfilms helped to stimulate the demand for having them published in book form. At the time of Richardson's death, it seemed that the best hope for this would be to find an American publisher, and there had already been some encouraging signs from the University of Chicago Press. In his efforts to help, Quincy Wright had in 1948 received enthusiastic support from Richardson's friend, Gregory Bateson ('the work is of first class importance') and from a professor at Yale, G Evelyn Hutchinson ('it takes courage to compose a book like this ... you may have hold of a trail blazer'). In contrast, a Chicago professor of sociology, W F Ogburn, had been unimpressed ('I hardly think it worth while to dig into or appraise the book carefully, since its exposition is wholly inadequate for publication'). There had been a rumour at Yale that, without the knowledge of the author, a photocopy had been made for a select group concerned with psychological warfare.

After 1948 there appears to have been a slackening in activity on the part of Wright and his friends until November 1953, when he received a letter from Dorothy Richardson informing him of Lewis' death and stressing the importance of publication. 'The problem', she wrote[33],

> is NOT *primarily* financial. It is that his writings appeal only to a very few trained in mathematics and sociology with some knowledge of psychology and the science of statistics. Consequently publishers are unwilling to risk an expensive book. But if we 'order' a publisher (or printer in this case?) to print the book and pay all expenses ourselves, that publisher (or printer) has no financial interest in advertising the book, and the copies go unsold. There has been, however, so much interest lately in different parts of America, that we wonder whether the book could be *published* (not only printed) there? The money, so laboriously saved, might pay half the cost.

Wright thereupon made a fresh approach to the University of Chicago Press and elicited the assistance of his Chicago colleagues N Rashevsky and L L Thurstone. In spite of this, he was unable to find a firm willing to undertake the work without a subsidy far in excess of what Dorothy could provide. He therefore sought financial backing from various societies and foundations, but without success.

Dorothy handed over the original manuscripts to Stephen Richardson when she visited him in Ithaca in 1954, and from then on he was actively involved in the search for a publisher. In so doing, he kept in close touch with Wright whom he had met the previous year. Others who played important roles at this stage included Richardson's admirer Carl Lienau and F Mosteller, under whom Stephen had studied at Harvard. Mosteller[34] told me how he first heard about Richardson when working as a consultant (round about 1950) for Project Rand before it turned into the Rand Corporation. Some members of the staff had come across Richardson's arms race model and found it to be of interest in some work related to the concept of *peacefare*. Mosteller was impressed by the pioneering nature of Richardon's methods and included some of the material in his courses on mathematical modelling for social scientists.

Many difficulties still had to be overcome and there were numerous disappointments. Finally, towards the end of 1957, Stephen was able to inform Wright that he had succeeded in concluding an agreement with the Boxwood Press with financial backing from the Lucius T. Littauer Foundation. The books came off the press in 1960. *Arms and Insecurity: A mathematical study of the causes and origins of war* was edited by Rashevsky with Ernesto Trucco, and *Statistics of Deadly Quarrels* by Quincy Wright and C C Lienau. The fears of so many publishers about the poor sales of such highly specialised books proved to be ill founded; they both had to be reprinted and the royalties were more than sufficient to repay the money advanced by the Foundation.

The main contents of the books have already been described in earlier chapters of this biography but there are several useful additions, such as prefaces by the editors, a biographical note by Ernest Gold (in fact, a reproduction of his Royal Society Memoir) and a preface by Richardson. There is also the following dedication:

> Written in memory of the insistence by Professor Karl Pearson, F.R.S., that popular beliefs ought to be tested by statistics. Also in memory of a comment by Dr. Charles Chree, F.R.S., on a controversy over which he presided, between two theorists: 'Ech, Richardson, if people would only look at the facts as they are given them, these troubles would not arise. But they *will* go wandering after their own ideas, which are mostly Will-o'-the-Wisps.'

The books were reviewed in a number of American journals of

sociology, politics and history. Some of the reviewers expressed reservations about the practical significance of Richardson's conclusions, but there was general appreciation of the originality of his methods and of his dedication. Probably the most widely read review was that by Sir Graham Sutton in the *Scientific American*. In view of his own expertise in these fields, it is not surprising that Sutton devoted nearly half of the three-page article to a survey of Richardson's contributions to mathematics, numerical weather prediction and atmospheric turbulence, preceded by a brief but pertinent account of his life. To appreciate the books it was, in his opinion, necessary to 'know something of the man who wrote them, and of the motives which led to their composition'. Sutton was not convinced about the validity of some of Richardson's deductions from his statistical findings; he clearly preferred the thinking which led to the arms race model. The construction of such a model was 'fraught with far greater danger than a statistical inquiry, but, on the other hand, if the model is valid, the gain in precision is immense'. Sutton ended by asking[34a]:

> What, then, are we to conclude about these remarkable studies? An intellectual exercise that leads nowhere, or the beginning of a new approach to the understanding of the most important problem of our day? The question is perhaps more appropriately addressed to a historian than to a mathematician, but for me the reading of these books has been more than a stimulating intellectual experience. Time has proved the worth of many of Richardson's contributions to physical science, eccentric though they must have appeared to his contemporaries. Will history be repeated?

In Britain, the most comprehensive review of the two books was in the *Journal of the Royal Statistical Society* by Ian Sutherland, whose father had been a good friend of Richardson. In just over ten pages he provides an excellent objective summary of Richardson's methods and of his main conclusions. His one criticism, fully justified, is that more could have been done by the author and the editors to indicate the relationships between the different sections of the books. In apologising—quite unnecessarily—for the length of his review, Sutherland expressed the view that the books 'represent a quite remarkable pioneer achievement in a field which many would previously have regarded as not amenable to mathematical treatment'. He continued: 'Richardson has shown unequivocally how much may be learnt from a mathematical and statistical approach to human conflict'. Richardson's chief hope had been that his work[35]

> might help people in authority to understand the circumstances in which the deterioration of international relations would be progressive, and to recognize that these had an emotional rather than a rational basis, so that they might in future act more rationally—a visionary hope, perhaps, but none the less laudable.

Writing in *The Friend*, Charles F Carter gave an equally sober appraisal of the books when he stated that Richardson's results 'are interesting, but do not of course yield conclusive guidance for statesmen'. He nevertheless felt it to be very desirable 'that others should take up and continue what Richardson so notably began'. He concluded[36]:

> ...it is impossible to resist the appeal of Richardson's vigour and enthusiasm, his Socratic dialogues, his immense range of scholarship. His successors will probably be dull in comparison, but they will have a great work to do.

If the publication of Richardson's books had been facilitated by the growing strength of the peace research movement in the United States, it is equally true that their appearance helped in turn to stimulate the growth of such movements, not only in the USA but also in other countries. In Richardson's native Britain, for example, there had been little enthusiasm for his ideas in the years immediately after his death; few had ever heard of his microfilms and fewer still had read them. There was, of course, an active peace movement, as witnessed by the strength of the Campaign for Nuclear Disarmament, but Richardson would probably have felt that the members were 'more concerned with propaganda, not research'. We could parody Sutcliffe by saying that his war studies had 'passed over with not a ripple' in the social sciences.

The situation began to change towards the end of the decade. Lentz's book made a profound impression on a number of serious-minded people in Britain, as in the United States. Among these was a Lancaster Councillor, Pat Deighan, who told me how Richardson had indirectly had an impact on his early career. In 1930, when he was only 16, the firm to which he was apprenticed as an electrical engineer and plumber enrolled him for a course at Paisley Technical College, where Richardson was, of course, Principal. He himself had other ideas; having just become fanatically keen on weightlifting and catch-as-catch-can wrestling, he preferred to spend his evenings at the local health club. One day he was summoned to his employer's office to explain why he never attended the classes at Paisley—an enquiry about his absences had been made by Richardson. When he replied that this was because he was training for the Scottish wrestling championships, he was immediately sacked[37].

In 1959, Deighan started a Peace Research Centre at Langthwaite House in Lancaster. This attracted a small group of enthusiastic young people to discuss Lentz's ideas, which although less mathematically oriented than those of Richardson would certainly have elicited his support.

In 1962, Lentz visited England and at Pat Deighan's suggestion a meeting was convened at the Friends International Centre in London to hear him. One of the organisers was a Quaker, Cedric Smith, Professor of Biometrics at University College, London—a disciple of Galton and Pearson. He gave me the following account of what happened[38].

> The meeting was attended by about a dozen people, including two Conservative Members of Parliament. Ted Lentz talked, the MPs suggested further meetings at the House of Commons—and elsewhere—but these never seemed to get very far; there was never enough unanimity. Meanwhile a "Round Table on Peace Studies" was set up at University College (but did not last very long). Finally, a Study Conference Weekend (28 June–1 July 1963) was held at Cumberland Lodge, Windsor Great Park; we invited a number of eminent academics in related fields, such as history, economics, law, politics and psychology, and we spent the weekend explaining to them the virtues of peace research. At the end of this meeting it was decided to form a Conflict Research Society.

By this time Richardson's books had been published and had attracted a certain amount of attention. One of the members of the Peace Research Centre, Paul Smoker, had been persuaded by Lentz to break off a university course in physics and instead to take up peace research. With his mathematical background, he was able to understand Richardson and he became very interested in his ideas. Charles Carter, who in his review of Richardson's books had advocated the continuance of such research, was appointed Vice-Chancellor of the new University of Lancaster and took the opportunity to promote the establishment of a research post in peace studies which was intended to carry on the same line of work as Richardson. The first appointee was Michael Nicholson, who had heard about Richardson through the writings of Boulding and Rapoport.

A year later, the Centre for the Analysis of Conflict, headed by an Australian, John Burton, was set up in the Law Faculty of University College (it is now located at the University of Kent in Canterbury). Its purpose is to conduct and publish social science research in the field of conflict analysis and resolution and it has tended to concentrate on finding practical solutions to specific problems rather than on the study of conflict in general. When Michael Nicholson came to London to join this Centre, his place at Lancaster University was taken by Paul Smoker.

Also in 1965 a symposium on 'Conflict in Society'[39] was held in London under the auspices of the CIBA Foundation. It was attended by many of the peace researchers whom we have been talking about—Boulding, Rapoport, Cedric Smith, Nicholson etc—and was organised by another Richardson 'fan', Tony de Reuck, who is now in the Department of Linguistic and International Studies at the University of Surrey. He had probably been aware of Richardson's work longer than any of the other participants; it was in 1939 that having just left school with a knowledge of calculus fresh in his mind he read one of Richardson's letters to *Nature* about his arms race theory. He also learnt something of Richardson's pioneering work in weather prediction while serving as a meteorologist during the Second World War. As a great admirer of Richardson, he was delighted when he heard about the formation of the Conflict Research Society and very happy to be associated with the CIBA symposium. The gathering was a great success, and helped the peace research movement to gain momentum.

The next development (from the point of view of the Richardson story) occurred in 1969 when it was decided to bring the Peace Research Centre from Lancaster to London and to attach it to the Conflict Research Society with Nicholson as director. At his suggestion, the Centre was renamed the Richardson Institute for Conflict and Peace Research.

Initially the Richardson Institute was funded largely from Quaker trusts. When this source of finance diminished in 1978, the University of Lancaster offered to take on the Institute, provided that external funds could be found for the director's salary. An agreement was reached whereby the Conflict Research Society retained some of its rights and the Richardson Institute became part of the Department of Politics at Lancaster, where Paul Smoker is now Reader. Michael Nicholson stayed on as director until 1982. The main Richardson souvenirs at the Institute are six volumes containing some of the original material for his microfilms, interspersed with a few letters and other notes, such as the diary of his visit to Danzig in 1939.

Another illustration of the growing interest in peace research in Britain was the establishment in 1975 of the School of Peace Studies at the University of Bradford. In his inaugural address as the first Professor of Peace Studies, Adam Curle referred to Richardson as one of those who had given intellectual standing to the subject. I am sure that Richardson would have shared his views on the interdisciplinary nature of peace research and on the distinction between peace researchers and peace activists[40]:

> The activists are concerned to change situations which they conceive to be liable to lead to war and violence, and if they do carry out research it is in order that they may be better informed and so act more efficiently. The 'pure' researchers tend to feel that it is enough to obtain the facts, that if only we were adequately informed we would avoid wars and violence.

The developments in Britain from the late 1950s onwards were paralleled in many other countries, so that by 1982, the Rector of the United Nations University, Soedjatmoko, estimated that there were at least 250 peace research institutions in the world[41]. What a contrast to the days, only 30 years earlier, when Richardson had been struggling to find a publisher for his books. After their publication in 1960, knowledge of Richardson's methods and results was no longer largely confined to the enthusiasts who had read the original microfilms—the members of the 'early Church' of the peace research movement, in the words of Boulding[42]. From then on, right up to the present day, there has been no lack of papers by researchers trying to carry on from where Richardson left off. Some have elaborated his list of fatal quarrels; others have refined his statistics. Above all, many have devoted their efforts to improving and updating his arms race model, as instanced by Paul Smoker's paper *A pilot study of the present arms race*. This appeared in the 1963 *General Systems Yearbook* as part of a 100-page section entitled *After Richardson*. The editors referred to Richardson as 'the first to pursue the systems approach on a large scale'.

In 1980, David Wilkinson of the University of California produced a major book *Lewis F. Richardson and the Statistical Study of War*. After commenting on the multitude of papers about *Statistics of Deadly Quarrels*, he addresses the question as to whether the contents of Richardson's book have now been assimilated into political science or whether they remain a provocative source of ideas. He concludes that whereas Richardson's analyses have been widely discussed, his annotated list of fatal quarrels has not been adequately exploited. To facilitate further use of this valuable data base, he presents a revised and somewhat simplified version of the list in machine-readable form—the original coding system was of course designed for manual use rather than for feeding into a computer.

In the same year, there appeared Carl Sagan's *Cosmos*, written in conjunction with his popular television series under the same name. Sagan uses Richardson's statistics of wars between 1820 and 1945 to produce a graph showing how many years one can expect to have to wait for wars of different magnitudes to occur. By extrapolating this curve he concludes that the waiting time until 'Doomsday'—a war of magnitude 10 which would effectively wipe out the whole population of the world—is about 1000 years. With a slightly different extrapolation, taking account of the proliferation of nuclear weapons since Richardson's day, Sagan points out that this waiting time may have been ominously shortened. To my mind, this is an over-simplification of Richardson's methods, but there is no denying its effectiveness for public presentation.

As an example of a recent development based on Richardson's arms race model, let us consider briefly Philip Schrodt's monograph *Preserving Arms Distribution in a Multi-Polar World* (1981). By linking the classical theory of the balance of power with Richardson's model, the author deduces conditions under which certain distributions of arms levels continue to be maintained even when the arms levels themselves are changing. In discussing the relevance of this kind of basic research to practical policy making, he argues that the basic research of today is the applied research of tomorrow. Although Schrodt's work is not aimed at answering specific questions facing mankind at present, he believes that it could well help policy makers to solve some of our future problems.

Richardson's model is frequently used in social science textbooks, at least in the United States, as for example Michael Olenick's *An Introduction to Mathematical Models in the Social and Life Sciences*, published in 1978. More than 20 pages of the book are devoted to Richardson's equations, with special emphasis on how they can be used to ascertain the conditions under which expenditures on armaments are likely to lead to stability or to a runaway arms race. Olenick includes a biographical sketch of Richardson, more accurate than most of the others.

It would be misleading to give the impression that the conclusions which Richardson deduced from his peace research have been universally accepted. There have been many adverse criticisms of his views on the pacifying influence of international trade. His analysis of the economic causes of war has likewise been seriously questioned. One of the most outspoken comments was in a recent paper *Lewis Fry Richardson: Apostle of Math* by William Eckhardt of the Canadian Peace Research Institute. While acknowledging his pioneering efforts, Eckhardt accuses Richardson not only of having an unconscious bias against economic causes of war but also of 'falling over himself backwards in order to avoid seeing any relationship between imperialism and war'[43]. In his view, the statistics which Richardson used to support his finding that common government seems to be associated with fewer civil wars of lower magnitudes could equally well justify the conclusion that imperialism is associated with more foreign wars of these magnitudes.

The question of objectivity is also discussed in David Eversley's Runnymede Lecture of 1972, *A Question of Numbers*. He stated the purpose of his talk at the outset[44]:

> I am less concerned with the question whether one can deduce from particular observations a set of generally valid laws, or whether such laws are in fact appropriate to social science, than with the question whether useful practical inferences may be drawn, and whether 'objectivity' is enough to achieve changes in policy and opinion.... The main question is: how useful is statistical evidence, however good, as an instrument for social policy making and opinion forming?

Eversley felt there was a danger of the computer becoming a substitute for the powers of observation and the application of common sense. Statistical methods can be efficacious, he argued, but they are not everything: we need also value judgments. This means that in interpreting the results of statistical analysis it is practically impossible to keep away from one's own personal beliefs; the important thing, therefore, is to make a personal avowal of one's subjective influences, not to pretend that they do not exist. In support of his thesis, Eversley cites a number of examples where the use of models has led to planning decisions of dubious value. He does not refer specifically to Richardson, but this may have been because he was not aware of any instance where Richardson's model had been applied in the solution of a practical problem. Eversley ended his lecture on a critical note[45]:

> The Social Scientist, as he has become more numerate, has also become more arrogant, more remote, and less able and willing to communicate. When he learns that quantitative methods are mostly a means to an end, not an end in themselves, he may perform a more useful role in society.

Richardson would have agreed with much of what Eversley had to say, especially regarding the difficulty in avoiding bias. In a personal avowal of his own prejudices, relevant to arms races, he wrote[46]:

> I have a prejudice that the moral evil in war outweighs the moral good, although the latter is conspicuous. This prejudice is derived from the Quakers, who brought me up. I am not ashamed of it; indeed it has been one of the two principal motives for writing this book [*Arms and Insecurity*]. The other principal motive is my prejudice that scientific method is more trustworthy than rhetoric.... I come from ancestry which the Nazis would call aryan, and the Communists would call bourgeois, but which I am happy to call English.... I am aware of nation prejudices and of the manner in which they are usually supported by facts selected, not by random sampling, but by pride and hate. Such bias I would fain avoid, if it were possible to do so.

The above-mentioned monograph by Schrodt came off the press in October 1981, an unplanned but fitting tribute to the centenary of Richardson's birth. The event was celebrated more deliberately at a meeting of the Royal Meteorological Society, when Henry Charnock spoke with characteristic charm and humour about Richardson's life and work. He was able to recount several personal incidents as he had been involved in the 'parsnip' experiment and had heard Richardson's lecture at the Eugenics Society. Charnock must be almost the only meteorologist to have read Richardson's microfilms.

A few months later, the Richardson Centenary Lectures were presented to a large gathering at the Paisley College of Technology. The first was by a French mathematician, now living in the United States as an IBM Fellow, Benoit B Mandelbrot. He described how Richardson's work on atmospheric diffusion and on the length of coastlines related to subjects which do not fit into the neat and tidy ideas of Euclidean geometry. Many other examples can be found in nature—the shapes of clouds, the structure of the human brain (the surface area of which is proportional to its volume!), the clustering of stars and nebulae, the Brownian motion of microscopic particles. These shapes are now being handled by what Mandelbrot calls 'fractals', a new branch of geometry to which he himself has made important contributions. In asking the question 'Does the wind possess a velocity?', Richardson had been bold enough to call attention to a mathematical function which was continuous but did not have a derivative. He was one of the few who dared to say that the Emperor has no clothes; without him, the world would have been much poorer.

Mandelbrot told me how he had 'discovered' Richardson, almost by chance. Before disposing of a series of journals which were rarely consulted, a librarian had asked him to glance through them. One was the issue of the *General Systems Yearbook* containing Richardson's full account of his work on contiguity, including his measurements of the lengths of coastlines. Mandelbrot added that this paper had been a revelation and had profoundly affected some of his subsequent researches. His book *The Fractal Geometry of Nature* (1982) has many appreciative remarks about Richardson and a three-page biographical sketch.

The second Centenary Lecture was by Michael Nicholson, Director of the Richardson Institute and well known writer on peace research. He referred to Richardson's war studies as having been premature—just as his work in weather prediction and other fields. Had it not been for a few perceptive people, his discoveries might have been completely ignored. The main problem was that, in his day, Richardson had no constituency—politicians and historians could not understand his differential equations while scientists were not involved in studying the causes of war. Things had changed after the Second World War, when the existence of nuclear weapons led some scientists to branch out into war studies. Furthermore, the successful application of mathematical analysis to economic theory had led social scientists to believe that they might also be able to contribute to the prevention of war. It was only then that Richardson's theories and methods had been found to be invaluable tools and that his outstanding achievements became recognised. His work was all the more remarkable for having been carried out in almost complete isolation—a tribute to his belief in himself and his sense of mission.

I asked Nicholson and other peace researchers whom I met in the course of writing this book to give me examples of how Richardson's findings had proved to be of practical value. There can be no doubt about the impact of his work on academics, but are the statesmen and politicians even aware of his achievements—or of those of his

Epilogue 261

successors? The most frequent reply was that, up to the present, the practical impact of Richardsonian peace research has been small, but evidence of the positive interest of governments has been provided by the establishment of the Arms Control and Disarmament Unit in the British Ministry of Defence and of similar bodies in the United States and elsewhere. Roy Dean, of the British unit, told me that, while he is familiar with the work of Richardson, he is not aware of any relevance in it to the problems faced in the current disarmament negotiations. He feels that much of the work of the peace research institutions is primarily theoretical and that, to achieve practical results, a bridge must be found between peace research and disarmament[47]. Boulding informed me that the most positive interest in some of his peace research had come from a military academy, while Richardson himself declined an offer by a military agency to publish his work.

In answering my questions, Adam Curle made an interesting analogy[48]:

> I don't think that peace research in the Richardson sense is any more relevant to the making or preservation of peace than is pure research in chemistry or physiology to curing a particular disease—although it may, subsequently, help to develop practical methods. I think that the actual process of healing fractured relationships has more affinities to psychotherapy; peacemaking is clinical rather than scientific although certain general principles may eventually emerge.

This analogy between peace research and medical research is taken a step further by Anatol Rapoport in his foreword to Mike Yarrow's stimulating book *Quaker Experiences in International Conciliation*. 'The analogy fails', he wrote[49],

> on the level of applying knowledge to organized action. In applying knowledge produced by medical research, there are no insuperable problems. Institutions empowered to translate such knowledge into action already exist: the medical profession, hospitals, a pharmaceutical industry, departments of public health, etc.... But where are the institutions ready, willing and (most importantly) *empowered* to translate new findings about the causes of war into appropriate action, no matter how strongly these findings may be supported by scientific evidence?

It seems probable that, at least in the early stages of his peace research, Richardson believed that his findings should be of interest to statesmen and politicians, if not in their day-to-day decision making, at any rate in their long-term planning; why else would he have sent them copies of his publications? In this connection, the following anecdote is related in the editors' preface to *Arms and Insecurity*[50]:

> In 1939 when the clouds of war were ominously gathering over Europe, an American journal received for publication a paper from Lewis F.

Richardson. In his letter of transmittal Richardson urged that the paper be published immediately because its publication might avert an impending war. The editors of the journal not only did not rush the paper but rejected it.

Nobody will of course entertain even for a moment the thought that the Second World War would not have occurred had the paper been published in time. However, the story, whether true or not, illustrates Richardson's strong belief that scientific knowledge may prevent disasters to humanity. We think that in this basic belief he was right. What he forgot is that even if his theory had been considerably more developed and immediately applicable, those individuals who are primarily responsible for decisions which precipitate wars would not have read his paper anyway; and even if they had attempted to read it, they could not have understood the mathematics of it.

No matter how true the above story may be, Richardson certainly displayed a more realistic attitude later on, as can be seen from the quotations on page 230 from his *Hints Concerning International Organizations*. Dorothy, on the other hand, never lost her conviction that her husband's work was of immediate practical value. She wrote, for instance, to L L Thurstone: 'The present state of world affairs cries out for this work "Arms and Insecurity". Our politicians do not see it, for they look only on one side behind a barrier of arms which blinds them utterly to the similarity of the people on the other side'[51]. She did not apparently appreciate the point made by Rapoport, namely that the politicians could not have understood the mathematics.

To Quincy Wright and the others who helped to get the books published, the important thing was not so much to have them read by statesmen as to make them more readily accessible to social scientists, so that Richardson's methods and results could be properly appraised by his peers. Only at a later stage, after the work had stood the test of time, would the essentials be conveyed in more intelligible language to those responsible for decisions about peace and war. James Joyce is said to have studied Norwegian so that he could read Ibsen in the original. Did Richardson really believe that the foreign office officials to whom he sent copies of his publications would be prepared to learn mathematics in order to understand him?

In asking this question I am perhaps not being entirely fair to Richardson, for, at least in the early phases of his peace research, his real problem was that he was unable to find people able and willing to assess and criticise his work—witness his complaint about the gap between the British Academy and the Royal Society. For his meteorological papers, the situation was quite different, in that there were plenty of experts capable of discussing his results both critically and authoritatively. The delay in recognising the full worth of his numerical method of weather prediction was not because his contemporaries could not understand him; it was simply that the technology necessary for applying the method was not available.

In mathematics, too, Richardson had no difficulty in obtaining criticism and help when required. This was all the more important in that he himself did not have a solid background of mathematical learning to fall back on. When he ran up against a mathematical problem, he frequently had to study what was for him a new branch of the subject in order to find a solution. Sometimes he even had to push forward the frontiers of knowledge in mathematics, as evidenced by his several mathematical papers. In contrast with his meteorological work, however, the originality of Richardson's contributions to mathematics has not been widely recognised; his papers are rarely mentioned in modern mathematical books. I have come across two notable exceptions. The first was Mandelbrot's book on fractals in which we find, as mentioned earlier, an enthusiastic appreciation of Richardson's mathematical insights. The second was *Methods of Numerical Mathematics* by the Russian Academician G I Marchuk (1982). This highly specialised 500-page book contains only ten chapters, but one of them is entitled *Richardson's method for increasing the accuracy of approximate solutions*. Is this a case of a prophet being not without honour, save in his own country?

As regards Richardson's papers in what we might call pure psychology, we have seen how his experiments on the quantitative estimation of sensation were the subject of great controversy during his lifetime. In this case, it was not the novelty of his results which prevented them from being immediately accepted in the psychological world; the problem stemmed rather from philosophical differences about what constitutes a measurement. Since that time there has been growing recognition of the great practical value of such measurements, for example in improving the quality of telephone circuits and in the development of colour photography. In the vast contemporary literature in this field, and especially in historic reviews of the subject, the pioneering nature of Richardson's work is often cited. A case in point is *Bias in Judgment* (now in the press) by E Christopher Poulton, who is related to Richardson through the Garnett family.

In these days, when scientists tend to specialise in one small branch of a subject so that they will ultimately know more about it than anybody else, one of the most impressive things about Richardson was the wide variety of topics to which he made original contributions—hydrology, meteorology, psychology, eugenics, mathematics, education and peace research. Was there some central theme to which all his works can be related, or does their diversity simply illustrate how he applied his creative energies to whatever happened to interest him at the moment or to whatever he was being paid to do? Some who knew him have wondered what he would have achieved if he had concentrated on one discipline: Gold suggested that he might have put Britain in the forefront of hydrological research, while N K Johnson wrote of the tragedy to meteorology when he resigned from the Meteorological Office. Rather than speculate on what would have happened if a director had not defalcated with the company's funds or if the Meteorological Office had not come under the Air Ministry, I myself prefer to reflect again on what Richardson might have done with his life if at the age of 26 he had been awarded the Fellowship at Cambridge. My feeling is that he would still have applied his talents in many different directions.

Some insight into Richardson's motivations is provided in the biographical notes which he wrote for the Royal Society. It was his strong curiosity which led him into research. His poor memory made it pleasanter for him to explore new subjects rather than to master any one of them. His solitary temperament enabled him to persist in unpopular researches. His switch from the physical sciences to the social sciences reflects his early resolve to eat the meal of life in the right order. His Quaker background and his family tradition of service to the community influenced his decision to apply the discipline he had acquired in mathematics and physics to the study of the causes of war. Richardson also gave a clue about the source of his originality when he wrote of the tendency of his mental machine almost to run of its own accord, a condition which he felt to be advantageous for creative thinking.

Richardson mentioned that in some ways his creative temperament was a nuisance—as when driving a car. Perhaps it was also responsible for his apparent difficulty in communicating his ideas to others. Nobody could claim that his book on weather prediction and his writings on the causes of war make easy reading; the narrative lacks continuity and the logical connection between one chapter and the next is not always evident. The trouble may have been that he could never allow himself sufficient time to organise his material into a coherent whole. Once he had solved one problem, his thoughts turned immediately to the next. Perhaps he was afraid to spend too much time in gathering together his previous results lest he might lose the inspiration to continue opening up new frontiers—and each new problem seemed to be more important than the last. As implied in his correspondence with Shaw (see p 141), he tended to lose interest in a subject once the main facts were known; he was happy to leave the details to others.

In spite of his determination to apply Occam's razor to his studies of war, I feel that Richardson sometimes let himself be carried away by the beauty of the mathematics. Some of his statistical elaborations may delight the mathematical reader, but it is difficult on occasion to justify such refined treatment of very crude original data. He once wrote to

Epilogue

Stephen: 'A beautiful theory, without any mercy, had me in thrall. It is now nearly finished. It is about the "latent roots of a matrix", a mathematical affair'[52]. He once remarked that he would like to have a matrix as an epitaph. But, of course, the real attraction to Richardson in applying mathematics to the social sciences was the way it can help in finding the truth. Verbal statements are often so vague that they can hardly be proved wrong, but once they have been expressed in the form of mathematical formulae, deductions can be made and tested by reliable techniques.

In the course of his life, Richardson was variously known as a chemist, physicist, mathematician, psychologist, meteorologist or just a plain scientist; I have even seen him referred to as an economist and as a biologist! He was nicknamed Doctor Dick, more frequently just the Doctor, and at one time Prophet—or Professor? He himself disliked labels, but if he been faced with the choice of only one I think it would have been 'searcher after truth'.

APPENDIX A

Published Works of L F Richardson

1905   (with C H Carpenter) Note on the structure of steel plates *Minutes Proc. Inst. Civ. Eng.* **155** 411

1906   *Biometrika* Index to Volumes I–V

1907a   On a freehand potential method *Rep. Br. Assoc. Adv. Sci.* (Leicester) 457

1907b   Pearson K and Pollard A F C, assisted by C W Wheen and L F Richardson An experimental study of the stresses in masonry dams *Drapers Co Research Memoirs (Technical Series)*

1908a   A freehand graphic way of determining stream lines and equipotentials *Proc. Phys. Soc.* **21** 88–124 (also *Phil. Mag.* S6 **15** 237–69)

1908b   The lines of flow in saturated soils *Proc. R. Dublin Soc.* **11** 295–316

1910   The approximate arithmetical solution by finite differences of physical problems involving differential equations, with an application to the stresses in a masonry dam *Phil. Trans.* A **210** 307–57

1911   The approximate solution of various boundary problems by surface integration combined with freehand graphs *Proc. Phys. Soc.* **23** 75–85

1913a   The measurement of mental "nature" and the study of adopted children *Eugen. Rev.* **4** 391–4

1913b   Review of magnetic disturbances at Eskdalemuir in the year 1913 *British meteorological and magnetic yearbook 1913* Pt IV 78–81

1914a   Annual review of magnetic disturbances at Eskdalemuir 1914 *British meteorological and magnetic yearbook 1914* Pt IV 75–8

1914b   Magnetic disturbances 1913 *Nature* **94** 450

1915a   Standardising a Milne-Shaw seismograph *Rep. Br. Assoc. Adv. Sci.* (Manchester) 74–9

1915b   The detection of distant thunderstorms by clicks in a telephone *Advisory Committee for Aeronautics Report T. 623*

1918   National voting in an international assembly *War and Peace* February 193–6

1919a	A form of Knudsen's vacuum manometer. *Proc. Phys. Soc.* **31** 270–6
1919b	Atmospheric stirring measured by precipitation *Proc. R. Soc.* A **96** 9–18
1919c	Measurement of water in clouds *Proc. R. Soc.* A **96** 19–31
1919d	The deflection of light during a solar eclipse *Nature* **104** 393–4
1919e	*The Mathematical Psychology of War* (Oxford: W Hunt)
1919f	Description of a line squall *Q. J. R. Meteorol. Soc.* **45** 70
1920a	Some measurements of atmospheric turbulence *Phil. Trans.* A 1–28
1920b	Convective cooling and theory of dimensions *Proc. Phys. Soc.* **32** 405–9
1920c	The supply of energy from and to atmospheric eddies *Proc. R. Soc.* A **97** 354–73
1920d	Experiments on numerical prediction *Annual Report of the Meteorological Committee for the year ending 21 March 1920* 79–80
1920e	Note on a theorem by Mr. G. I. Taylor on curves which oscillate regularly *Proc. Lond. Math. Soc.* **20** 211–12
1920f	Sun-flash balloons for continuous signalling *Q. J. R. Meteorol. Soc.* **46** 293–4
1921a	Lizard balloons for signalling the ratio of pressure to temperature *Prof. Notes Meteorol. Off.* No 18
1921b	Cracker balloons for signalling temperature *Prof. Notes Meteorol. Off.* No 18
1921c	Memorandum on the upper air works, by V Bjerknes, in collaboration with L F Richardson *International Commission for the Investigation of the Upper Air Proceedings of the seventh meeting Bergen July 1921* Appendix II
1921d	Tables of 0.288^{th} powers (T N Doerr) Introduction by L F Richardson *Q. J.R. Meteorol. Soc.* **47** 196
1922a	*Weather Prediction by Numerical Process* (London: Cambridge University Press)
1922b	*Computing forms for numerical calculations described in "Weather Prediction by Numerical Process"* (London: Cambridge University Press)
1922c	Review of *On the dynamics of the circular vortex* by J Bjerknes *Q. J. R. Meteorol. Soc.* **48** 375–6
1922d	(with A Wagner and R Dietzius) An observational test of geostrophic approximation in the stratosphere *Q. J. R. Meteorol. Soc.* **48** 328–41
1923a	Theory of the measurement of wind by shooting spheres upward *Phil. Trans.* A **223** 345–82
1923b	Wind above the night-calm at Benson at 7 a.m. *Q. J. R. Meteorol. Soc.* **49** 34
1923c	Demonstration of an electromagnetic inductor *Proc. Phys. Soc.* **35** 213
1924a	The aerodynamic resistance of spheres shot upward to measure the wind *Proc. Phys. Soc.* **36** 67–80
1924b	How to observe the wind by shooting spheres upward *Prof. Notes. Meteorol. Off.* No 34
1924c	Attempts to measure air-temperature by shooting spheres upward *Q. J. R. Meteorol. Soc.* **50** 19–22
1924d	Review of *The calculus of observations* by E T Whittaker and G Robinson *Q. J. R. Meteorol. Soc.* **50** 163
1924e	A holiday resort for geophysicists? *Q. J. R. Meteorol. Soc.* **50** 381
1924f	*Hints to Meteorological Observers* 8th edn by W Marriot (London: Stanford). Section on turbulence pp 29–30 by LFR

1925a How to solve differential equations approximately by arithmetic *Math. Gaz.* 415–521
1925b Turbulence and vertical temperature difference near trees *Phil. Mag.* **49** 81–90
1925c The brown corona and the diameters of particles *Q. J. R. Meteorol. Soc.* **51** 1–6
1925d Photometric observations on clouds and clear skies *Q. J. R. Meteorol. Soc.* **51** 7–23
1926a (with Denis Proctor) Diffusion over distances ranging from 3 km to 86 km *Mem. R. Meteorol. Soc.* No 1
1926b (with Russell E Munday) The single-layer problem in the atmosphere and the height-integral of pressure *Mem. R. Meteorol. Soc.* No 2
1926c (with Denis Proctor and Robert C Smith) The variance of upper wind and the accumulation of mass *Mem. R. Meteorol. Soc.* No 4
1926d Springs for vertical seismographs *Mon. Not. R. Astron. Soc. Geophys. Suppl.* **1** 403–11
1926e Atmospheric diffusion shown on a distance–neighbour graph *Proc. R. Soc.* A **110** 709–37
1926f Power in the League of Nations *World Outlook* 7 May 1926
1927a Report on photometers for a survey of the reflectivity of the earth's surface *International Union of Geodesy and Geophysics Meteorology Section Prague 1927*
1927b (with J Arthur Gaunt) The deferred approach to the limit *Phil. Trans.* A **226** 299–361
1927c (As secretary to a committee) Meteorological studies in connection with the Toronto meeting of the British Association *Q. J. R. Meteorol. Soc.* **53** 295–300
1928a Thresholds when sensation is regarded as quantitative *Br. J. Psychol.* **19** 158–66
1928b The amount of uniformly-diffused light that will go in series through two apertures forming opposite faces of a cube *Phil. Mag.* **6** 1019–23
1928c (with students of Westminster Training College) Contact potential in the Dolezalek electrometer connected idiostatically *Proc. Phys. Soc.* **40** 234–9
1928d (with V Stanyon) An absolute current balance having a simple approximate theory *Proc. Phys. Soc.* **41** 36–42
1929a A search for the law of atmospheric diffusion *Beitr. Phys. Frei. Atmos.* 24–9
1929b Quantitative mental estimates of light and colour *Br. J. Psychol.* **20** 27–37
1929c Imagery, conation and cerebral conductance *J. Gen. Psychol.* **2** 324–52
1930a (with R S Maxwell) The quantitative mental estimation of hue, brightness or saturation *Br. J. Psychol.* **20** 365–7
1930b (with J S Ross) Loudness and telephone current *J. Gen. Psychol.* **3** 288–306
1930c (with J Arthur Gaunt) Diffusion regarded as a compensation for smoothing *Mem. R. Meteorol. Soc.* No 30
1930d (with R S Maxwell) A new type of practical examination designed to be a fair competition *Proc. Phys. Soc.* **42** 108–25

1930e The analogy between mental images and sparks *Psychol. Rev.* **37** 214–27
1930f Reflectivity of woodland, fields and suburbs between London and St Albans *Q. J. R. Meteorol. Soc.* **61** 31–8
1932a The measureability of sensations of hue, brightness or saturation, in *Report of a joint discussion on vision held on June 3rd 1932 at Imperial College of Science by the Physical and Optical Societies* (London: Physical Society) pp 112–16
1932b Part of the discussion on a paper by W A Leyshon on: 'Periodic movements of the negative glow in discharge tubes' *Proc. Phys. Soc.* **44** 189
1932c Review of *Manual of Meteorology* by Sir Napier Shaw *Nature* **129** 220–1
1932d Determinism *Nature* **129** 316
1933a A quantitative view of pain *Br. J. Psychol.* **23** 401–3
1933b Third order aberrations of seismograph springs *Mon. Not. R. Astron. Soc. Geophys. Suppl.* **3** 125–31
1933c Photography of faint transient light-spots *Nature* **131** 401–3
1933d Time-marking a cathode-ray oscillograph *Proc. Phys. Soc.* **45** 135–41
1933e Part of the discussion on a paper by N R Campbell on: 'The measurement of visual sensations: a criticism of Dr. L.F. Richardson's proposed method of measuring sensations by "mental estimates"' *Proc. Phys. Soc.* **45** 585
1935a Mathematical psychology of war *Nature* **135** 830–1
1935b Mathematical psychology of war *Nature* **136** 1025
1935c Time-marking a cathode-ray oscilloscope by harmonics *Proc. Phys. Soc.* **47** 258–62
1937a Hints from physics and meteorology as to mental periodicities *Br. J. Psychol. (General Section)* **28** 212–5
1937b The behaviour of an osglim lamp Parts I and II *Proc. R. Soc.* A **162** 293–335
1937c The behaviour of an osglim lamp Part III *Proc. R. Soc.* A **163** 380–90
1938a A psychology class at an evening institute *Education for commerce* Jan 1938 103–24
1938b The arms race of 1909–13 *Nature* **142** 792–3
1939a Generalized Foreign Politics *Br. J. Psychol. Monograph Suppl.* No 23
1939b A visit to Danzig and neighbourhood *Paisley Daily Express* 21 and 23 August 1939
1939c A visit to Danzig *Northern Echo* 28 August 1939
1940   Quantitative estimates of sensory events (Final report of committee) Appendix I E by L.F. Richardson *Rep. Br. Assoc. Adv. Sci.* 339–40
1941a Frequency of occurrence of wars and other fatal quarrels *Nature* **148** 598
1941b Mathematical theory of population movement *Nature* **148** 784
1942   Comments on M.R.P. 43 *Meteorological Research Paper* 55
1944   Stability after the war *Nature* **154** 240
1945a The distribution of wars in time *J. R. Stat. Soc.* **107** 242–50
1945b The distribution of wars in time *Nature* **155** 610
1946a Chaos, international and inter-molecular *Nature* **158** 135
1946b The probability of encounters between gas molecules *Proc. R. Soc.* A **186** 422–31

1947a The number of nations on each side of a war *J. R. Stat. Soc.* **109** 130–56
1947b Social science in the gap between the Royal Society and the British Academy *Nature* **159** 269
1947c *Arms and Insecurity* (A book on microfilm, privately produced) revised 1949
1948a Variation of the frequency of fatal quarrels with magnitude *J. Am. Stat. Assoc.* **43** 523–46
1948b (with Henry Stommel) Note on eddy-diffusion in the sea *J. Meteorol.* **5** 238–40
1948c War moods *Psychometrika* **13** 147–74 and 197–232
1949a The persistence of national hatred and the changeability of its objects *Br. J. Med. Psychol.* **22** 166–8
1949b Meteorological publications by L.F. Richardson as they appear to him in October 1948 *Weather* **4** 6–9
1950a *Statistics of Deadly Quarrels* (A book on microfilm, privately produced)
1950b A method for computing principal axes *Br. J. Psychol. (Statistical Section)* **3** 16–20
1950c War and eugenics *Eugen. Rev.* **42** 25–36
1950d A purification method for computing the latent columns of numerical matrices and some integrals of differential equations. *Phil. Trans.* A **242** 439–91
1950e *Psychological Factors of Peace and War* ed. T H Pear (London: Hutchinson) Chapter X *Threats and Security* and Chapter XI *Statistics of Fatal Quarrels*
1950f Gilbert Hancock Richardson (Obituary) *The Friend* **108** 583
1951a Spectrum of atmospheric turbulence *Nature* **167** 318
1951b Could an arms-race end without fighting? *Nature* **168** 567
1952a Dr. S.J.F. Philpott's wave-theory *Br. J. Psychol. (General Section)* **43** 169–99
1952b Is it possible to prove any general statements about historical fact? *Br. J. Sociol.* **3** 77–84
1952c Contiguity and deadly quarrels: the local pacifying influence *J.R. Stat. Soc.* A **115** 219–31
1952d Transforms for the eddy-diffusion of clusters *Proc. R. Soc.* A **214** 1–20
1952e The criterion of atmospheric turbulence *Q. J. R. Meteorol. Soc.* **78** 422–5
1953a The submissiveness of nations *Br. J. Stat. Psychol.* **4** 77–90
1953b Three arms-races and two disarmaments *Sankhyā: Indian J. Stat.* **12** 205–28
1960a *Arms and Insecurity: a mathematical study of the causes of war* ed. N Rashevsky and E Trucco (Pittsburgh: Boxwood and Chicago: Quadrangle)
1960b *Statistics of Deadly Quarrels* ed. Quincy Wright and C C Lienau (Pittsburgh: Boxwood and Chicago: Quadrangle)
1960c *The World of Mathematics* ed. J R Newman (London: Allen and Unwin) Volume II; Chapter 6 *Mathematics of war and foreign politics* Chapter 7 *Statistics of Deadly Quarrels* (Reprinted, with some abridgment, from Richardson 1950e)
1961 The problem of contiguity: an appendix of statistics of deadly quarrels *General Systems Yearbook* **6** 139–87
1965 *Weather Prediction by Numerical Process* reprint of Richardson (1922a) with a new introduction by Sydney Chapman (New York: Dover)

Appendix B

Unpublished Papers by L F Richardson

1	1898	*A North Durham Colliery*	BS
2	1915	*Conditions of a Lasting Peace in Europe*	OMA
3	1923	*Barometric Recurrences* (with S P Peters and G A Wright)	MO
4	1936	*Quanta and Diffusion*	OMA
5	Late 1930s	*Gregariousness and its Opposite* (incomplete)	OMA
6	Late 1930s	*An Abstract Formulation of Fashions*	OMA
7	1939	*Some Biological Waves and the Equation* $d^2 \log y/dx^2 = 1 - y$ (by J Carson and L F Richardson)	OMA
8	1945	*An Impartial Selection of Fatal Quarrels*	RI
9	1951	*Hints concerning International Organizations*	CCL
10	1953	*Voting in an International Organization*	OMA

Note: The last column indicates where the original (or the best) copy is held. The abbreviations are explained in Appendix E.

APPENDIX C

Published works about L F Richardson

Obituaries
1. Anon 1954 *Bull. Br. Psychol. Soc.* **22**
2. Anon 1953 *Eugen. Rev.* **55** 210–11
3. Anon 1953 *The Friend* **111** 981
4. Anon 1953 *Dunoon Observer and Argyllshire Standard* 3 October 1953
5. Anon 1954 *Meteorol. Mag.* **83** 26–7
6. Anon 1953 *Paisley and Renfrewshire Gazette* 3 October 1953
7. Anon 1953 *The Times* 3 October 1953
8. Anon 1954 *Westminster Club Bulletin* February 1954 p 6
9. Ashford O M 1953 *The Friend* **111** 1066
10. Brunt D 1954 *Q. J. R. Meteorol. Soc.* **80** 127–8
11. Ellis T 1953 *The Friend* **111** 1014
12. Gold E 1953 *The Times* 19 October 1953
13. Gold E 1954 *Obituary Notices of Fellows of the Royal Society* **9** 217–35
14. Richardson D 1953 *The Friend* **111** 995
15. Richardson D 1954 *A Quaker Scientist: In thanksgiving for the life of Lewis Fry Richardson* (Published privately)
16. Sheppard J T 1953 *Annual Report of the Council of King's College, Cambridge*
17. Sheppard P A 1953 *Nature* **172** 1127
18. Todd J 1954 *Mathematical Tables and Other Aids to Computation* **8** 242–5

Biographical Articles and Essays
19. Ashford O M 1949 *Weather* **4** 9–10
20. Ashford O M 1973 *The Friend* **131** 19–21
21. Ashford O M 1981 *The Friend* **139** 1281–4
22. Ashford O M 1981 *Weather* **36** 323–5
23. Chapman S 1965 Introduction to the Dover Edition of *Weather Prediction by Numerical Process* v–x

24 Charnock H 1981 *Weather* **4** 316–22
25 Eames C and R 1973 *A Computer Perspective* (Cambridge MA: Harvard University Press) 80–1
26 Eckhardt W 1980 *Pioneers of Peace Research III. Lewis Fry Richardson: Apostle of Math* (St Louis: Peace Research Lab)
27 Gleiser M 1980 *Datamation* 181–4
28 Gold E 1961 *Dictionary of National Biography 1951–1960* (Oxford: Oxford University Press) 837–9
29 Olenick M 1978 *An Introduction to Mathematical Models in the Social and Life Sciences* (Reading MA: Addison-Wesley) 41–2
30 Platzman G W 1967 *Bull. Am. Meteorol. Soc.* **48** No 8 514–50
31 Platzman G W 1968 *Bull. Am. Meteorol. Soc.* **49** No 5 496–500
32 Rapoport A 1957 *J. Conflict Resolut.* **1** 249–99
33 Rapaport A 1968 *International Encyclopedia of the Social Sciences* **13** 513–7
34 Richardson S 1957 *J. Conflict Resolut.* **1** 300–4
35 Van Den Dongen 1980 *Foundations of Peace Research* (London: Housmans)

Reviews
Weather Prediction by Numerical Process (1922)
36 Anon 9 February 1922 *Methodist Recorder*
37 Anon 28 March 1922 *Manchester Guardian*
38 Chapman S 1922 *Q. J. R. Meteorol. Soc.* **48** 282–4
39 Exner F 1923 *Meteorol. Z.* **40** 189–91
40 Jeffreys H 1922 *Phil. Mag.* **44** 285–6
41 Johnson N K 1922 *Math. Gaz.* **11** 125–7
42 McAdie A 1923 *Geogr. Rev.* **13** 324–5
43 Shaw Napier 1922 *Nature* **110** 762–5
44 Whipple F J W 1922 *Meteorol. Mag.* **57** 61–3
45 Woolard E W 1922 *Mon. Weather Rev.* **50** 72–4

Weather Prediction by Numerical Process (Dover reprint 1965) (selection only)
46 Knighting E 1966 *Weather* **XXI** No 7 267
47 Phillips N A 1966 *Mathematics of Computation* **20** 633
48 Platzman G W 1967 see item 30 above
49 Van Mieghem J 1967 *WMO Bull.* **XVI** 60

Generalized Foreign Politics (1939)
50 Forder H G 1940 *Math. Gaz.* **29** 227–8
51 Piaggio H T H 1939 *Nature* **144** 692

Arms and Insecurity and *Statistics of Deadly Quarrels* (1960) (selection only)
52 Carter C F 1961 *The Friend* **119** 453–4
53 Kecskemeti P 1960 *Science* **132** 1931–2
54 Sutherland I 1962 *J.R. Stat. Soc.* A **125** 473–83
55 Sutton O G 1961 *Sci. Am.* **204** 193–200

APPENDIX D

Other References

Allen B W 1933 *William Garnett. A Memoir* (Cambridge: Heffers)
Allport B M 1938 *Personality* (London: Constable)
Anon 1914 *The International Industry of War* (London: Union of Democratic Control)
Anon 1956 Dorothy Richardson (Obituary) *The Friend* **114** 1188
Anon 1922 *Meteorol. Mag.* **57** 7
Baker P J 1914 *The Friend* **LIV** 626
Batchelor G K 1950 *Q. J. R. Meteorol. Soc.* **76** 133
—— 1976 Geoffrey Ingram Taylor 1886–1975 *Biographical Memoirs of Fellows of the Royal Society* **22** 565–633
Bateson Gregory 1935 *Naven* (Cambridge: Cambridge University Press)
Binyon L 1918 *For Dauntless France* (London: Hodder and Stoughton)
Bjerknes V 1904 *Meteorol. Z.* **21** 1–7
—— 1914 *Mon. Weather Rev.* **42** 11–14
—— 1920 *Q. J. R. Meteorol. Soc.* **46** 119–38
Bjerknes V, Bjerknes J, Solberg H, and Bergeron T 1934 *Hydrodynamique Physique* (Paris: Les Presses Universitaires de France)
Blackwell M J 1958 *Meteorol. Mag.* **87** 129–32
Boulding K 1962 *Conflict and Defense* (New York: Harper)
—— 1975 *Collected Papers* vol 5 (Colorado: Associated University Press)
Boyce A O 1889 *The Richardsons of Cleveland* (London: Samuel Harris)
Brayshaw A N 1941 *Memoir and Selected Writings* (Birmingham: Society of Friends)
Brown S K 1973 *Bootham School York 1823–1973* (York: published privately)
Brunt D 1934 *Physical and Dynamical Meteorology* (Cambridge: Cambridge University Press)
Calder R 1950 Olaf Stapledon: Father of Super Man (Obituary) *Manchester Guardian* 7 September 1950
Campbell N R 1933 *Proc. Phys. Soc.* **45** 565–90
Cardwell D S L 1974 *Artisan to Graduate* (Manchester: Manchester University Press)

Catchpool T C 1919 *On Two Fronts* (London: Headley)
Cave C J P 1920 *Q. J. R. Meteorol. Soc.* **46** 155
—— 1925 *Q. J. R. Meteorol. Soc.* **51** 67–76
Christopherson D G 1972 Richard Vynne Southwell 1888–1970 *Biographical Notices of Fellows of the Royal Society* **18** 549–64
Chromov S P 1940 *Einführing in die synoptische Wetteranlage* (German transl. G Swoboda) (Vienna: Springer)
Clark J E 1925 *The Friend* **83** 822
Corless R 1933 *Q. J. R. Meteorol. Soc.* **59** 89–93
Crichton J 1950 *Meteorol. Mag.* **79** 337–40
Curle Adam 1975 *The Scope and Dilemma of Peace Studies* Inaugural Lecture (Bradford: University of Bradford)
Dean Roy 1981 *Disarmament* **4** 33–43
De Morgan A 1872 *A Budget of Paradoxes* (London: Longman and Green)
—— 1915 *A Budget of Paradoxes* new edition ed. D E Smith (London: Open Court)
Dines W H 1920 *Q. J. R. Meteorol. Soc.* **46** 163–71
—— 1931 *Collected Scientific Papers* (London: Royal Meteorological Society)
Douglas C K M 1931 *Q. J. R. Meteorol. Soc.* **57** 245–53
Ellis E L 1972 *The University College of Wales, Aberystwyth 1872–1972* (Cardiff: University of Wales Press)
Eversley D 1973 *A Question of Numbers?* (London: Runnymede Trust)
Exner F 1917 *Dynamische Meteorologie* (Leipzig: B G Teuber) (Second edition 1925 (Vienna: Springer))
Eysenck H J 1979 *The Structure and Measurement of Intelligence* (Berlin: Springer)
Fiedler A F 1983 *Olaf Stapledon: A Man Divided* (Oxford: Oxford University Press)
Frith E A and Smeal R W (undated) *The Little Grey Book* (published privately)
Gandin L S 1965 *Machines forecast the Weather* (Leningrad: Gidrometeoizdat)
Garnett R 1937 *Kenneth Gordon Garnett (1892–1917)* (Rochester: Stanhope Press)
Garnett W and Campbell L 1882 *The Life of James Clerk Maxwell* (London: Macmillan)
Gaunt T 1950 *John Arthur Gaunt. A Memoir by his Father* (published privately)
Gold E 1945 William Napier Shaw 1854–1945 *Obituary Notices of Fellows of the Royal Society* **5** 203–30
Graham J W 1922 *Conscience and Conscription* (London: Allen and Unwin)
Graves R 1929 *Goodbye to all that* (London: Jonathan Cape)
Gregg R B 1936 *The Power of Non-Violence* (London: Routledge)
Hughes W R 1956 *Indomitable Friend. The Life of Corder Catchpool* (London: Allen and Unwin)
Humphreys W J 1929 *Physics of the Air* 2nd edn (Philadelphia: Lippincott)
James W 1901 *Principles of Psychology* (London: Macmillan)
Johnson N K 1950 *Weather* **5** 87–90
Kolmogoroff A N 1941 *C.R. Acad. Sci. URSS* **30** 301
Koschmieder H 1933 *Dynamische Meteorologie* (Leipzig: Akademische)
Lentz T 1955 *Towards a Science of Peace* (New York: Bookman Associates)
Mandelbrot B B 1982 *The Fractal Geometry of Nature* (San Francisco: Freeman)
Marchuk G I 1982 *Methods of Numerical Mathematics* (2nd edn) (New York: Springer)
Mason B J 1966 *Weather* **21** 382–93

Maxwell R S 1930 *Br. J. Psychol.* **20** 181–9
McDonald J E 1957 *Weather* **12** 107–14
McDougall W 1901 *Brain* **24** 577
Munday R 1932 *Br. J. Educ. Psychol.* **2** 1–7
Munsinger H 1975 *Psychol. Bull.* **82** 625–59
Obukhov A M 1941 *Bull. Acad. Sci. URSS, Geogr. Geofiz. Ser.* No 4–5
O'Connor G 1982 *Ralph Richardson: An Actor's Life* (London: Hodder and Stoughton)
Pearson K 1892 *A Grammar of Science* (London: Black)
—— 1904 *Report on Drapers' Company Grant*
Philpott S J F 1932 Fluctuations in Human Output *Br. J. Psychol. Monogr. Suppl.* No 17
Platzman G 1979 *Bull. Am. Meteorol. Soc.* **60** 302–12
Poulton E C (in the press) *Bias in Judgment*
Pritchard F C 1951 *The Story of Westminster College* (London: Epworth Press)
Pyatt E 1983 *The National Physical Laboratory* (Bristol: Adam Hilger)
Rashevsky N 1947 *Mathematical Theory of Human Relations* (Bloomington: Principia)
Reuck A de 1966 *Conflict in Society* (London: Churchill)
Richardson E 1909 *Doors* (London: Headley)
—— 1914 *The Dim Divine* (London: Fifield)
Richardson G 1850 *Annals of the Cleveland Richardsons* (published privately)
—— 1864 *Journal of George Richardson* (London: Bennett)
Richardson H and Simmons A T 1905 *An Introduction to Practical Geography* (London: Macmillan)
Richardson J 1757 *Journal: An Account of the Life of that Ancient Servant of Christ* (London: Luke Hende)
Richardson L 1977 *Lawrence Richardson: Selected Correspondence 1902–1903)* (Cape Town: Van Riebeck Society)
Russell B 1914 *War—The Offspring of Fear* (London: Union of Democratic Control)
Russell L 1938 *My Monkey Friends* (Bristol: Arrowsmith)
Sagan C 1980 *Cosmos* (New York: Random House)
Schrodt P A 1981 *Preserving Arms Distributions in a Multi-Polar World* (Denver: University of Denver)
Shaw N 1920 *Meteorol. Mag.* **55** 166
—— 1923a *Forecasting Weather* (3rd ed) (London: Constable)
—— 1923b *The Air and its Ways* (Cambridge: Cambridge University Press)
—— 1931 *Manual of Meteorology* vol IV (Cambridge: Cambridge University Press)
Smagorinsky J 1972 *The General Circulation of the Atmosphere* in *Meteorological Challenges* (Ottawa: Information Canada)
Smith T 1930 *Br. J. Psychol.* **20** 362–4
Soedjatmoko 1982 *Development Forum* **10** 1 and 10
Sorokin P A 1937 *Social and Cultural Dynamics* (New York: American Book Co)
Southwell R V *et al* 1937–8 *Proc. R. Soc.* A **159** 315; **161** 155; **164** 447; **168** 317
Southwell R V 1946 and 1956 *Relaxation Methods in Theoretical Physics* vol I and II (Oxford: Clarendon)
Stommel H 1949 *J. Mar. Res.* **8** 199–225

Sutcliffe R C 1982 *Q. J. R. Meteorol. Soc.* **108** 996–7
Sutton O G 1942 Note on the Use of the Richardson Number in Meteorological Problems *Meteorological Research Paper 43*
Swift J 1733 *On Poetry a Rapsody* (Dublin)
Tatham M and Miller J E 1920 *The Friends Ambulance Unit 1914–1919* (London: Swarthmore)
Taylor G I 1921 *Proc. Lond. Math. Soc.* **20** 196–212
—— 1959 *The Present Position in the Theory of Turbulent Diffusion* in *Advances in Geophysics* vol 6 (New York: Academic)
Thompson P D 1983 *Bull. Am. Meteorol. Soc.* **64** 755-69
Thomson J J 1936 *Recollections and Reflections* (London: Bell)
West Ranyard 1945 *Psychology and World Order* (London: Pelican)
Whittaker J M 1968 A. C. Aitken *Biographical Memoirs of Fellows of the Royal Society* **14**
Wilkinson D 1980 *Deadly Quarrels: Lewis F. Richardson and the Statistical Study of War* (Berkeley: University of California Press)
Wilkinson L P 1980 *Kingsmen of a Century 1873–1972* (Cambridge: King's College)
Wright Q 1942 *A Study of War* (Chicago: University of Chicago Press)
Yarrow C H M 1978 *Quaker Experiences in International Conciliation* (New Haven: Yale University Press)

APPENDIX E

Archives Containing Material Relating to L F Richardson

BL   *Bodleian Library, Oxford.* The archives of the British Association for the Advancement of Science have several files on activities in which Richardson was involved.

BR   *Bertrand Russell Archives, McMaster University, Hamilton.* Letter from Bertrand Russell re Richardson.

BS   *Bootham School, York.* Essay and Nature Diary by Richardson.

CU   *Cambridge University Library.* Napier Shaw archives. Files of Cambridge University Press.

FH   *Friends House Library, London.* The Balkwill Genealogy (typescript) deals with the families of Richardson's father and mother.

LU   *University College Library, London.* Letters from Richardson to Karl Pearson.

MO   *Meteorological Office Library, Bracknell.* Napier Shaw correspondence. Minutes of Meteorological Committee. Eskdalemuir files.

PC   *Paisley College of Technology, Paisley.* About 30 files mainly of lecture notes, but with some correspondence and miscellaneous notes on experiments. About 60 textbooks *ex libris* Richardson, with marginal notes in his handwriting. Details are given in a catalogue *Lewis Fry Richardson (1881–1953): Manuscripts and documents held in the Library of Paisley College of Technology* (1983).

PR   *Public Records Office, Kew.* Files of National Peat Industries and Sunbeam Lamp Co. Air Ministry files on transfer of Meteorological Office to Air Ministry.

RI   *Richardson Institute, University of Lancaster.* Six loose-leaf bound

volumes, mainly working material for books on peace research, but with some correspondence and other notes. Two files on languages and religion, and on mapping of populations.
RS *Royal Society Library, London.* Personal Record of L F Richardson, with Supplement. Correspondence with Sir Joseph Larmor. Minutes of Gassiot Committee. Minutes of British National Committee on Geodesy and Geophysics.
TC *Trinity College, Cambridge.* Correspondence with Sir Geoffrey Taylor. Memoir of J Arthur Gaunt.
UC *University of Chicago, Joseph Regenstein Library.* Correspondence with Quincy Wright.
UO *University of Oslo. University Library.* Bjerknes Papers.
WC *Westminster College Library, Oxford.* Minutes of Governing Body of Westminster and Southlands Colleges. Staff Register.
WI *Contemporary Medical Archives Centre, Wellcome Institute for the History of Medicine, London.* Files of the Eugenics Society.

Personal collections
CCL Carl C Lienau, New York
ET   Elaine Traylen, Norfolk
RW   Ranyard West, Kirkcudbright
OMA Oliver M Ashford, Oxford
GP   George Platzman, Chicago
GR   Gordon Rogers, Devon
RF   Richardson family, Northumberland

APPENDIX F

Notes, Including Sources of Quotations

Full references for items LFR with date are given in Appendix A, and for items with other authors and date in Appendix D. Abbreviations, additional to those in Appendix E:

AI	*Arms and Insecurity*
GFP	*Generalized Foreign Politics*
MPW	*The Mathematical Psychology of War*
PFPW	*Psychological Factors of Peace and War*
RP	*Personal Record of L F Richardson (RS)*
SDQ	*Statistics of Deadly Quarrels*
WPNP	*Weather Prediction by Numerical Process*

Chapter 1
1. Quoted from Richardson's Personal Record (RP)
2. This paragraph is mainly based on the Balkwill Genealogy (FH)
3. In the possession of the Richardson family (RF)
4. For more information on Sir Ralph Richardson's family, see O'Connor (1982)
5. From Edith Richardson (1914)

Chapter 2
1. From Catherine Richardson's diary (RF)
2. RP Supplement 1–2
3. RP
4. See Brayshaw (1941)
5. RP
6. See Brown (1973)
7. This diary is held by Bootham School (BS). Unfortunately, his later diary, for which he was awarded a prize, is missing.

Notes, Including Sources of Quotations 281

8 RP Supplement 4
9 RP
10 See Wilkinson L P (1980)
11 RP
12 RP Supplement 5
13 Based on information in Appendix C Item 14

Chapter 3
1 RP Supplement 3
2 LFR (1903)
3 Ellis (1972)
4 Original held by OMA
5 Appendix C Item 10
6 Letter from Mansel Davies to OMA dated 9 October 1982 (OMA)
7 Most of the information about National Peat Industries is from the company files (PR)
8 Appendix C Item 13 220
9 LFR (1908b)
10 LFR (1908a)
11 LFR (1910)
12 Letter from LFR to A C Aitken dated 31 December 1952 (OMA)
13 LFR (1907b)
14 RP Supplement 6
15 Pearson (1892)
16 LFR (1906)
17 Letter from LFR to K Pearson dated 6 April 1907 (LU: Pearson Papers)
18 Appendix C Item 16
19 For more information on William Garnett see Allen (1933)
20 Pearson (1904)
21 For more information on Kenneth Garnett see Garnett (1937)
22 Letter from Olaf Richardson to OMA dated 26 September 1981 (OMA)
23 Most of the information about the Sunbeam Lamp Company is from the company files (PR)
24 David Richardson's diary (RF)
25 Letter from J Edmundson to OMA dated 16 December 1982 (OMA)
26 Letter from J J Thomson to LFR dated 20 May 1912 (OMA)
27 See Cardwell (1950)
28 The quotations in the following paragraphs are from this notebook which is now in my possession (OMA)
29 The quotations in this paragraph are from Richardson (1913a)
30 From *Eugenics Review* 1913 67
31 Munsinger (1975) 623
32 Appendix C Item 24
33 Appendix C Item 34
34 Letter from Keith Bovey to OMA dated 3 November 1981 (OMA)
35 Letter from Maurice Wilson to LFR dated September 1914 (OMA)
36 These experiments are mentioned in Thomson (1936)

Chapter 4

1. RP Supplement 4
2. Letter from Shaw to LFR dated 8 November 1913 (MO)
3. RP Supplement 3
4. Appendix C Item 5 29
5. Correspondence between LFR and the Office of Works, Edinburgh, 3 November to 17 December 1915 (file at Eskdalemuir)
6. Correspondence between LFR and Shaw, 2 July 1915 to 11 March 1916 (file at Eskdalemuir)
7. Letter LFR to J Larmor dated 9 August 1921 (RS)
8. WPNP ix
9. Correspondence between LFR and Shaw, 25 August to 2 October 1915 in File C.E. 90 (MO)
10. Letter from Shaw to Munitions Invention Dept. dated 15 November 1916 (MO)
11. RP Supplement 6
12. Graham (1922) 155–6
13. Tatham and Miller (1920) viii
14. Hughes (1956) 33
15. Tatham and Miller (1920) 95
16. Minutes of the Meteorological Committee 28 October 1914 (MO)
17. RP Supplement 2
18. Letter to OMA from G Hutchinson dated 5 October 1982 (OMA)
19. Based on Tatham and Miller (1920)
20. From Robert Charnley's Memoirs (unpublished) (FH)
21. Binyon (1918) 105. Reprinted by permission of Hodder and Stoughton Ltd
22. Frith and Smeal (undated)
23. Letter to OMA from A Molyneux dated 7 October 1981 (OMA)
24. Letter to OMA from R Charnley dated 7 October 1981 (OMA)
25. Appendix C Item 11
26. Letter to OMA from H Morrell dated 15 September 1982 (OMA)
27. Fiedler (1983)
28. Calder (1950)
29. In file *LFR's reflections on the war* (OMA)
30. Appendix B Item 2 3–4
31. Ibid. 8
32. Ibid. 9–10
33. MPW 3
34. Ibid. 2
35. Ibid. 3
36. GFP v
37. MPW 7–8
38. Ibid. 33
39. MPW 1
40. Ibid. 50
41. LFR (1953b) 207
42. Russell (1914) 11–2
43. Appendix D Anon (1914) 1

44 MPW 15–16
45 RP Supplement 3
46 Ibid. 6
47 Letter to George Allen and Unwin from B Russell dated 29 December 1918 (BR)
48 RP Supplement 7
49 GFP v
50 WPNP 211
51 Ibid. ix
52 File *Ideas for new meteorological instruments* (OMA)
53 Appendix C Item 34 301
54 A copy of the citation is in Robert Charnley's Memoirs (FH)

Chapter 5
1 Dines (1931) 2
2 Ibid. 6
3 Letter from LFR to Shaw dated 28 October 1918 (MO)
4 Letter from Shaw to LFR dated 5 November 1918 (MO)
5 Minutes of Meteorological Committee 26 March 1919 (MO)
6 LFR (1921b)
7 LFR (1921a)
8 Ibid.
9 Dines (1920) 378
10 WPNP ix
11 Ibid. vii
12 Bjerknes (1904) (Engl. transl. by Y Mintz (MO))
13 Bjerknes (1914)
14 WPNP viii
15 Letter from Shaw to V Bjerknes dated 6 September 1913 (UO)
16 WPNP 4
17 Ibid. 19
18 Ibid. 65
19 Ibid. 214
20 Ibid. 66
21 Swift (1733) lines 339 *et seq.*
22 De Morgan (1872) 377
23 WPNP 49
24 Ibid. 50
25 Ibid. 106
26 Ibid. 156
27 Ibid. vii
28 Ibid. 212
29 Ibid. 211
30 Ibid. 217
31 Mason (1966) 388
32 WPNP vii–viii
33 Ibid. 219
34 Appendix C Item 23 viii

35 WPNP 220
36 Ibid. 223
37 Ibid. 229
38 Ibid. vii
39 Letter from LFR to Cambridge University Press dated end of July 1921 (CU)
40 Letter from LFR to Cambridge University Press dated 16 August 1921 (CU)
41 Appendix C Item 30 545
42 Appendix C Item 36
43 Appendix C Item 37
44 Appendix C Item 45 73–4
45 Appendix C Item 42
46 Appendix C Item 43 763
47 Ibid. 764
48 Appendix C Item 41 126
49 Appendix C Item 40 286
50 Appendix C Item 38 282
51 Ibid. 284
52 Appendix C Item 44 62
53 Appendix C Item 39 English translation from Appendix C Item 30 520
54 Letter from Exner to LFR dated 8 May 1923 (OMA)
55 Sutcliffe (1982) 997
56 Douglas (1931)
57 Chromov (1940) 12 (Engl. transl. by OMA)
58 Shaw (1923a) 545
59 Bjerknes *et al* (1934) vol 3 848 (Engl. transl. by OMA)
60 Letter from LFR to V Bjerknes dated 10 March 1934 (OMA)
61 Letter from Dorothy Richardson to Peggy Garnett dated 25 May 1932 (OMA)
62 Bjerknes (1920)
63 Letter from V Bjerknes to his wife dated 7 November 1919 (UO)
64 Letter from V Bjerknes to R K Wenger dated 26 November 1919 (UO)
65 Letter from LFR to V Bjerknes dated 4 January 1920
66 The information about the transfer of the Meteorological Office to the Air Ministry has been obtained from File A 18550/19 (PR), the Annual Reports of the Meteorological Committee (MO) and Corless (1933)
67 PR Supplement 1
68 Johnson (1950)
69 Letter from Sir Charles Normand to OMA dated 11 August 1982 (OMA)
70 Cave (1925)
71 Letter from Sir Harold Jeffreys to J Burton dated 6 July 1983 (in Burton's possession)
72 Annual Report of the Meteorological Committee for year ending 31 March 1924, page 10
73 Dines (1931)

Chapter 6
1 Letter from V Bjerknes to Napier Shaw dated 13 May 1920 (UO)
2 Letter from V Bjerknes to W Ekman dated 9 June 1920 (UO)

3   Shaw (1923b) 74–90. There is also a more detailed unpublished account of the proceedings of the meeting by Shaw (MO)
4   Proceedings of the meeting (unpublished), page 33 (MO)
5   Minutes of Meeting of Governing Body of Westminster and Southlands Colleges 18 May 1920 (WC)
6   Pritchard (1951) 144
7   *Westminsterian* Winter 1929 10
8   Pritchard (1951) 145
9   PR Supplement 6
10 Dines (1931) 2
11 Letter to OMA from University of London Library dated 26 January 1982 (OMA)
12 *Westminsterian* Easter 1926
13 Appendix C Item 8
14 LFR (1923c)
15 LFR (1928d)
16 LFR (1930d)
17 Letter to OMA from D Browne dated 31 March 1982 (OMA)
18 Letter to OMA from Ralph Richardson dated 4 June 1982 (OMA)
19 Southwell *et al* (1937–38)
20 Christopherson (1972)
21 Southwell (1946) vol I 36
22 Letter to OMA from H Lamb dated 1 January 1982 (OMA)
23 Suggested by F Horner in letter to OMA dated 20 October 1982 (OMA)
24 Letter from LFR to J Jeans dated 22 November 1922 (PC)
25 File of LFR correspondence and notes (PC)
26 RP Supplement 4
27 LFR (1919c)
28 LFR (1920a)
29 Appendix D Anon (1922)
30 According to LFR (1951b), the phrase *Richardson number* was first used by W Paeschke
31 Letter reproduced in LFR (1942)
32 LFR (1952d)
33 Note in WPNP revision file (GP—photocopy)
34 Taylor (1959) 105–6
35 Letter from LFR to Taylor dated 17 March 1919 (St Johns College, Cambridge)
36 Letter from LFR to Taylor dated 19 November 1933 (St Johns College, Cambridge)
37 Gaunt (1950) 4
38 Appendix C Item 13 221
39 LFR (1929a)
40 Gaunt (1950) 24
41 Appendix C Item 15
42 Appendix C Item 31 499–500
43 LFR (1923a, 1924a, b and c)
44 LFR (1924b)

45 *Meteorol. Mag.* 1923 **58** 251
46 RP Supplement 4 and Allport (1938)
47 LFR (1929c) 345
48 Ibid. 329
49 LFR (1928a) 165
50 Summary based on LFR (1929b)
51 Smith (1930)
52 LFR (1930a) 367
53 Letter to OMA from H Jeffreys dated 9 January 1982 (OMA)
54 Letter from LFR to Shaw dated 4 June 1927 (OMA)
55 Letter from Shaw to Cave dated 6 August 1928 (CU)
56 Letter to LFR from V Bjerknes dated 4 April 1921 (UO)
57 Letter to LFR to V Bjerknes dated 9 April 1921 (UO)
58 Ibid. 23 July 1921 (UO)
59 File of miscellaneous notes by LFR (OMA)
60 LFR (1921c)
61 Appendix C Item 31 497–8
62 International Commission for the Exploration of the Upper Air: Report of the Meeting in London 1925
63 LFR (1949b)
64 LFR (1919c) 25
65 LFR (1927a) and (1930f)
66 Minutes of the board of Governors of Paisley Technical College (PC)
67 Letter to LFR from Shaw dated 14 July 1929 (OMA—photocopy)
68 Letter from LFR to Shaw dated 20 July 1929 (OMA—photocopy)

Chapter 7
1 LFR (1930e) 215
2 Ibid. 226
3 LFR (1937b) 315
4 Appendix B Item 7 2B
5 Letter from LFR to J Carson dated 21 April 1940 (OMA)
6 LFR (1937a) 214
7 Letter from LFR to OMA dated 20 December 1940 (OMA)
8 Letter to OMA from C A Oakley dated 13 December 1982 (OMA)
9 LFR (1938a) 124
10 LFR (1933a) 402
11 Ibid. 402
12 LFR (1932a) 113
13 LFR (1933e)
14 *Rep. Br. Assoc. Adv. Sci.* 1940 334
15 Minutes of the Board of Governors of Paisley Technical College (PC)
16 RP Supplement 3
17 Letter from LFR to OMA dated 29 March 1934 (OMA)
18 GFP 48
19 From lecture notes on *Optics* (PC)
20 Appendix D Anon (1956)

21 Letter from Dorothy Richardson to Peggy Garnett dated 25 August 1931 (OMA)
22 Letter from Dorothy Richardson to Peggy Garnett dated 1 February 1932 (OMA)
23 Letter from Keith Bovey to OMA dated 3 November 1981 (OMA)
24 Letter from Dorothy Richardson to Peggy Garnett dated 1 February 1932 (OMA)
25 Letter from LFR to V Bjerknes dated June 1933 (UO)
26 Appendix C Item 30 547
27 Letter from V Bjerknes to C W Oséen dated 29 September 1936 (BU)
28 Letter from LFR to Cambridge University Press dated 26 September 1936 (CU)
29 Letter from V Bjerknes to C W Oséen dated 17 January 1938 (BU)
30 Letter to OMA from Alison Douglas dated 13 July 1981 (OMA)
31 RP Supplement 2
32 Ibid. 1
33 GFP v
34 LFR (1935b)
35 LFR (1953a) 78
36 Letter from LFR to Psychometric Corporation dated 25 August 1938 (CU)
37 Letter to LFR from Cambridge University Press dated 4 October 1938 (CU)
38 Letter from LFR to Cambridge University Press dated 5 October 1938 (CU)
39 Ibid. 7 October 1938 (CU)
40 GFP 1
41 Ibid. 2
42 Ibid. 3
43 Boulding (1962) 30
44 GFP 18
45 Ibid. 22
46 Ibid. 83–5
47 Appendix C Item 51
48 Appendix C Item 50 228
49 LFR (1939c)
50 LFR (1939b)
51 File entitled *Separate Fatal Quarrels VOL I* (RI)
52 Letter from LFR to Board of Governors dated 16 February 1940 (PC)
53 Minutes of the Board of Governors of Paisley Technical College meeting on 28 February 1940 (PC)
54 RP Supplement 3
55 File of miscellaneous notes (PC)
56 Appendix C Item 34 303
57 Letter to OMA from Olaf Richardson dated 30 January 1983 (OMA)
58 Appendix C Item 15

Chapter 8
1 SDQ 6
2 Ibid. 8
3 PFPW 240

4   SDQ 7
5   PFPW 246
6   LFR (1941a)
7   LFR (1948a) 543
8   Ibid. 544
9   Appendix C Item 34 303
10 These quotations are from Appendix B Item 5
11 Letter to LFR from Gilbert Richardson dated 28 June 1941 (OMA)
12 These quotations are from Appendix B Item 6
13 SDQ 20
14 Appendix C Item 54 474
15 SDQ 184
16 Ibid. 128
17 PFPW 242
18 SDQ 131
19 Ibid. 137
20 PFPW 248
21 SDQ 172
22 Ibid. 212
23 Ibid. 231–2
24 Ibid. 237
25 Ibid. 242
26 Ibid. 206
27 Ibid. 210
28 AI 5
29 Ibid. 62
30 Ibid. 74
31 Ibid. 75
32 Ibid. 77–8
33 Ibid. 109–10
34 Ibid. 135
35 Ibid. 208
36 Ibid. 148
37 Ibid. 145–6
38 Ibid. 230–1
39 Ibid. 231

Chapter 9
1   File of miscellaneous notes (PC)
2   File *Physics Instruments and Techniques I* (PC)
3   Letter to OMA from David Eversley dated 8 July 1981 (OMA)
4   LFR (1947a) 138
5   Letter to OMA from Ranyard West dated 15 September 1983 (OMA)
6   Letter to OMA from Ranyard West dated 11 September 1981 (OMA)
7   Letter to Ranyard West from LFR dated 13 November 1943 (RW)
8   Letter to Ranyard West from LFR dated 6 April 1944 (RW)
9   West (1945) 29
10 Letter to Ranyard West from LFR dated 23 July 1946 (RW)

Notes, Including Sources of Quotations 289

11 Letter to Ranyard West from LFR dated 7 April 1952 (RW)
12 Letter from LFR to C P Blacker dated 23 April 1946 (WI)
13 Letter from D W Harding to C P Blacker dated 6 May 1946 (WI)
14 Letter from Godfrey Thomson to C P Blacker dated 3 May 1946 (WI)
15 Letter from LFR to C P Blacker dated 17 June 1946 (WI)
16 Letter from LFR to C P Blacker dated 23 August 1946 (WI)
17 Letter from LFR to C P Blacker dated 4 October 1946 (WI)
18 Letter from LFR to Ranyard West dated 6 April 1948 (RW)
19 Letter from LFR to Ranyard West dated 27 January 1947 (RW)
20 Letter from LFR to Ranyard West dated 16 April 1947 (RW)
21 Whittaker (1968) 7
22 Letter to LFR from A C Aitken dated 27 December 1952 (OMA)
23 Letter from LFR to Quincy Wright dated 21 December 1943 (UC)
24 Letter from LFR to Quincy Wright dated 6 September 1944 (UC)
25 Letter to LFR from Quincy Wright dated 29 November 1947 (UC)
26 Letter from LFR to Quincy Wright dated 31 May 1951 (UC)
27 Letter to LFR from Quincy Wright dated 29 November 1947 (UC)
28 Letter from LFR to Quincy Wright dated 8 December 1947 (UC)
29 Letter from LFR to Quincy Wright dated 1 December 1952 (UC)
30 Letter from LFR to Quincy Wright dated 17 November 1948 (UC)
31 Letter from LFR to OMA dated 9 November 1945 (OMA)
32 SDQ 258
33 Ibid. 285
34 Ibid. 263
35 Letter to LFR from Quincy Wright dated 29 November 1947 (UC)
36 Letter from LFR to Quincy Wright dated 8 December 1947 (UC)
37 LFR (1948c) 147
38 Ibid. 155
39 Ibid. 156
40 Ibid. 219
41 Ibid. 220
42 Ibid. 230
43 LFR (1949a) 167
44 Ibid. 168
45 Letter to OMA from Ralph Pickford dated 17 November 1982 (OMA)
46 Mandelbrot (1982) 403
47 Stommel (1949)
48 Batchelor (1950)
49 Letter from LFR to OMA dated 11 April 1952 (OMA)
50 Letter from LFR to OMA dated 3 September 1948 (OMA)
51 Letter from LFR to OMA dated 2 October 1948 (OMA)
52 Letter from LFR to Sutcliffe dated 2 February 1936 (GP—photocopy)
53 Letter from LFR to Pasquill dated 30 January 1949 (in Pasquill's possession)
54 Letter to OMA from Pasquill dated 22 October 1983 (OMA)
55 Letter from Blacker to LFR dated 22 June 1949 (WI)
56 LFR (1950c) 27
57 Ibid. 35
58 Letter from LFR to Blacker dated 2 January 1950 (WI)

59 Letter to LFR from Blacker dated 4 January 1950 (WI)
60 Letter from O L Zangwill to OMA dated 4 January 1982 (OMA)
61 Letter from LFR to Stephen Richardson dated 28 November 1948 (OMA)
62 Letters from LFR to Rogers dated September 1950 to January 1951 (GR)
63 AI xiv
64 SDQ xxxvi
65 Appendix B Item 8
66 PFPW 15
67 LFR (1960c) 1238–9
68 GFP 22
69 LFR (1953a) 89
70 SDQ 189
71 Ibid. 1
72 Quotations from Appendix B Item 10
73 Letter from C C Lienau to A W Richardson dated 19 December 1943 (CCL)
74 Letter from LFR to C C Lienau dated 12 February 1951 (CCL)
75 Letter from LFR to C C Lienau dated 16 March 1951 (CCL). I have not found the article which Richardson published under his pseudonym 'Dafry'.
76 Letter from LFR to CCL C C Lienau dated 10 April 1951 (OMA)
77 Letter to OMA from C C Lienau dated 27 July 1983 (OMA)
78 Quotations from Appendix B Item 9
79 Letter from LFR to C C Lienau dated 22 June 1951 (CCL)
80 Philpott (1932)
81 Letter to OMA from D W Harding dated 4 October 1982 (OMA)
82 LFR (1952a) 174
83 Ibid. 177
84 Ibid. 194
85 Ibid. 199
86 Letter to OMA from D W Harding dated 16 October 1981 (OMA)
87 Letter from LFR to Sir John Sheppard dated 31 August 1953 (CU)
88 LFR's House and Garden Diary (OMA)
89 Letter from LFR to Quincy Wright dated 27 September 1953 (UC)
90 Letter from Dorothy Richardson to Sir John Sheppard dated 2 October 1953 (CU)
91 Quoted in letter from Margaret Peacock to OMA dated 1 May 1982 (OMA)

Chapter 10
1 Appendix C Item 7
2 Ibid. Item 12
3 Ibid. Item 13
4 Ibid. Item 6
5 Ibid. Item 4
6 Ibid. Item 16
7 Ibid. Item 3
8 Ibid. Item 9
9 Ibid. Item 14
10 Ibid. Item 11
11 Ibid. Item 1

Notes, Including Sources of Quotations

12 Ibid. Item 8
13 Ibid. Item 5
14 Ibid. Item 10
15 Ibid. Item 17
16 Ibid. Item 34 303
17 Recorded on tape (ET)
18 Letter to OMA from Dorothy Richardson dated 5 December 1955 (OMA)
19 Letter to OMA from Dorothy Richardson undated (early 1954) (OMA)
20 WPNP (1965 reprint) xi
21 Platzman (1979) 304
22 Appendix C Item 31 497–8
23 Letter from J Charney to Stephen Richardson dated 16 October 1953 (OMA)
24 Smagorinsky (1972) 5
25 McDonald (1957) 107
26 Appendix C Item 31 496
27 Appendix C Item 30 514
28 Speech by Sir John Mason 6 October 1972 (MO)
29 Speech by Edward Heath 6 October 1972 (MO)
30 Boulding (1975) viii
31 Appendix C Item 32 249
32 Ibid. 297–8
33 Letter from Dorothy Richardson to Quincy Wright dated 4 November 1953 (UC)
34 Letter from F Mosteller to OMA dated 16 April 1984
34a Appendix C Item 55 200
35 Appendix C Item 54 482
36 Appendix C Item 52 454
37 Letter from P Deighan to OMA dated 13 July 1983 (OMA)
38 Letter from Cedric Smith to OMA dated 13 May 1982 (OMA)
39 de Reuck (1966)
40 Curle (1975)
41 Soedjatmoko (1982)
42 Appendix C Item 35 36
43 Appendix C Item 26
44 Eversley (1973) 5
45 Ibid. 27
46 AI xv
47 Dean (1981)
48 Letter to OMA from Adam Curle dated 4 September 1981 (OMA)
49 Yarrow (1978) xiii
50 AI ix
51 Letter from Dorothy Richardson to L L Thurstone dated 9 March 1955 (UC)
52 Letter from LFR to Stephen Richardson dated 28 November 1948 (OMA)

APPENDIX G

Acknowledgments

I am most grateful to the following members of the Garnett and Richardson families for lending me letters and other manuscripts, for reminiscing about Lewis and Dorothy Richardson, for answering my questions and for checking parts of my typescript: John Garnett, Pauline Hunt, Julian Hunt, Arthur Hunt Cooke, Peggy Jay, Mary Philipson, Dan Piggott, Christopher Poulton, Constance Richardson, Herbert Richardson, Merl Richardson, the late Olaf Richardson, the late Sir Ralph Richardson, Helen Thomson, Elaine and Michael Traylen,

I owe my thanks to the following for providing information and for reading parts of my draft: George Batchelor, Donal Browne, James Carson, Pat Deighan, Carl Lienau, Cedric Smith, Ranyard West.

Original letters were shared with me by Elaine Austin, Frank Pasquill, Gordon Rogers and Sylvia Ross.

Personal recollections of the Richardsons were provided by: Robert Charnley, George Hutchinson, Arthur Molyneux and Herbert Morrell (who served with Richardson in the FAU); Tom Selwood (Benson); V A Carpenter, W J Fewings and N C Patten (Westminster College); J Carson, J Denholm and J McLean (Paisley College of Technology); R Alcorn, Jessie Burns, Dr and Mrs D Walker (Kilmun); Gwen Ashford, Richard Ashford, Robert Barbour, Keith Bovey, Alison Douglas, David Eversley, Hubert Lamb, Kenneth Laurie, Kathleen Lewis, Gordon Rogers, Sylvia Ross (family friends); Sir Harold Jeffreys (Cambridge).

Many other people helped in my research: John Aldridge (Benson); E J C Sackett (Westminster College); J Huppert (Cambridge); T M Howie and S James (Paisley College of Technology); A C J Greenwood and Bill Young (Eskdalemuir); Margaret Peacock (Glasgow Friends Meeting); Clifford Smith (Bootham); J Burton, R Hide, F Horner, K Langlo, R W Lewis, Sir John Mason, W J Maunder, the late Sir Charles Normand, R S Ratcliffe, J S Sawyer, R S Scorer, H Stommel, R C Sutcliffe, P Thompson and D Wheeler (Meteorology); H J Eysenck, D W Harding, T McLardy, C A Oakley, R Pickford and O L Zangwill (Psychology); R Burnside, P G Drazin, M Gibson, B B Mandelbrot and

Acknowledgments

R P Pearce (Mathematics); K Boulding, A Curle, C R Dean, J Mongar, F Mosteller, W Polachek, A de Reuck, J D Singer, P Smoker and I Sutherland (Peace research).

I am most grateful for the help provided by the following libraries, organisations, publishers, societies and universities: Bodleian Library, Boxwood Press, British Association for the Advancement of Science, British Library, British Psychological Society, Cambridge University Library, Cambridge University Press, Cavendish Laboratory, Dover Publications, Dunoon Observer and Argyllshire Standard, Eskdalemuir Observatory, Eugenics Society, Friends House Library, Gateshead Public Library, General Electric Company, The Guardian, The Institute of Physics, Jesus College (Oxford), King's College (Cambridge), Lancaster University (Richardson Institute), London University, McMaster University (Mills Memorial Library), Meteorological Office, Methodist Recorder, National Physical Laboratory, Oslo University, Paisley College of Technology, Public Record Office, Punch, Royal Commission on Historical Manuscripts, Royal Historical Society, Royal Meteorological Society, Royal Society Library, Royal Statistical Society, St John's College (Cambridge), Science Museum, Science Reference Library, Scout Association, Trinity College (Cambridge), Tyne and Wear County Council, University of Chicago (Joseph Regenstein Library), University of Manchester, University of Newcastle, University College of Wales (Aberystwyth), Wellcome Contemporary Archives, Westminster College (Oxford), Westfield College (London).

Copyright material is reproduced by permission, as appropriate, of several of the above bodies. Quotations and photographs from Meteorological Office publications and records are reproduced by permission of the Controller of Her Majesty's Stationery Office. Quotations from *Nature* are reproduced by permission of Macmillan Journals Limited. Permission for reproduction of other copyright material was granted by American Meteorological Society, Constable and Co Ltd, Epworth Press, Hutchinson Publishing Group Ltd, Mathematical Association, Royal Society, University College (London), School of Peace Studies (University of Bradford), Yale University Library. Figure 9.6 is from the BBC Hulton Picture Library.

Finally, I should like to thank Neville Hankins and Ian Kingston of Adam Hilger Ltd for their friendly efficiency in converting my manuscript into a book.

Index†

Aberystwyth, University College, 20–1
Accuracy, fictitious, 151, 173, 189
Agar W E, 18, 25
Aggression, 181
Aitken A C, 205–6
Albedo, 70, 138
Anecdotes about LFR, warning on accuracy of, 13
Anti-gregariousness, 176
Appeasement, 187
Applications, warlike, of LFR's work, 41, 51, 127–9, 251, 261
Arms and Insecurity (LFR), 204, 252–3
Arms race, 64, 161, 164, 167, 185–6, 222–3
Arms race model, LFR's, 61, 161, 163, 165, 221–3, 257
Atmospheric diffusion, *see* Diffusion, atmospheric
Aurora borealis, 14, 145–6

Balance of power, 161
Balloons, experiments with, 76, 78, 105, 123, 134
Barbour, Robert, 151
Batchelor G K, 214
Bateson, Gregory, 204, 221, 251
Belligerency, 181

Benson Observatory, 72, 73, 76–7, 104–5, 107
Bergen, 76, 80, 108, 128–9, 134–6
Bergen School (of meteorology), 80, 104, 108, 158, 258, 263
Bibliographies of LFR's works, 19, 25, 251
Binyon, Laurence, 55
Biometrika, 24, 144
Bjerknes J, 109, 119, 130
Bjerknes, Vilhelm:
  associations with LFR, 80, 101–4, 108, 157
  career, 74–5, 79, 80
  honours, 156, 158, 214, 244
  research activities, 74, 78–9, 80–1, 90, 101, 108, 134
  works, 79, 80, 101
Blacker C P, 201–4, 216
Bootham School, York, 4, 7, 11–16, 34–5, 191
Boulding, Kenneth, 164, 249, 257, 261
Boyce, Anne Ogden, 2–4
Boy Scouts, 31, 41, 102, 117
Bradford, University of, 256
Brain, human, 143–5
Brayshaw, A Neave, 11–12
Bright, John, 3

†NB LFR = L F Richardson

Index

British Acadamy, 219–20
British Association, 48, 122, 133, 149, 162, 238
British Journal of Sociology, 220
British Journal of Statistical Psychology, 223
British Psychological Society, 147, 162, 200–1, 217, 221, 238
Browne, Donal and family, 112, 115–16
Brunt, Sir David, 20, 100, 238–9
Bulletin of the American Meteorological Society, 245–6, 250
Bulletin of the British Psychological Society, 238
Burton, John, 255

Calwagen E, 135–6
Cambridge University, 17, 35, 233, *see also* Cavendish Laboratory, Cambridge *and* King's College, Cambridge
Cambridge University Press, 94, 157–8, 162
Campaign for Nuclear Disarmament, 254
Campbell N R, 148
Carpenter, Sir Harold, 19, 25
Carson, James, 144–5
Carter, Charles F, 254–5
Carter, Roger, 67
Casualties, number of war, 172, 185, 207
Catalogue of fatal quarrels, 173
Catchpool, T Corder, 52, 158, 167
Cave C J P, 106, 119–20, 133–4
Cavendish Laboratory, Cambridge, 17, 26, 34
Centenary Lectures, Richardson, 259–60
Centre for the Analysis of Conflict, 255
Champagne, Battle of, 54 5, 66, 71
Chapman, Sidney, 98, 245
Charney, Jule, 242–4
Charnley, Robert, 54, 57
Charnock, Henry, 39, 259
Chree, Charles, 20, 42, 49, 252

Christopherson D G, 117
Chromov S P, 100
CIBA Foundation, 255–6
Civil defence, 169–70
Clark, James Edmund, 11–12, 34, 102, 119
Class struggle, 184
Colour, quantitative estimation of, 131–3, 148
Commission for Scientific Aeronautics, 79
Committee for Investigation of the Upper Atmosphere, 133–4
Computer Perspective, A (exhibition), 246
Computers, electronic, 242
Computing forms, 88, 157
Conflict in Society (symposium), 255–6
Conflict Research Society, 255–6
Contiguity, 200, 225–6
Cosmic rays, 145
Cosmos (Sagan), 257
Cotterell, Arthur N, 58, 70
Crafoord Prize, 213–14
Curle, Adam, 256, 261

Dafry (LFR pseudonym), 228
Dalton Hall, Manchester, 42
Dean, Roy, 261
Defence, 181, 187, 223, 230
Deighan, Pat, 254
De Morgan, Augustus, 85–6
Determinism, 163–4, 210, 250
Deterrence, 224
Differential equations, approximate solution of, 22–3, 25, 45, 49, 77, 81, 83, 116, 144, 146
Diffusion, atmospheric, 123–5, 127
Diffusion by continuous movements (Taylor), 125
Dines, William Henry, 72–3, 76–7, 107, 111
Disarmament, unilateral, 161
Discipline in schools and colleges, 34–7
Douglas C K M, 100, 109
Dover Publications, 245

Dream, Richardson's, 94, 248–9
Dreams, analysis of, 176–8
Dunoon Observer and Argyllshire Standard, 237
Durham College of Science, 10, 16, 26, 35

Echo-sounding, 39
Eckhardt, William, 258
Eddy diffusion, 212–14 (*see also* Diffusion, atmospheric)
Eddy motion, 83–4, 124, 164
Ellis, Tom, 57, 238
Eskdalemuir Observatory, 20, 42–5, 81
Esperanto, 8, 57, 181
Eugenics, 24, 38, 59, 102–3
Eugenics Review, 38
Eugenics Society, 38, 201, 216
Eversley, David, 198, 258–9
Exchange rates, variability in international, 185–6
Exner, Felix M, 99–100
Eysenck H J, 39

Faith, definition of, 218
Fantasy, Richardson's, 45, 77, 81, 91, 93, 96
Fashions, formulation of, 177
Fatal quarrels, statistical analysis of, 171–4, 178–9, 220
Finite differences, see Differential equations
For Dauntless France (Binyon), 55
Forder H G, 167, 186
Forecasting Weather (Shaw), 101–2
Forecasting, weather
  empirical methods of, 77–8
  Richardson's method of, 81–9, 99–102
  use of computer for, 242–3
Fractals, 260
Friends Ambulance Unit (FAU), 52–8, 60, 71
Friends International Centre, 254
Friends, Society of, see Quakers
Friend, The, 52, 238, 254
Frontiers
  importance of, 181
  length of, 200, 226
Front, meteorological, 80, 109
Fry family (LFR's mother), 4

Galton, Sir Francis, 24, 38
Gandhi, 223, 230, 250
Gandin L S, 93
Garnett family, 26–7, 152
Garnett, Hilda (sister-in-law), 28, 240–1
Garnett, John (nephew), 160, 196–7
Garnett, Kenneth (brother-in-law), 27, 103
Garnett, Margaret (Poulton), 27
Garnett, J C Maxwell (brother-in-law) 27, 34, 37, 45, 114, 123, 202, 217–8
Garnett, Peggy (Mrs Peggy Jay), 155–6
Garnett, Rebecca (mother-in-law), 10, 44–5, 118
Garnett, Stuart (brother-in-law), 18, 26–7
Garnett, William (father-in-law), 10, 16, 26, 30, 44–5
Gassiot Committee of the Royal Society, 42
Gaunt, J Arthur, 125–7
Generalized Foreign Politics (LFR), 162–7, 189–190, 209, 222
General Systems Yearbook, 226, 257, 260
Geographical Review, 96
Giblett M A, 133
Glazebrook R T (Sir Richard), 20, 23, 26, 42, 45
God, definition of, 183
Godske C L, 156–8
Gold, Ernest, 22–3, 25, 127, 135, 237, 252
Golders Green, Richardson's home in, 114
Government, common, 224–5, 258
Government, world, 225–6, 230
Graves, Robert, 62
Gregariousness, 175, 211
Guild J, 148
Guns, location of, 49–51

Harding D W, 202, 232–3
Harker J A, 20, 31, 44–5
Hatred, irreducible minimum of, 211
Heath, Edward, 247–9
Helmholtz, Hermann von, 18, 45, 109, 180, 244
Heredity, 25, 38
Hergesell, Hugo, 79, 88
Hewson, E Wendell, 245
History, quantitative, 219
Hitler A, 161, 167–8, 190, 245
Houstoun R A, 148–9
Howard, Luke, 3
Hunt Cooke, Arthur, 118–19
Hutchinson, George, 57, 60, 122
Hutchinson, G Evelyn, 251

Ido (international language), 8, 68, 93, 95
Imperialism, 258
Institute of Physics, The, 119
Instruments, meteorological, *see* Observations, meteorological
Intelligence tests, 159
Intermarriage (mixed marriages), 59, 216, 230
International Commission for the Investigation of the Upper Air, 134–7
International Industry of War, The, 64
International Meteorological Organization 79, 134, *see also* World Meteorological Organization
International organizations, voting in, 67, 226–7
International Union of Geodesy and Geophysics (IUGG) 137–8, 156
Introduction to Mathematical Models in the Social and Life Sciences, An (Olenick), 258
Isle of Wight, (Seaview, holiday home of the Garnett family), 26, 28, 66, 102, 116, 118, 126, 153

James, William, 132–3
Jeans, Sir James, 120, 175, 183
Jeffreys, Sir Harold, 97, 107, 134

Johnson N K (Sir Nelson), 97, 106, 127
Journal of the American Statistical Association, 174
Journal of Conflict Resolution, 249
Journal of Meteorology, 212
Journal of the Royal Statistical Society 253

Kilmun, (Hillside House, Richardson home), 192–4, 205, 211–13, 233
King's College, Cambridge, 16, 18, 234–5, 237
Kites, meteorological, 72, 121
Knight, Francis H, 15,37
Kolmogoroff A N, 124
Koschmieder H H, 100

Lamb, Hubert H, 117
Lancaster, University of, 255–6
Language (*see also* Esperanto *and* Ido)
  diversity of, 182
  international, 8, 59, 138, 181
League of Nations, 225
League of Nations Union, 27, 123, 155–6, 161, 221
Lentz, Theodore (Ted), 249, 254
Lewis Fry Richardson: Apostle of Math (Eckhardt), 258
Lewis Fry Richardson and the Statistical Study of War (Wilkinson), 257
Lienau, Carl C, 227–31, 252
Line squall, 70
Little Grey Book, The, 56, 58
Loch Long, (Knap Cottage, Richardson summer home), 154, 192
London University, 24–5, 106, 111–13, 139, 255
Loudness, measurement of, 131–2
Lucius T Littauer Foundation, 252
Ludlam, Ernest B, 52–3, 139, 211

McAdie, Alexander, 96
McDonald, James E, 245
McDougall W, 60, 130, 143, 159, 176
MacLean, Angus, 142

Manchester College of Technology
  (School of Technology), 27, 34–7
Manchester Guardian, 95
Mandelbrot, Benoit B, 259–60, 263
Mansel Davies, 21
Marchuk G I, 263
Martin C E and K, 199, 226
Marx, Karl, 184, 250
Mason B J (Sir John), 89–90, 247–8
Mathematical Gazette, 97, 167
Mathematical Psychology of War, The
  (LFR), 60–5, 160
Mathematics (*see also* Differential
  equations)
  applications to war studies, 60–1,
    188, 221, 250
  beauty of, 264–5
  checking of LFR's, 205
Matrix, 179–80, 205, 265
Maxwell, James Clerk, 24, 26, 44, 101
Maxwell R S, 132–3, 140, 148–9
Mental imagery, estimating intensity
  of, 130–1
Meteorology as an exact science, 80
Meteorological Committee (Council),
  45–6, 75, 105–6
Meteorological Magazine, 98, 238
Meteorological Office, 20, 73, 94,
  105–7, 146, 246–9
Meteorological Society *see* Royal
  Meteorological Society
Meteorological stations, distribution
  of, 82, 90–1, 135, 242
Meteorological textbooks, treatment
  of LFR's method, 100–2
Meteorologische Zeitschrift, 98
Meteorological Section of IUGG,
  137–8
Methodist Recorder, 94–5
Methods of Numerical Mathematics
  (Marchuk), 263
Michigan, University of, peace
  research at, 249–50
Microfilm, publication of LFR war
  books on, 204, 249, 251
Models, idealised, justification for
  the use of, 177, 188–9
Molyneux, Arthur, 56, 70

Monthly Weather Review, 95–6
Morrell, Herbert, 57
Mosteller, Frederick, 252
Munday, Russell E, 102
Munsinger, Harry, 39

Nansen, Fridtjof, 25
National Peat Industries, 21–3, 25
National Physical Laboratory (NPL),
  19, 26, 133
Natural History Journal, (Quaker
  schools magazine), 12–14
Nature
  letters from LFR to, 161, 174–5,
    208, 219, 223
  obituary, 239
  review of LFR books, 96–7, 166
Nature–nurture controversy, 38–9
Neon lamps, analogy with human
  brain, 143–5
Newcastle upon Tyne (The Gables,
  Richardson family home), 4, 9,
  31
Newman, James R, 221–2
Nicholson, Michael, 255–6, 260
Nobel Prize, 52, 158
Noel Baker, Lord (J Philip Baker),
  52–3
Northern Echo, 167
Nuclear weapons, 224
von Neumann, John, 242–4

Oakley C A, 147
Observations, meteorological
  upper-air, 70, 72–3, 78–9, 89, 91,
  128–30, 136–7 *see also*
  Meteorological stations,
  distribution of
Obukhov A M, 124
Occam's razor, 61, 87, 131, 264
Ogburn W F, 251
Olenick, Michael, 258
Oséen C W, 156, 158

Pacification *see* Peace, maintenance of
Pacifism (conversation with Roger
  Carter), 67

Pain, assessment of, 147–8
Paisley and Renfrewshire Gazette, 237
Paisley Daily Express, 167
Paisley Technical College (now College of Technology)
 archives, 240–1
 LFR's activities at, 16, 139, 147, 149, 168, 199, 254, 259
 Principal's residence at Castlehead, 142, 144, 168
Pantopiric examinations, 114
Parcel, air (concept introduced by LFR), 86
Parsnips, diffusion experiment with, 211–13, 259
Pasquill F, 214–15
Patents, LFR's, 32, 39–41
Patten N C, 113, 140
Peacefare (project at Rand Corporation), 252
Peace, maintenance of
 and armaments, 184–5, 190, 222
 and common government, 225, 258
 and common language, 59, 181
 and intermarriage, 59, 216, 230
 and international law, 59
 and international trade, 164, 187–8, 258
 and justice, 166
 LFR's precepts for, 230
 and world governments, 225, 230
Peace movements, attitude to research, 254
Peace research
 different approaches, 171, 173, 209
 growth of, 168, 249–57
 summary of results of LFR's work, 216, 229
 value of, 258, 261–2
Peace research institutions, 245, 256
Pear T H, 221
Pearson, Karl, 19, 23–5, 27, 144, 252
Pease, Edward, 3
Petterssen, Sverre, 136
Phenology, 12, 119
Phenomena biological, wavelike behaviour of, 144
Philosophical Magazine, 97–8

Philpott S J F, 113, 231–3
Photometer, 138
Physical and Dynamical Meteorology (Brunt), 100
Physical Society of London, 113, 119–20. 148
Physics of the Air (Humphreys), 100
Piaggio H T H, 166, 186
Platzman, George, 245–6, 250
Pocock, Michael, 154, 160
Poison gas, meteorological applications to distribution of, 97, 127, 169–70
Poisson distribution, 180, 207, 229
Potted History Book, 179–80
Poulton, E Christopher, 263
Poulton, Margaret (Mrs Maxwell Garnett), 27
Prediction, weather *see* Forecasting, weather *and Weather Prediction by Numerical Process*
Prejudice, 201, 259
Preserving Arms Distribution in a Multi-Polar World (Schrodt), 257
Proceedings of the Royal Society, 123, 208, 214
Proctor, Henry Richardson, 11
Propaganda, anti-war, 65, 203
Psychological Factors of Peace and War (ed. Pear), 221
Psychology, 37, 51, 112, 147, 159
Psychometrika, 162, 209
Public opinion, assessment of, 210
Publisher, difficulty in finding a, 65, 162, 200–4, 207, 227
Pyatt E, 19

Quaker Experiences in International Conciliation (Yarrow), 261
Quaker Meetings, 119, 142, 158, 198, 233
Quaker schools, 3, 8, 12, 19, *see also* Bootham School, York
Quaker traditions, 1 2, 6, 8 9, 51
Quakers (Society of Friends), 7, 16, 25, 28, 42, 51, 117, 133, 236, 256, 259, 264

Quarterly Journal of the Royal Meteorological Society, 98, 119, 214, 238
Question of Numbers, A, (Everstey) 158–9

Radiation, research on atmospheric and solar, 70, 73, 76, 86–7, 138
Rand Corporation, 252
Rapoport, Anatol, 250–1, 261
Rashevsky, Nicholas, 249, 252
Relaxation Methods in Theoretical Physics (Southwell), 117
Religion, 69, 183–4, 218
de Reuck, Tony, 255
Reviews of LFR books, 94–9, 252–4
Richardson, Arthur (brother), 6
Richardson, Catherine (mother—Catherine Fry) 4, 9–10
Richardson, Catherine (sister), 8, 10
Richardson Centenary Lectures, 259–60
Richardson, David (father), 1, 4–6, 9, 18, 51
Richardson, Dorothy (wife—Dorothy Garnett)
  activities after death of LFR, 238, 240–1, 251
  collaboration in work of LFR, 76, 132, 191, 239–40
  correspondence, 127, 155–6, 240, 251–2, 262
  engagement and marriage, 26, 29–30
  miscarriages, 31, 66, 102, 114
  social work, 66, 114–5, 154–6, 170
  as Quaker, 28, 118
Richardson, Edith (sister), 6–7, 175
Richardson, Elaine (daughter—Mrs Michael Traylen), 103, 160–1, 191, 195, 233, 239, 242
Richardson, George, 3–4
Richardson, Gilbert (brother), 7, 11, 31, 68, 177
Richardson, Hannah, 2–3
Richardson, Hugh (brother), 6, 11, 13, 31, 37
Richardson, Institute, 255–6
Richardson, John, 2

Richardson, Lawrence (brother), 7, 11, 31, 202
Richardson, Lewis Fry
  academic qualifications and honours, 15, 17, 111–13, 120, 141, 244–5
  acknowledgement of other people's ideas, 114, 116, 125, 144, 207
  adoption of children, 103
  appointments, 19, 20, 21, 26, 30, 34, 42, 110, 139
  author's recollections of, 143, 150
  biographical notes for Royal Society, 24, 169, 264
  biographical sketches, 258, 260, *see also* Gold, Ernest *and* Richardson, Stephen
  birth, 9
  break with meteorology, 130, 133, 214
  celebration of centenary, 259–60
  childhood, 9–11
  coining of new words, 93, 114, 121, 187
  collaboration, scientific, 19, 24–5, 76–7, 126, 136, 144–5, 148–9, 212–3
  controversy with Southwell. 116–7, 244
  correspondence, quotations from, 50, 73–4, 102, 105, 122, 133–4, 141, 144, 157–8, 201–9, 214–15, 216–19, 227–8, 234–5, 243–4
  curiosity and doubt, 10, 16, 106, 149, 159, 218
  death, 234–5
  diaries, 13–14, 196–7
  education, 10–17
  experiments, 70–1, 92, 120, 122, 129–31, 136, 143–5, 195–6
  fallibility, 92, 120, 122
  handwriting, 16, 128
  humour, 13, 20, 32–3, 62, 68, 104, 115, 180, 216–17, 228
  illness, 10, 18, 147, 235–6
  interest
    in eugenics, 25, 38, 59, 103

in international languages, 59, 68, 95, 138, 181
in meteorology, 13, 25, 45, 77, 120
in music, 160–1
in phenology, 119
in psychology, 37, 51, 112, 130, 147, 159
international activities, 135–8
lecture on relations between physical and other sciences, 151–2
marriage, 29–30
memory, poor, 15, 64, 111, 217, 234, 264
nicknames, 55–7, 140, 143, 265
obituary notices, 237–9
opinions and attitudes, 19, 48, 59, 64–5, 69, 124, 154, 187, 218–20, 224
parody, use of, 62, 68, 85, 149, 214
practical work, 195–6
professorship, proposal and refusal, 170, 245
Quakers, influence of and work for 119, 128, 264
recognition, posthumous, 242, 244–9, 250–1, 255–9, 263
social life, 31, 102, 114, 193–5
solitude, love of, 43, 114, 175, 264
teacher, success as, 37–8, 113, 150–1
temper, loss of, 21, 88–9, 92, 152, 199
travels abroad, 26, 31, 43, 76, 106, 109, 133–6, 141, 167–8
works *see* Index of Published and Unpublished Works
Richardson number, 122, 237
Richardson, Olaf (son), 103, 145, 152, 191, 240–1
Richardson prize (proposal), 245
Richardson, Sir Ralph (nephew), 6, 116, 132
Richardson, Stephen (son)
life of, 103, 152, 191, 241
memoir about LFR, 40, 71, 239, 251
recollections of LFR, 115, 170, 195

role in publication of LFR's war books, 249, 252
Richardson Wing (Meteorological Office), 246–9
Richardsons of Cleveland, The, (Boyce) 2–4
Rogers, Gordon, 218–19
Ross J S, 131–2, 140
Rowling, Cecil, 118
Royal Historical Society, 220
Royal Meteorological Society, 12, 103, 106, 112, 119–20, 129, 138–9, 245–6, 259
Royal Society, 23–4, 94, 112, 169, 208, 219–20, 262
Royal Statistical Society, 204, 208
Runnymede Lecture (David Eversley), 258–9
Russell, Bertrand, 63–5
Russell, Lilian, 194
Rutherford, Ernest (Lord Rutherford), 17, 37, 43

Sagan, Carl, 257
St Michael's School Limpsfield, 66, 118, 125
Schott G A, 20–1
Schrodt, Philip, 257–9
Schuster, Sir Arthur, 46–7
Scientific American, 253
Searle G F C, 16
Section Sanitaire Anglaise (SSA 13) *see* Friends Ambulance Unit
Seismographs, 46, 48, 50–1
Sensation, quantitative estimation of, 131–2, 147–9, 263
Shaw W N (Sir Napier)
attitude to administration, 73–4, 108
and V Bjerknes, 81, 109, 158
career, 26, 44–5, 106
correspondence, quotations from, 50–1, 73–5, 81, 140–1
support for LFR, 24, 44, 47, 49, 103, 133, 138
works, 96–7, 100–1, 111
Sheppard, Sir John, 237
Sheppard, Peter A, 215, 239

Shrimpton A G, 49
Simplified spelling 8, 177–8
Simpson C G (Sir George), 94, 106–7
Smith, Cedric, 255
Smith T, 133
Smoker, Paul, 255–7
Society of Friends *see* Quakers
Socratic dialogue, examples of, 162–6, 183, 188–90, 225
Soedjamoto, 256
Solutions, approximate of differential equations *see* Differential equations
Sorokin, Pitrim A, 173
Southwell R V (Sir Richard), 116–7, 244
Stanford University (seminar), 249
Stapledon, Olaf, 57–9, 122
Statics and Kinematics (Bjerknes), 80
Stations, meteorological *see* Meteorological stations
Statistics of Deadly Quarrels (LFR), 204, 252–3
Stephenson, William, 204
Stocksfield (Wheelbirks, Richardson estate), 5, 115,
Stommel, Henry, 211–13, 231
Study of War, A, (Wright), 172, 206
Submissiveness, 222–3
Sunbeam Lamp Company, 30, 33–4
Sutcliffe R C, 100, 214, 254
Sutherland, Ian, 253
Sutton O G (Sir Graham), 122, 127–8, 253
Swan, Ellen, 114–5, 156, 191
Swift, Jonathan, 85

Taylor G I (Sir Geoffrey), 121–2, 124–5, 214
Telephone, noises on, 48–9, 120
Thompson, Godfrey, 202–3
Thomson, Sir J J, 16, 22, 34–5, 41, 45
Thought, analogy to flow of water, 62
Threats, effectiveness of, 222
Thunderstorms, location of, 48–9
Thurstone L L, 252, 262
The Times, 106, 237

Touch, quantitative estimation of sensation of, 132
Towards a Science of Peace (Lentz), 249, 254
Toynbee A, 220, 250
Trade, international, 164, 187–8, 230, 258
Traylen, Michael (son-in-law), 233
Turbulence, atmospheric, 83–4, 86, 89, 121–3, 127, 169–70
Turbulivity, 93, 121

Unesco, 227
United Nations, 223, 226–7
United Nations Association, 221
University of Chicago Press, 251–2

Voting in international organizations 67, 226–7

War
 causes of, 59, 161, 179, 182, 184–5, 200, 258
 classification and definition of, 172, 206, 220
War–the offspring of fear (Russell), 63–4
Warfinpersal, 187
War moods, 178, 209–11, 221
Wars
 distribution of in time, 179, 204
 frequency of, 181
 list of past,
  LFR, 179
  Wright, 172–3, 180
  Sorokin, 172
Watson, Robert Spence, 10, 30
Watson Watt R A (Sir Robert), 119
Weather, 214
Weather maps, Bjerknes', 79, 88
Weather Prediction by Numerical Process (LFR)
 computing forms, 88, 157
 Dover reprint, 245
 first draft, 49
 index, 133
 loss of working copy, 66
 notes for second edition, 157–8

preface, 77–8, 80–1, 86, 90, 93–4
publication of, 49, 94, 121, 162, 245
reviews, 94–9, 245
summary of contents, 81–94
Westminster Club Bulletin, 238
Westminster Training College (now Westminster College), 110–11, 125, 138–40, 240
West, Ranyard, 200–5, 217, 219
Whipple F J W, 98
Whirls (ditty by LFR), 85, 245
Wilkinson, David, 257
Wilson, Maurice, 40
Wind, measurement of, *see* Observations, meteorological
Wind stress, LFR's apparatus for measuring, 215

Woods Hole, 144–5, 211, 213
Woolard, Edgar W, 95
Workman H B, 110–11, 139
Works by LFR, published and unpublished *see* below
World of Mathematics, The (ed. Newman), 221
World Meteorological Organization (WMO), 91
World, ideal, 229
Wright, Quincy, 172, 206–9, 234–5, 249, 251–2, 262

Yarrow, Mike, 261

Zamenhof L L, 181–2

Index of Published Works†
An absolute current balance having a simple accurate theory (1928), 113
The analogy between mental images and sparks (1930), 143
The approximate solution by finite differences of physical problems involving differential equations (1910), 23
Arms and Insecurity (1960), 252–3
Atmospheric diffusion shown in a distance–neighbour graph (1926), 123
Atmospheric stirring measured by precipitation (1919), 121
Contiguity and deadly quarrels: the local pacifying influence (1952), 224
Could an arms-race end without fighting? (1951) 222–3
The deferred approach to the limit (1927), 126
Description of a line squall (1919), 70
The detection of distant thunderstorms by clicks in a telephone (1915), 49
Diffusion as a compensation for smoothing (1930), 126
Diffusion over distances ranging from 3 km to 86 km (1926), 123
A holiday resort for geophysicists? (1924), 144
Imagery, conation and cerebral conductance (1929), 131
Loudness and telephone current (1930), 131
The Mathematical Psychology of War (1919), 60–5
The measurability of sensations of hue, brightness or saturation (1932), 148
Measurement of water in clouds (1919), 121
Memorandum on the upper air works: Appendix (1921), 136
Meteorological publications by L.F. Richardson as they appear to him in October 1948 (1949), 137

†A complete bibliography is given in Appendix A. This index lists only those works mentioned specifically in the text.

National voting in an international assembly (1918), 67
The number of nations on each side of a war (1947), 207
Dr. S.J.F. Philpott's wave theory (1952), 230
Is it possible to prove any general statements about historical fact? (1952), 220
The problem of contiguity (1961), 226
A psychology class at an evening institute (1938), 147
Quantitative estimates of sensory events: Appendix (1940), 149
A quantitative view of pain (1933), 147–8
The single-layer problem in the atmosphere and the height-integral of pressure (1926), 102
Standardising a Milne–Shaw seismograph (1915), 48
Statistics of Deadly Quarrels (1950), 204, (1960), 252–3
The supply of energy from and to atmospheric eddies (1920), 122
The submissiveness of nations (1953), 223
Threats and security (1950), 221
Thresholds when sensation is required as quantitatiave (1928), 132
Transforms for the eddy-diffusion of clusters (1952), 214
Turbulence and vertical temperature difference near trees (1925), 122
War and eugenics (1950), 216, 224
(see also *Weather Prediction by Numerical Process* in Index)

Index of Unpublished Works†
An abstract formulation of fashions, 177
Conditions of a lasting peace in Europe, 59
Gregariousness and its opposite, 175–7
Hints concerning international organisations, 227–31
An impartial selection of fatal quarrels, 220
A North Durham Colliery, 15
Quanta and diffusion, 146
Some biological waves, 144–5

†A complete list of LFR's unpublished works is given in Appendix B, This index lists only those works specifically mentioned in the text.